POPULAR PHYSICS.

PECK'S GANOT.

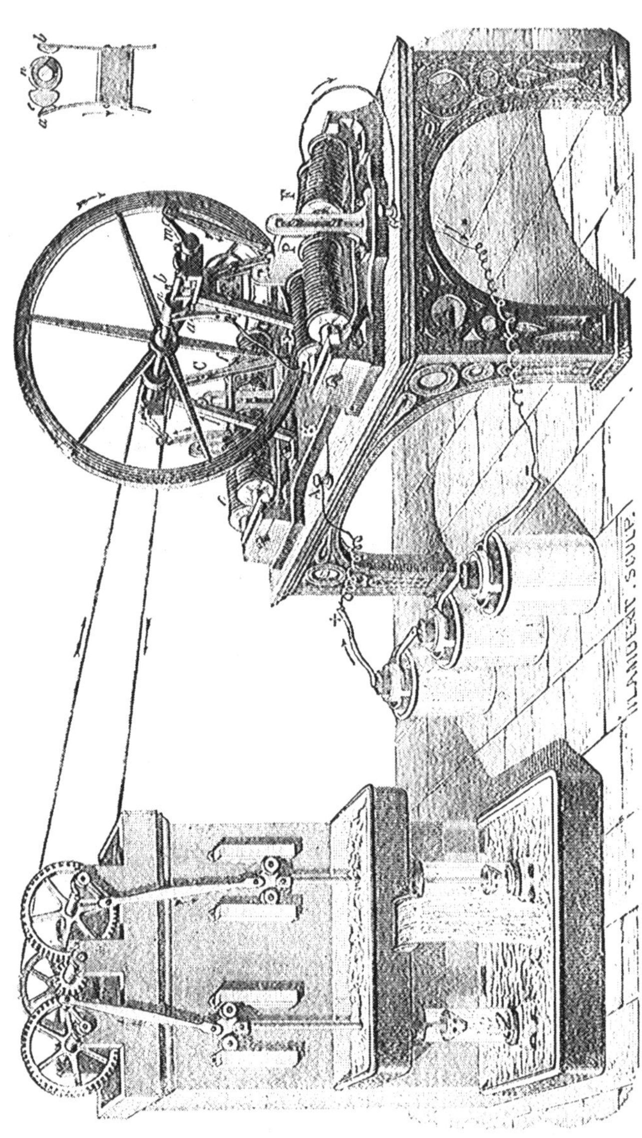

INTRODUCTORY COURSE

OF

NATURAL PHILOSOPHY

FOR THE USE OF

SCHOOLS AND ACADEMIES.

EDITED FROM

GANOT'S POPULAR PHYSICS,

BY

WILLIAM G. PECK, LL.D.,

PROFESSOR OF MATHEMATICS AND ASTRONOMY, COLUMBIA COLLEGE, AND OF MECHANICS IN THE SCHOOL OF MINES.

A. S. BARNES & CO.,

NEW YORK AND CHICAGO.

1871.

THE NATIONAL SERIES

OF

STANDARD TEXTS IN THE SCIENCES.

I.—NORTON'S FIRST BOOK IN PHILOSOPHY...$1 00
PECK'S GANOT'S NATURAL PHILOSOPHY.. 1 75

II.—PORTER'S FIRST BOOK IN CHEMISTRY.... 1 00
PORTER'S PRINCIPLES OF CHEMISTRY.... 2 00

III.—JARVIS' PRIMARY PHYSIOLOGY 75
JARVIS' PHYSIOL. AND LAWS OF HEALTH.. 1 75

IV.—WOOD'S OBJECT LESSONS IN BOTANY..... 1 40
WOOD'S CLASS-BOOK OF BOTANY 3 50

V.—STEELE'S 14 WEEKS IN ASTRONOMY...... 1 50

VI.—PAGE'S ELEMENTS OF GEOLOGY.......... 1 25

VII.—CHAMBERS' ELEMENTS OF ZOOLOGY.......... 1 50

PREFACE.

THE rapid spread of scientific knowledge, and the continually widening field of its application to the useful arts, have created an increased demand for new and improved text-books on the various branches of NATURAL PHILOSOPHY.

Of the elementary works that have appeared within a few years, those of M. GANOT stand preëminent, not only as popular treatises, but as thoroughly scientific expositions of the principles of PHYSICS. His "Traité de Physique" has not only met with unprecedented success in France, but has been extensively used in the preparation of the best works on Physics that have been issued from the American press.

In addition to the "Traité de Physique," which is intended for the use of Colleges and higher institutions of learning, M. GANOT has recently published a more elementary work, adapted to the use of schools and academies, in which he has faithfully preserved the prominent features and all the scientific accuracy of the larger work. It is characterized by a well-balanced distribution of subjects, a logical development of scientific principles, and a remarkable clearness of definition and explanation. In addition, it is profusely illustrated with beautifully executed engravings, admirably calculated to convey to the mind of the student a clear conception of the principles unfolded. Their completeness and accuracy are such as to enable the teacher to dispense with much of the apparatus usually employed in teaching the elements of Physical Science.

In preparing an American edition of this work on POPULAR PHYSICS, it has not been the aim of the editor to produce a strict translation. No effort, however, has been spared to preserve throughout, the spirit and method of the original work. No changes have been made, except such as have seemed calculated to harmonize it with the system of instruction pursued in the schools of our country.

By a special arrangement with M. GANOT, the American publishers are enabled to present *fac-simile* copies of all the original engravings.

NEW YORK, *June 1st*, 1860.

NOTE.

At the request of many teachers, a chapter has been prepared on the Application of Physical Principles to Machines. In the revised edition several cuts, with their corresponding numbers, have been omitted.

NEW YORK, June 1st, 1871.

CONTENTS.

INTRODUCTION.

CLASSIFICATION OF THE SCIENCES.

CHAPTER I.

PRELIMINARY PRINCIPLES AND MECHANICS OF SOLIDS.

CHAPTER II.

MECHANICS OF LIQUIDS.

CHAPTER III.

MECHANICS OF GASES AND VAPORS.

CHAPTER IV.

ACOUSTICS.

CHAPTER V.

HEAT.

CHAPTER VI.

OPTICS.

CHAPTER VII.

MAGNETISM.

CHAPTER VIII.

STATICAL ELECTRICITY.

CHAPTER IX.

DYNAMICAL ELECTRICITY.

CHAPTER X.

ELECTRO-MAGNETISM.

CHAPTER XI.

APPLICATIONS TO MACHINES.

POPULAR PHYSICS.

INTRODUCTION.

CLASSIFICATION OF THE SCIENCES.

SCIENCE is a knowledge of the laws that govern the Universe.

A LAW is a necessary relation between cause and effect. It is assumed as the foundation of all Science, that *like causes produce like effects.* This principle is an inductive truth, founded upon universal experience.

By the UNIVERSE we mean all that has been created, whether material or immaterial. The Universe may be regarded as made up of *mind* and *matter.* MIND is that which thinks and wills; MATTER is that of which we become cognizant through the medium of the senses. Science admits of two corresponding divisions, *Science of Mind*, or METAPHYSICS, and *Science of Matter*, or NATURAL PHILOSOPHY.

NATURAL PHILOSOPHY is that branch of science which treats of the laws that govern the material Universe.

Matter exists in two states, *organized* and *unorganized;* it is organized when its particles are aggregated into organs adapted to the support of life; in all other cases it is un-

---

What is Science? What is a Law? Define the Universe. Mind. Matter. What are the two divisions of Science? What is Natural Philosophy? In what two states may Matter exist? Illustrate.

organized. Natural Philosophy admits of two corresponding divisions: *Science of Organized Matter*, or PHYSIOLOGY, and *Science of Unorganized Matter*, or GENERAL PHYSICS.

Physiology, which treats of the laws of matter as modified by the principle of vitality, is divided into two principal branches: *Animal Physiology*, or ZOÖLOGY, and *Vegetable Physiology*, or BOTANY. Both of these branches, with their various subdivisions, belong to the domain of NATURAL HISTORY.

All unorganized matter may be divided into two classes, *Celestial* and *Terrestrial*. General Physics admits of two corresponding divisions. That branch which treats of celestial bodies, including the earth as a whole, is called ASTRONOMY; that which treats of terrestrial bodies, is called TERRESTRIAL PHYSICS.

TERRESTRIAL PHYSICS is again subdivided into two branches. The first is called *Physics Proper*, or simply PHYSICS; it treats of the general properties of bodies. The second is called CHEMISTRY; it treats of the nature of the ultimate particles of bodies and of their laws of combination. The first of these branches, or PHYSICS, is the subject treated of in the following pages.

Besides the branches above enumerated, and which may be called *Pure Sciences*, there are others that depend upon, or are applications of, two or more of them. Such, for example, are the sciences of GEOLOGY, MINERALOGY, PHYSICAL GEOGRAPHY, &c. These are called *Mixed Sciences*.

Into what may Natural Philosophy be divided? What is Physiology, and what are its branches? How may Unorganized Matter be divided? What are the corresponding divisions of General Physics? Define them. How is Terrestrial Physics divided? What is Physics Proper? Chemistry? What are the Pure Sciences, and what are some of the Mixed Sciences?

CHAPTER I.

PRELIMINARY PRINCIPLES AND MECHANICS OF SOLIDS.

I.—DEFINITIONS AND GENERAL PROPERTIES OF MATTER.

Definition of Physics—Physical Agents.

1. PHYSICS is that branch of Natural Philosophy which treats of the general properties of bodies, and of the causes that modify these properties.

The principal causes that modify the properties of bodies are: *Gravitation*, *Heat*, *Light*, *Magnetism*, and *Electricity*. These causes are called *Physical Agents.*

Definition of a Body.

2. A BODY is a collection of material particles; as a stone, or a block of wood. A body which is exceedingly small is called a *Material Point.*

Bodies are made up of small particles, called *Molecules*, and these again are composed of still smaller elements, called *Atoms.* These atoms are inconceivably small, and are held in their places by the action of two opposing systems of forces, called *Molecular Forces.* Those which tend to draw atoms together are called *Attractive Forces*, and those which tend to push them asunder are called *Repellent Forces.* Heat is the principal if not the only repellent force in Nature.

(**1.**) What is Physics? What are Physical Agents? Name them. (**2.**) Define a Body. A Material Point. An Atom. A Molecule. What are Molecular Forces? Define Attractive and Repellent Forces.

Mass and Density.

3. The MASS of a body is the quantity of matter which it contains.

Different bodies, having the same volume, contain very different quantities of matter; for example, a cubic inch of lead contains nearly eleven times as much matter as a cubic inch of water. The masses of bodies are proportional to their weights.

The DENSITY of a body is the degree of closeness of its particles.

Those bodies in which the particles are close together are said to be *dense*; thus, platinum and mercury are *dense* bodies. Those in which the particles are not close together are said to be *rare*; thus, steam and air are *rare* bodies. The densities of bodies having the same bulk are proportional to their weights.

Classification of Bodies.

4. Bodies may exist in two different states, the *solid* and the *fluid*.

SOLIDS are those which tend to retain a permanent form; as stones, metals, and the like. The particles of such bodies adhere to each other with considerable energy, and this adhesion can be overcome only by the exertion of some effort.

FLUIDS are those whose particles move freely amongst each other; as water, alcohol, and air. Such bodies have no tendency to retain a permanent form, but assume at once the form of the containing vessel.

Fluids are divided into *Liquids* and *Gases* or *Vapors*. *Liquids* are sensibly incompressible; as water, wine, and milk. *Gases* and *vapors* are highly compressible; as atmospheric air and steam.

(**3.**) What is the mass of a body? Density? Give examples of dense and rare bodies. (**4.**) How are bodies divided? Define solids and fluids. How are fluids divided? Define liquids, and gases or vapors.

In solids, the molecular forces of attraction are greater than the repellent forces, hence the difficulty of separating their molecules; in liquids, the attractive and repellent forces are sensibly balanced; in gases, the repellent are more powerful than the attractive forces.

Many bodies may exist in each of the three states in succession. Thus, if ice be heated until the repellent forces balance those of attraction, it passes into the liquid state and becomes water; if still more heat be applied, the repellent forces prevail over those of attraction, and it passes into the state of vapor and becomes steam.

General Properties of Bodies.

5. All bodies possess certain properties, the most important of which are: *Magnitude, Form, Impenetrability, Inertia, Porosity, Divisibility, Compressibility, Dilatability, and Elasticity.*

Magnitude and Form.

6. The MAGNITUDE of a body is its bulk, or the portion of space that it fills. It is evident that a body can not exist without possessing the three attributes of length, breadth, and thickness.

The FORM of a body is its external shape. Bodies may have the same magnitude and be very different in shape; they may likewise be of the same form and yet be of very different magnitudes.

Impenetrability.

7. IMPENETRABILITY is that property by virtue of which no two bodies can occupy the same place at the same time. This property is self-evident, although phenomena are observed which would seem to conflict with it. Thus, when a pint of alcohol is mixed with a pint of water, the volume of the resulting mixture is less than a quart. This diminu-

Illustrate. (5.) What properties belong to all bodies? (6.) What is Magnitude? Form? (7.) What is Impenetrability? Illustrate.

tion of volume arises from the particles of one of the fluids insinuating themselves between those of the other; but it is clear that where a particle of alcohol is, there a particle of water can not be. In like manner when a nail is driven into a board, the particles of the latter are thrust aside and compressed to make room for those of the former.

Inertia.

8. INERTIA is the tendency which a body has to maintain its state of rest or motion. If a body is at rest it has no power to set itself in motion, or if it is in motion it has no power to change either its rate of motion or the direction in which it is moving. Hence, if a body is at rest, it will remain at rest, or if in motion, it will move on uniformly in a straight line until acted upon by some force.

The reason why we do not see bodies continue to move on uniformly in straight lines, when set in motion, is that they are continually acted upon, by forces which change their state of motion. Thus, a ball thrown from the hand, besides meeting with the resistance of the air, is continually drawn downwards by the attraction of the earth, till at last it is brought to rest.

Many familiar phenomena are explained by the principle of inertia. For example, when a vehicle in motion is suddenly arrested, loose articles in it are thrown to the front, because they tend to keep the motion which they had acquired. When a man in running strikes his foot against an obstacle, the inertia of the upper part of his body carries it forward, and he falls to the ground. For the same reason, when a man jumps from a car in motion, he will be in danger of falling in the direction of the moving car. It is inertia which renders accidents upon railroads so terrible. When from any cause the locomotive is suddenly arrested, the inertia of the entire train acts to pile the cars together in one general wreck. It is the inertia of the hammer that enables it to overcome the resistance

---

Give examples of apparent penetrability. (8.) What is Inertia? Illustrate. *Why do we not see bodies conform to the law of inertia? Give examples of the principle of inertia.*

which the wood offers to the entering nail; and in driving piles, the principal effect is due to the inertia of the descending ram.

Porosity.

9. Porosity is the degree of separation between the molecules of a body. The intervals between the molecules are called *pores*. When these intervals are very great, the body is said to be *porous*, as in steam, air, and gases. When the intervals are very small, the body is said to be *dense*, as in gold, platinum, and mercury. *Pores* must not be confounded with *cells*, as in sponge, light bread, and the like.

All bodies are more or less porous.

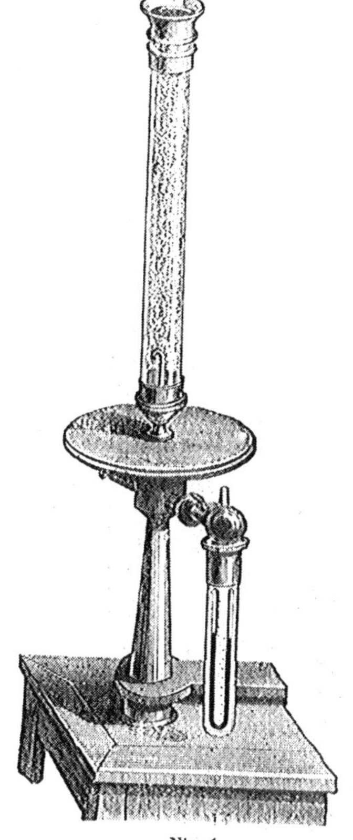
Fig. 1.

The following experiment shows the porosity of leather. A long glass tube (Fig. 1) is surmounted by a brass cup, with a thick leather bottom, fitting the tube air-tight. The lower end of the tube terminates in a brass cap, which is attached to a machine for exhausting the air from the tube, called an air-pump.

If a quantity of mercury is

(9.) What is Porosity? When are bodies porous? When dense? *Explain the experiment showing the porosity of leather.*

poured into the upper cup, and the air exhausted from the tube, the mercury, being pressed down by the external air, is seen falling through the leather in small drops like rain.

Gold was shown to be porous by some Florentine philosophers in the following manner. A hollow sphere of gold was filled with water and tightly closed, after which it was subjected to great pressure. The water was seen to issue from the globe and form on its surface like dew. The experiment has since been repeated with other metals, and with like results.

Gases are shown to be porous by their enormous reduction in volume when compressed; if a gas be introduced into a jar, it will spread by its expansive force and completely fill the vessel; if a second gas be introduced into the same vessel, it likewise expands and fills the vessel as though the first gas did not exist. This proves that the molecules of the second gas arrange themselves in the pores of the first.

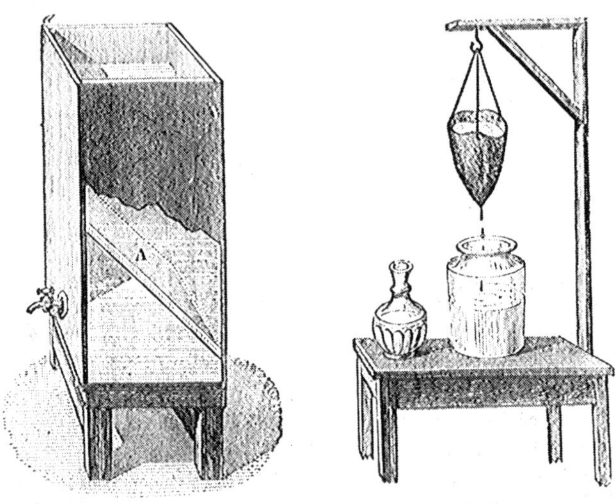

Fig. 2. Fig. 3.

The property of porosity finds an important application in the process of filtering, that is, in separating foreign particles from liquids.

Explain the Florentine experiment. What are filters?

Fig. 2 represents a filter for purifying water; it is simply a box divided into two parts by a partition of porous stone, *A*. The water to be filtered is placed in the upper part, from which it passes slowly into the lower part through the pores of the stone. In one corner of the box is a tube, *a*, which permits the air to escape as the lower part of the box fills with water. The purified water is drawn off by means of a faucet near the bottom of the box.

Fig. 3 represents a filter used by chemists. It consists of a pocket of some porous material, as felt, for example, suspended by cords. The substance to be filtered is poured into the pocket, from which the liquid escapes slowly through the pores, leaving the solid parts behind.

Filters are also formed by layers of powdered charcoal, or finely ground quartz, through the pores of which the liquids pass. It is to a natural filtration through sand that many kinds of spring water owe their purity.

It is in consequence of porosity, that burning coals covered up with ashes continue to burn slowly. The air which is necessary to combustion penetrates through the pores of the ashes, in sufficient quantity to keep the fire from being entirely extinguished.

Finally, it is in consequence of their porosity, that many kinds of wood absorb moisture from the air, and tend to swell and crack; this difficulty is remedied by applying oils and varnishes, which close the pores and exclude the moisture.

Divisibility.

10. Divisibility is that property by virtue of which a body may be divided into parts. All bodies are capable of subdivision, and in many cases the parts that may be obtained are of almost inconceivable minuteness.

The following examples serve to show the extreme smallness of the molecules of matter. A single grain of carmine imparts a sensible color to a gallon of water; this gallon of water may be separated into a million of drops, and if we suppose each drop to contain ten particles of carmine, which is a low estimate, we shall have

Explain the water filter. Explain the chemist's filter. Other applications of porosity. (**10.**) What is Divisibility? *Give examples of divisibility by solution.*

divided the grain of carmine into ten million of molecules, each of which is visible to the naked eye.

The microscope reveals to us, in certain vegetable infusions, *animalculæ* so small that several hundred of them can swim in a drop of water that adheres to the point of a needle. These little animals are capable of motion, and even of preying upon each other; they therefore possess organs of motion, digestion, and the like. How minute, then, must be the molecules which go to make up these organs.

A grain of musk is capable of diffusing its odor through an apartment for years, with scarcely an appreciable diminution of its weight. This shows that the molecules of musk continually given off to replenish the odor, are of inconceivable smallness.

The blood of animals consists of minute red globules swimming in a serous fluid; these globules are so small that a drop of human blood, no larger than the head of a small pin, contains at least 50,000 of them. In many animals these globules are still smaller; in the musk deer, for example, a single drop of blood of the size of a pin's head contains at least a million of them.

Compressibility.

11. Compressibility is the property of being reduced to a smaller space by pressure. This property is a consequence of porosity, and the change of bulk comes from the particles being brought nearer together by the pressure. Sponge, india-rubber, cork, and elder pith, are examples of compressible bodies; they may be sensibly diminished in volume by the pressure of the fingers. Gases are, however, the best examples of compressible bodies.

Fig. 4 represents an apparatus by means of which the compressibility of gases may be shown. It consists of a tube of glass, with metallic caps, completely closed at its lower end. An air-tight piston is introduced at the upper end, and on being pushed down we see the inclosed air reduced to the half, fourth, and even the hundredth part of its original bulk.

Examples of minute animals. Examples of odoriferous bodies. Blood globules.
(**11.**) What is Compressibility? Examples. *Explain the experiment.*

Liquids are only slightly compressible, nevertheless nice experiments show that even they can be somewhat reduced in bulk by pressure.

Fig 4.

Metals are compressible, as is shown in the process of stamping coins, metals, and the like.

Dilatability.

12. DILATABILITY is the property that a body possesses of assuming a greater bulk under certain circumstances.

In the experiment upon air, explained in the last article,

Are liquids compressible? Are metals compressible? How shown? (**12.**) What is Dilatability?

if the piston be raised after the air has been compressed, it will expand and fill the tube. Almost all bodies expand on being heated. It is on this principle that thermometers are constructed. In cooling, bodies contract.

A familiar example of dilatability and contractibility is shown in the process of fitting the tire upon a carriage wheel. The tire is made a little smaller than the wheel, but on being heated it expands so as to embrace it; on cooling it contracts again and draws the parts of the wheel tightly together.

The same property of metals has been used for producing great pressures, and even for restoring inclined walls to an erect position.

Elasticity.

13. Elasticity is the property which bodies possess of recovering their original shape and size after having been either compressed or extended.

Bodies differ in their degree of elasticity, yet all are more or less elastic. India-rubber, ivory, and whalebone are examples of highly elastic bodies. Putty and clay are examples of those which are only slightly elastic.

If air be compressed, its elasticity tends to restore it to its original bulk; this property has been utilized in making air-beds, air-cushions, and even in forming car-springs. If a spring of steel be bent, its elasticity tends to unbend it; this principle is employed in giving motion to watches, clocks, and the like. If a body be twisted, its elasticity tends to untwist it, as is observed in the tendency of yarn and thread to untwist; this principle, under the name of *torsion*, is used to measure the deflective force of magnetism. If a body be stretched, its elasticity tends to reduce it to its original length, as is shown by stretching a piece of india-rubber, and then allowing it to contract.

We see that the elasticity of a body may be brought into play by four different methods: by *pressure*, by *flexure* or bending, by *torsion*

Example. *Application in putting tire upon a wheel. Example of restoring walls.* (**13.**) What is Elasticity? Give examples of highly and slightly elastic bodies. *Give examples of the applications of elasticity. How may elasticity be brought into play? Examples.*

or twisting, and by *tension* or stretching. In whatever way it may be developed, it is the result of molecular displacement. Thus, when air is compressed, the repulsions between the molecules tend to expand it. Again, when a spring is bent, the particles on the outside are drawn asunder, whilst those on the inside are pressed together; the attractions of the former and the repulsions of the latter tend to restore the spring to its original shape.

The most elastic bodies are gases; after them come tempered steel, whalebone, india-rubber, ivory, glass, &c.

Fig. 5 illustrates the method of showing that ivory is elastic, and at the same time that the cause of its elasticity is molecular displacement. It consists of a polished plate of marble, over which is spread a thin layer of oil. If a ball of ivory be let fall upon it from different heights, it will at each time rebound, leaving a circular impression on the plate, which is the larger as the ball falls from a greater height. This experiment shows that the ball is flattened each time by the fall, that the flattening increases as the height increases, and that the action of the compressed molecules causes it to rebound.

Fig. 5.

The property of elasticity is utilized in the arts in a great variety of ways. When a cork is forced into the mouth of a bottle, its elasticity causes it to expand and fill the neck so as to render it both water and air-tight. It is the elasticity of air that causes india-rubber balls, filled with air, to rebound when thrown upon hard substances. It is the elasticity of steel that renders it of use in

What bodies are most elastic? How is it shown that ivory is elastic? Explain the experiment. Explain some of the applications of elasticity. Corking bottles. Springs.

springs for moving machinery, as well as for easing the motion of carriages over rough roads. It is the elasticity of cords that renders them applicable to musical instruments. It is the elasticity of air that renders it a fit vehicle for transmitting sound. It is the elasticity of the etherial medium pervading space that renders it capable of transmitting light.

II.—MECHANICAL PRINCIPLES

Definition of Mechanics.

14. MECHANICS is that branch of Physics which treats of the laws of rest and motion. It also treats of the action of forces upon bodies.

Rest and Motion.

15. A body is at REST when it retains its position in space. It is in MOTION when it continually changes its position in space.

A body is at rest with respect to surrounding bodies, when it retains the same relative position with respect to them, and it is in motion with respect to surrounding objects when it continually changes its relative position with respect to them. These states of rest and motion are called *Relative Rest* and *Relative Motion*, to distinguish them from *Absolute Rest* and *Absolute Motion.*

When a body remains fixed on the deck of a moving vessel or boat, it is at rest with respect to the parts of the vessel, although it partakes with them in the common motion of the vessel. When a man walks about the deck of a vessel, he is in motion with respect to the parts of the vessel, but he may be at rest with respect to objects on shore; this will be the case when he travels as fast as the vessel sails, but in an opposite direction. In consequence of the earth's motion around its axis and about the sun, together with the motion

---

Stringed instruments. Transmission of light. (**14.**) What is Mechanics? (**15.**) When is a body at rest? When in motion? Explain relative and absolute rest and motion. *Illustrate by examples.*

of the whole solar system through space, it is not likely that any part of our system is in a state of absolute rest for any appreciable length of time.

Different kinds of Motion.

16. MOTION may be *rectilinear* or *curvilinear;* it is rectilinear when the path of the moving body is a straight line, and it is curvilinear when this path is a curved line. The motion of a train of cars along a straight track is an example of rectilinear motion; the motion of the same train in passing round a curve is an example of curvilinear motion.

Uniform Motion—Velocity.

17. UNIFORM MOTION is that in which a body passes over equal spaces in equal times. Thus, every point on the surface of the earth is, by its revolution, carried around the axis with a uniform motion.

In this kind of motion the space passed over in one second of time is called the *velocity*. Thus, if a train of cars travel uniformly at the rate of 20 miles per hour, its velocity is 29.3 feet. Instead of taking a second as the unit of time, we might adopt a minute, or an hour. In the same case as before we might say, that the velocity of the train is one third of a mile per minute, or twenty miles per hour.

Varied Motion—Accelerated and Retarded Motion.

18. VARIED MOTION is that in which a body passes over unequal spaces in equal times. If the spaces passed over in equal times go on increasing, the motion is *accelerated;* such is the motion of a train of cars when starting, or that of a body falling towards the surface of the earth. If the spaces passed over go on decreasing, the motion is

(**16.**) What is Rectilinear Motion? Curvilinear Motion? Examples.
(**17**) What is Uniform Motion? Example. What is meant by velocity? Example.
(**18.**) What is Varied Motion? When is it accelerated and when retarded?

retarded; such is the motion of a train of cars when coming to rest, or that of a body thrown vertically upwards.

When the spaces passed over in equal times are continually increased or decreased by the same quantity, the motion is *uniformly accelerated*, or *uniformly retarded*. The motion of a body falling in a vacuum, is *uniformly accelerated;* that of a body shot vertically upwards in a vacuum, is *uniformly retarded*.

The velocity of a body having varied motion at any time, is the rate of the body's motion at that time. In varied motion the velocity is continually changing.

Forces, Powers, and Resistances.

19. If a body is at rest, any cause that tends to set it in motion, is called a *Force;* if a body is in motion, any cause that tends to make it move faster, or slower, or to change its direction, is called a *Force*.

A Force, then, is any cause that tends to change the state of a body, with respect to rest or motion.

The attractions and repulsions between the molecules of bodies are forces: the muscular efforts of men or animals, employed in accomplishing any kind of work, are forces; the elastic efforts of gases and vapors are forces.

Forces which act to produce motion are called *Powers;* those which act to prevent or destroy motion are called *Resistances*. The effort of steam employed in moving a train of cars is a *power*, whilst friction and the inertia of the air, which tend to retard the motion, are *resistances*. Powers tend to accelerate motion, and are for that reason called *Accelerating Forces*. Resistances, on the contrary, tend to retard motion, and are for that reason called *Retarding Forces*.

Examples. Define uniformly accelerated and uniformly retarded motion. Examples. (**19.**) What is a Force? Examples. Define Powers and Resistances. Examples. By what other names may they be called?

Distinctive Characteristics of Forces.

20. In order that the effect of any force may be completely understood, three things must be known: its *point of application*, its *direction*, and its *intensity*.

The *point of application* of a force is the point where it exerts its action. Thus, in Fig. 6, which represents a child drawing a wagon, the force exerted by the child has its point of application at *A*.

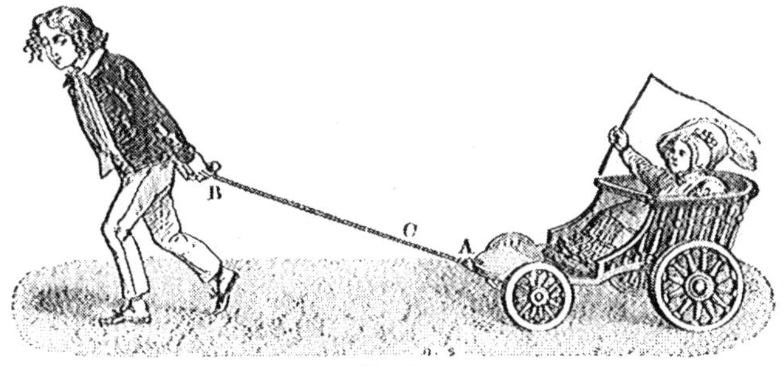

Fig. 6.

The *direction* of a force is the line along which it acts; thus, in Fig. 6, the line *AB* is the direction of the force exerted by the child.

The *intensity* of a force is the energy with which it acts; thus, in the same example as before, the intensity of the force exerted is the energy which the child exerts in overcoming the resistance of the wagon.

The intensity of a force is measured in pounds; thus, a force of fifty pounds is a force necessary to sustain a weight of fifty pounds. The intensity of a force may be represented by a distance which is usually laid off on the line of direc-

(**20**) What three elements determine a force? Define the point of application. The line of direction. The intensity. How is the intensity measured? How represented? Example.

tion of the force. Having assumed some unit of length, say one tenth of an inch, to represent one pound, this is set off as many times as the force contains pounds. In the example taken, if we suppose the force exerted to be seven pounds, and lay off from *A* to *C* seven tenths of an inch, then will *AC* represent the force both in direction and intensity.

Resultant and Component Forces.

21. When a body is solicited by a single force, it is evident, if no obstacle intervene, that it will move in the direction of that force; but if it is solicited at the same time by

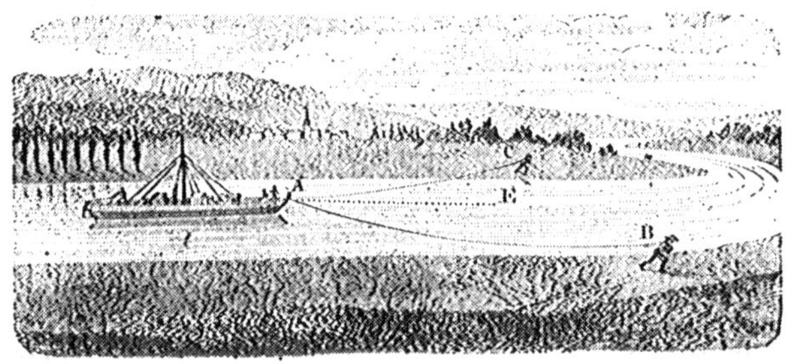

Fig. 7.

several forces acting in different directions, it will not, in general, move in the direction of any one of them. For example, if two men on opposite sides of a river tow a boat by means of a rope, as represented in Fig. 7, the boat will not move either in the direction *AB*, or *AC*, but it will move in some intermediate direction, as *AE*; that is, it will advance as though it were solicited by a single force directed from *A* towards *E*. This single force, which would produce the same effect as the two separate forces, is called

(**21.**) What is a Resultant of several forces?

their *Resultant.* The separate forces are called *Components* of the resultant.

In general the *resultant* of any number of forces is a single force whose effect is equivalent to that of the whole group. The individual forces of the group are called *Components.*

Parallelogram of Forces.

22. It is shown in Mechanics (Peck's Mechanics, Art. 25), that if *AB* and *AD*, Fig. 8, represent two forces acting at *A*, their resultant will be represented by *AC*. That is, *if two forces are represented in direction and intensity by the adjacent sides of a parallelogram, their resultant will be represented in direction and intensity by that diagonal which passes through their point of intersection.*

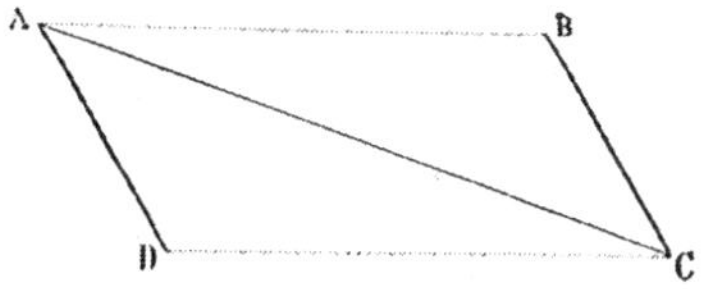

Fig. 8.

This principle is called the *Parallelogram of Forces.*

The operation of finding the resultant when the components are given is called *Composition of Forces;* the reverse operation is called *Resolution of Forces.*

Fig. 9.

When two forces are applied at the same point, as shown in Fig. 9, we lay off distances *AB*

What are Components? Illustrate. (**22.**) Enunciate the parallelogram of forces.

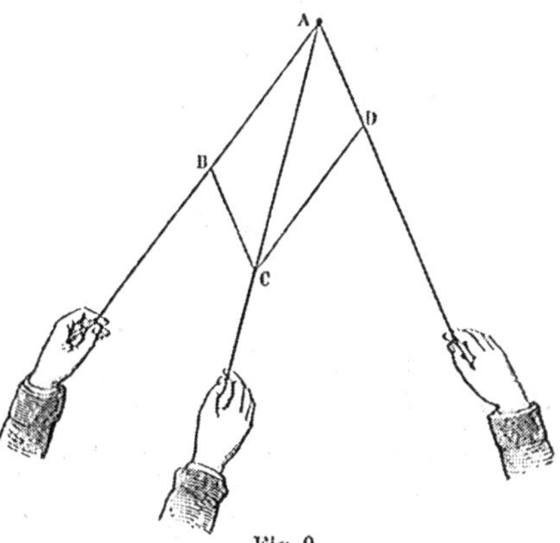

and AD to represent the forces, and having completed the parallelogram, we draw its diagonal AC; this will be their resultant. If the resultant AC is known, and the directions of its components are given, we draw through C the lines CD and CB parallel to their directions; then will the intercepted lines AD and AB be components of the force AC.

Practical Example of Composition of Forces.

23. A bird, in flying, strikes the air with both wings, and the latter offers a resistance which propels him forward.

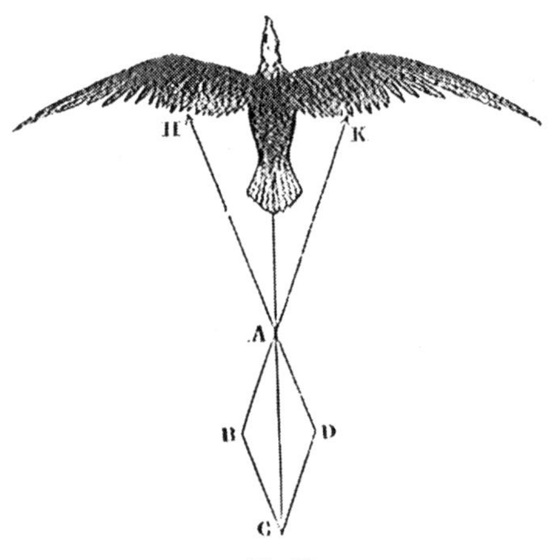

Fig. 10.

Let AK and AH, in Fig. 10, represent these resistances. Draw AB and AD equal to each other, and complete the parallelogram AC; draw also the diagonal AC. Then will AC represent the resultant of the two forces, and the bird will move exactly as though impelled by the single force CA.

How is the resultant found when the components are known? How are the components found? (**23.**) Explain the flight of a bird.

Practical Example of Resolution of Forces.

24. When a sail-boat is propelled by a breeze acting on the quarter in the direction *va* (Fig. 11), we may, by the rule in Art. 22, resolve the intensity of the wind into two components, one, *ca*, in the direction of the keel, and the

Fig. 11.

other, *ba*, at right angles to it. The first component alone is effective in giving a forward motion to the boat, whilst the second is partly destroyed by the resistance which the water offers to the keel, and partly employed in giving a lateral motion to the boat. This lateral motion is called *leeway*.

Resultant of Parallel Forces.

25. When two forces act in the same direction, as when two horses pull at the ends of a whiffle-tree to draw a wagon, *their resultant is equal to the sum of the forces.* When they act in a contrary direction, as in the case of a steamboat ascending a river, where the force of the engine acts to propel the boat forward, whilst the current acts to

(**24**.) Explain the sailing of a boat. (**25**.) What is the resultant of parallel forces when they act in the same direction? When they act in opposite directions? Examples.

retard its progress, *their resultant is equal to the difference of the forces.*

Equilibrium of Forces.

26 When several forces acting upon a body exactly balance each other, they are said to be *in equilibrium.*

Fig. 12.

The simplest case of equilibrium is that of two equal forces acting against each other, as in the case where two men of equal strength pull at the two ends of a rope, as shown in Fig. 12.

In the same manner, if two buckets of equal weight are suspended in a well from the ends of a rope passing over a pulley, they will be in equilibrium.

When a body rests upon a table, there is an equilibrium between the weight of the body which urges it downwards, and the resistance of the table which prevents it from falling. If the weight becomes greater than the resistance, the table breaks and the body falls.

Centrifugal and Centripetal Forces.

27. The CENTRIFUGAL FORCE is the resistance which a body offers to a force which tends to deflect it from its course.

In consequence of its inertia, a body always tends to move in a straight line, and if we see it move in a curved line it is because some force is acting to turn it from its path. This deflecting force is called the *Centripetal Force*, and

(**26.**) When are forces in equilibrium? Illustrate by examples. (**27.**) What is the Centrifugal Force? Centripetal Force?

because action and reaction are always equal, *the centripetal and centrifugal forces are always opposed and equal to each other.* If a ball is whirled about the hand, being retained by a string, it has a continual tendency to fly off, which tendency is resisted by the strength of the string; the tendency to fly off is due to the centrifugal force, and the force which resists this tendency is the centripetal force.

The curved path in which a body moves may be regarded as made up of short straight lines, and if at any instant the centripetal

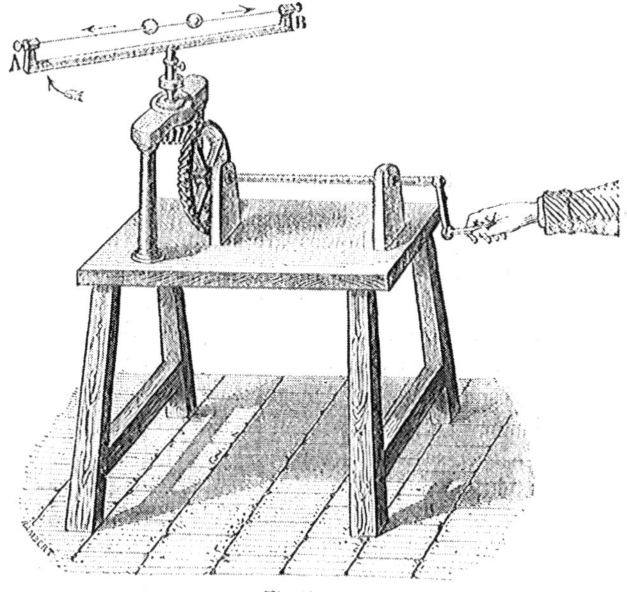

Fig. 13.

force were destroyed, the body would continue to move along that line on which it was situated; that is, its new path would be tangent to its old one.

The existence of the centrifugal force may be shown experiment-

Example. *How does a body move when the centripetal force is destroyed?*

ally by the apparatus represented in Fig. 13. It consists of a bar, *AB*, having its ends bent up so as to hold a wire which is stretched between them. On this wire two ivory balls are strung so as to slide along it, and the whole bar is made to turn about an axis at right angles to it by means of a crank and two bevelled wheels. When the bar is made to revolve about the axis, the balls, acted upon by the centrifugal force, are thrown against the ends of the bar with an energy which becomes greater as the motion of revolution becomes more rapid.

Some Effects of the Centrifugal Force.

28. When a train of cars turns round a curve in the road, the centrifugal force tends to throw the train off the track, a tendency which is resisted by raising the outer rail and by making the wheels conical.

It is in consequence of the centrifugal force, that the mud adhering to the tire of a carriage-wheel is thrown off in all directions.

In the circus, where horses are made to travel rapidly around in a curved path, the centrifugal force tends to overturn them outwards, which tendency is partly overcome by making the outside of the track higher than the inside, and partly by both horse and rider inclining inwards, so as to make the resultant of their weight and the centrifugal force perpendicular to the path.

When a sponge filled with water and held by a string is whirled rapidly around, the centrifugal force throws off the water and leaves the sponge dry. This principle has been used for drying clothes in the laundry.

A very remarkable effect of the centrifugal force is the flattening of our earth at the poles. The earth turns on its axis every twenty-four hours, which rotation gives rise to a centrifugal force at every point of its surface. At the

---

Explain the experiment. (**28**) Give examples of the action of the centrifugal force. Cars on a curve. Mud from wheel. Circus. Sponge. Effect on the form of the earth.

equator the centrifugal force is greatest, because the velocity is there the greatest, and from the equator it grows feebler towards each pole, where it is zero. The centrifugal force at every point is perpendicular to the axis, and may be resolved into two components, one directed outwards from the centre, and the other perpendicular to this. The former component lessens the weight of bodies, and the latter acts to heap the particles up towards the equator. It has been found that the earth is a spheroid, flattened at the poles. The polar diameter is about twenty-six miles shorter than the equatorial diameter. Observations upon the heavenly bodies show that other planets are in like manner flattened at their poles.

The manner in which the centrifugal force acts to flatten a sphere, is shown experimentally by an apparatus, represented in Fig. 14. This apparatus consists of a vertical rod to which a motion of rotation may be imparted, as shown in Fig. 13. At the lower part of this rod four strips of brass are firmly fastened and bent into circles, as shown by the dotted lines; their upper ends are fastened to a ring which is free to slide up and down the rod. When the axis is made to revolve rapidly, the centrifugal force causes the ring to slide down the rod, the hoops become more curved, as shown in the figure, and the whole assumes the appearance of a flattened sphere.

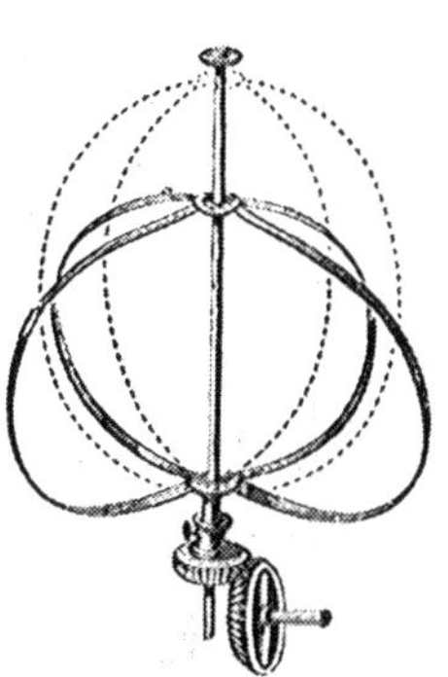

Fig. 14.

Machines.

20. A MACHINE is any contrivance by means of which a force acting at one point is made to produce an effect at some other point.

Effect on the weight. *Explain the experiment.* (**20.**) What is a Machine?

According to this definition every tool used in the arts is a machine; in common language, however, the term is only applied to more complex combinations. In this sense, a machine consists of a collection of moving pieces called *elements*, kept in position by a *frame*. The piece to which the motive power is applied is called the *recipient*, the piece that performs the work is called the *tool*, and these, with their connecting pieces, make up a *train of mechanism*. The elements of a train are called *Elementary Machines*, or *mechanical powers*, and are seven in number (Art. 449).

Of these the *lever* is most important to the student of Physics. The others are fully discussed in Chapter XI.

The Lever.

30. A LEVER is an inflexible bar free to turn about a fixed point, called the *Fulcrum*, and acted upon by two forces which tend to turn it in opposite directions. The force which acts as a motor, is called the *Power*, the other one is called the *Resistance*.

Levers are of three classes, according to the position of the fulcrum with respect to the power and resistance.

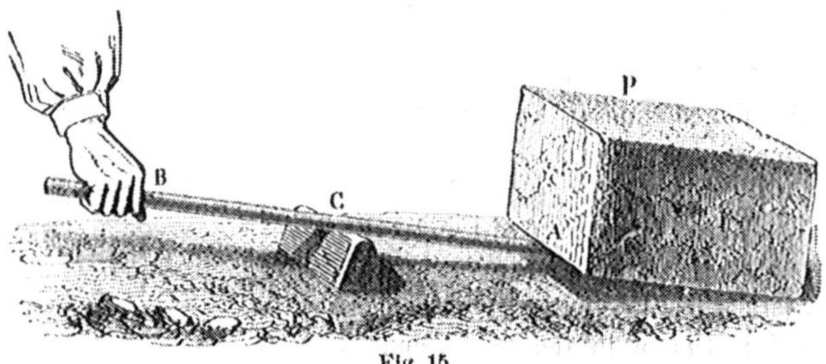

Fig. 15.

Lever of the first class.—In this class the fulcrum is between the power and the resistance. Such a lever is represented in Fig. 15. The hand is the power, the weight *P* is the resistance, and the fixed point *C* is the fulcrum.

What is a train of mechanism? What is the recipient? Tool? (**30.**) What is a lever? How many classes are there? Examples of each.

Lever of the second class.—In this class the fulcrum is beyond both the power and resistance, and nearest the resistance. Such a lever is shown in Fig. 16. The power is applied at *B*, the resistance at *A*, and the fulcrum is at *C*.

Fig 16.

Lever of the third class.—In this class the fulcrum is beyond both the power and the resistance, and nearest the power, as shown in Fig. 17.

In every class of lever, the distances from the fulcrum, to the power and resistance, are called *Lever Arms.* In each of the figures in this article, *CB* is the lever arm of the power, and *CA* the lever arm of the resistance.

Conditions of Equilibrium of the Lever.

31. It is demonstrated in Mechanics (Art. 78), that the effect of a force produced by the aid of a lever increases as its lever arm increases, so that, if the lever arm be doubled or tripled, the effect of the force is always doubled or tripled.

What are the Lever Arms? (**31.**) What is the relation between the power and resistance?

Hence it was that ARCHIMEDES was able to say, that he could lift the world if he had a place on which to rest his lever.

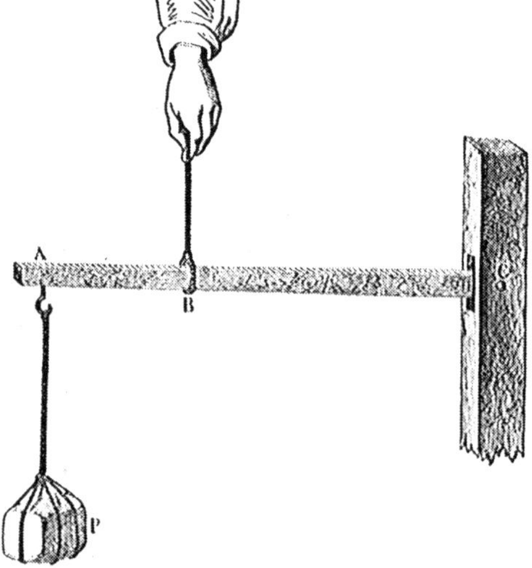

Fig. 17.

Since the effect of a force increases with its arm of lever it is necessary, in order that the power and resistance may be in equilibrium, that they should be to each other inversely as their lever arms. That is, if the power is three times the resistance, the lever arm of the former should only be one third as long as that of the latter, and so on. If the power is equal to the resistance, they will be in equilibrium when their lever arms are equal.

From what has been said, it follows, that the power is always greater than the resistance in the third class of levers, and less than it, in the second class. In the first class the power may be either greater or less than the resistance. We say in common language

---

Between the power, and velocity?

that there is a loss of power in using a lever of the third class, and a gain of power in using one of the second class.

In performing any work with a lever, the paths passed over by the points of application of the power and resistance are proportional to their lever arms; that is, the longer the lever arm the greater the path passed over, and the greater its velocity. This is expressed by saying, that *what is gained in power is lost in velocity*. It is for this reason that we say there is no real gain of power in the employment of a lever.

Examples of Levers.

32. Levers are of continual use in the arts, forming component parts of nearly every machine.

Fig. 18.

A pair of scissors affords an example of the first class of levers. The fulcrum is at *C*, Fig. 18, the hand furnishes the power, and the substance to be cut the resistance.

The common balance, yet to be described, is a lever of this class as is also the handle of a pump.

The ordinary nut-cracker is an example of levers of the

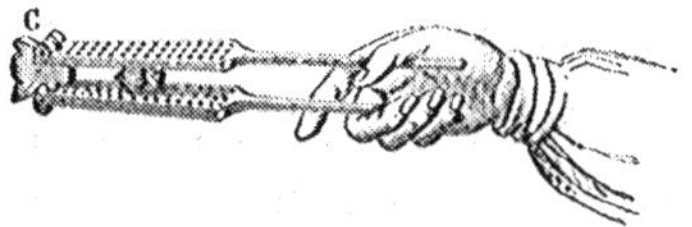

Fig. 19.

second class. The fulcrum is at *C*, Fig. 19; the power is the hand, and the resistance is the nut to be cracked.

Is there any gain of power in using a lever? (**32**) Applications. Explain the scissors. The nut-cracker.

The oars of a boat are levers of the second class. The end of the oar in the water is the fulcrum, the hand is the power, and the boat, or rather the resistance of the water which it has to overcome, is the resistance. The shears employed for cutting metals belong to this class of levers.

The treadle of a flax-spinner, or of a lathe, is an example of a lever of the third kind. The fulcrum is at *C*, Fig. 20, the foot is the power, and the work to be done is the resistance.

Fig. 20.

The bones of the animal frame are many of them levers of this class. Thus, in the bone of the forearm in man, the elbow joint is the fulcrum, the muscle attached just below the joint is the power, and a weight to be raised is the resistance.

Other Machines.

33. Besides the lever there are two other simple machines, the *cord* and the *inclined plane*. The former re-

Oars of a boat. Treadle of a spinner. *Bone of the forearm.* (**33.**) What are the other simple machines?

quires no description, and the latter will be explained further on. From these machines, as elements, are formed by combination, *the pulley*, *the wheel and axle*, *the screw*, and *the wedge*. These seven make up what are commonly called the *Mechanical Powers*, and from them may be constructed every machine, however complicated. For a more detailed account of the general principles of Mechanism and Machines, the reader is referred to Chapter XI.

III.—PRINCIPLES DEPENDENT ON THE ATTRACTION OF GRAVITATION.

Universal Gravitation.

34. THE earth exerts a force of attraction upon all bodies near it, tending to draw them towards its centre. This force, called the *Force of Gravity*, when unresisted imparts motion, and the body is said *to fall;* when resisted it gives rise to pressure, which is called *Weight*.

NEWTON showed that the force of gravity, as exhibited at the earth's surface, is only a particular case of a general attraction extending throughout the Universe, and continually tending to draw bodies together. This general attraction he called *Universal Gravitation*. It is mutually exerted between any two bodies whatever, and it is by virtue of it that the heavenly bodies are retained in their orbits.

The law of universal gravitation may be easily explained. If we take the mutual attraction of two units of mass, at a unit's distance from each other, as 1, then will their mutual attraction at any other distance be equal to 1 divided by the square of that distance; thus, if the distance is 2, their attraction will be $\frac{1}{4}$ of what it was at the

---

What machines are formed by combinations of simple machines? Name the seven mechanical powers. (34.) What is the Force of Gravity? What is its effect when unresisted? When resisted? What is Universal Gravitation? *Explain the law of Universal Gravitation.*

distance 1; if their distance is 3, their attraction will be $\frac{1}{9}$ of what it was at the distance 1, and so on. If one of the masses contains m units of mass, and the other one unit, the force will be m times as great as though they were both units of mass; that is, the attraction will be equal to m, divided by the square of the distance between the bodies. If the second body contain n units of mass, the attraction will be n times as great as before; that is, it will be mn, divided by the square of the distance between the bodies.

This law, discovered by NEWTON, may be expressed as follows: *Any two bodies exert upon each other a mutual attraction, which varies directly as the product of their masses, and inversely as the square of their distance apart.*

Effect of Gravitation on the Planets.

35. It is by the influence of gravitation that the planets are retained in their orbits. Their motion is the same as though they had been projected into space with an impulse, and then continually drawn from the right lines along which inertia tends to carry them, by the attraction of the sun. The planets also attract the sun, but their masses being exceedingly small in comparison with that of the sun, their effects in disturbing its position are exceedingly small. The orbits of the planets are ellipses differing but little from circles.

Force of Gravity.

36. The FORCE OF GRAVITY is that force of attraction which the earth exerts upon all bodies, tending to draw them towards its centre.

As has been stated, it is only a particular case of Universal Gravitation. It is, therefore, subject to the same law, that is, it varies directly as the mass of the body acted

Enunciate NEWTON'S law. (**35.**) What is the effect of gravitation on the planets? What are the orbits of planets? (**36.**) What is the Force of Gravity? How does it vary?

upon, and inversely as the square of its distance from the centre of the earth.

The shape of the earth has been shown by careful measurement to be that of a spheroid; that is, of a sphere slightly flattened at the poles. The mean radius is a little less than 4000 miles. On account of the flattening of the earth at the poles, different points are at slightly different distances from the centre, and consequently the force of gravity varies slightly at different places on the surface. For ordinary purposes, however, we may regard the earth as a perfect sphere, and the force of gravity as constant all over its surface.

Vertical and Horizontal Lines.

37. A Vertical Line is a line along which a body falls freely. All vertical lines are directed towards the centre of the earth, but for places near together they may be regarded as parallel.

In Fig. 21, the lines *ao* and *bo* are verticals, but if they are not far apart, their convergence is so small that they may be taken as

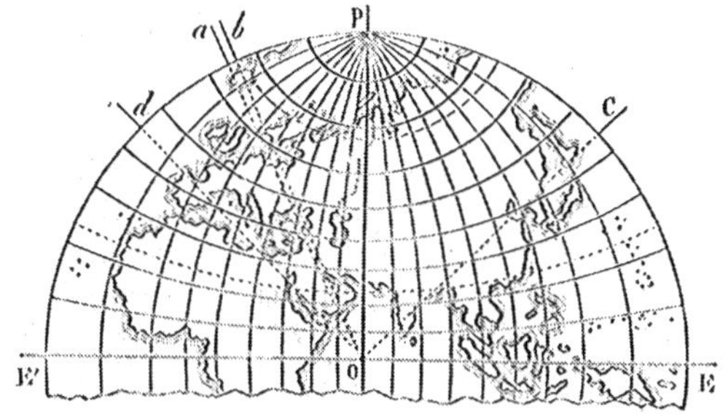

Fig. 21.

parallel. If, however, their distance apart is considerable, they can not be regarded as parallel. A man standing erect has his body in

What is the shape of the earth? (37.) What is a Vertical Line? Where do verticals meet? When may they be considered parallel? *When not parallel? Illustrate.*

a vertical, and it may happen that two persons on opposite sides of the globe, as at E and E', may both stand erect, and yet their heads be turned in exactly opposite directions, their feet being turned towards each other. Points where this may happen are said to be *antipodes*.

A HORIZONTAL LINE, or PLANE, at any place is one which is perpendicular to a vertical line at that place. The surface of still water is horizontal, or *level*. For small areas this surface may be regarded as a plane, but when a large surface is considered, as the ocean, it must be considered as curved, conforming to the general outline of the earth's surface.

Upon the principle of verticals and horizontals, all of our instruments for levelling and making astronomical observations are constructed.

The Plumb-Line.

38. A PLUMB-LINE, is a line having a heavy body, usually of lead, suspended at one of its ends. When the other end is held in the hand, the lead, tending towards the centre of the earth, stretches the string in the direction of the force of gravity.

It is used for indicating a vertical line. In Engineering and Architecture it is of continual use. For determining whether a wall is vertical, it is accompanied by a square plate, whose length is just equal to the diameter of the cylindrical leaden weight, and which has a hole at its middle point, just large enough to admit the passage of the string. The edge of the plate is applied to the masonry, as shown in Fig. 22, and if the *plumb-bob* just touches the wall, it must be vertical.

Weight.

39. The WEIGHT of a body is the pressure which it

What are antipodes? What is a Horizontal Line, or Plane? Level? Applications to instruments. (**38.**) What is a Plumb-Line? Describe it and its use. (**39.**) What is Weight?

exerts upon any body that prevents it from falling towards the earth.

Fig. 22.

The weight of a body is due to the force of gravity, acting upon all of its particles, but it must not be confounded with the force of gravity. Weight is only the effect of gravity when resisted; when gravity is unresisted it produces quite another effect, that is, motion.

At the same place the weights of bodies are proportional to their masses, or the quantities of matter which they contain We shall see hereafter that the weight of bodies may be determined by means of the balance; the force of gravity is determined by the velocity which it can impart to a body in a certain time, as will be shown more fully hereafter.

Centre of Gravity.

40. The Centre of Gravity of a body is that point through which the direction of its weight always passes.

We have seen that the weight of a body is the resultant of the action of gravity upon all of its particles. It is shown

---

Is weight the same as gravity? How is weight determined? How is gravity measured? (**40**.) What is the Centre of Gravity?

in Mechanics, that whatever may be the form of a body, or whatever may be its position, the direction of its weight always passes through a single point. This point is the *centre of gravity*.

The determination of the centre of gravity in the general case requires the aid of mathematics, but in many cases its position is evident. In a uniform straight bar, it is at the middle point. In a square, or a rectangular, or a circular, or an elliptical disk it is at the centre, or middle point.

Equilibrium of heavy Bodies.

41. The centre of gravity being the point at which the weight is applied, it follows that, if this point is held fast by any support whatever, the effect of the weight is completely counteracted, and the body will be in a state of equilibrium.

If a body has but a single point of support, it can be in equilibrium only when its centre of gravity lies somewhere on a vertical through that point. An example is shown in Fig. 23, which represents a boy balancing a cane upon his finger. In the figure, g is the centre of gravity, and that point must be kept exactly over the point of support. This is a case of *unstable equilibrium*.

Fig. 23.

If a body has but two points of support, it can

Where is the centre of gravity of a straight line? Of a square? Rectangle? Circle? Ellipse? (41.) When is a body in equilibrium? When a body rests on a point, where must the centre of gravity be? Example.

be in equilibrium only when its centre of gravity lies in a vertical drawn through some point of the line joining these two points. An example is shown in Fig. 24, which represents a man standing on stilts. To be in equilibrium, his centre of gravity must be exactly over the line joining the feet of his stilts. This is also a case of *unstable equilibrium.*

Fig. 24.

Fig. 25.

The art of balancing, in which circus-riders and rope-dancers are so expert, consists in skillfully keeping the centre of gravity supported.

If a body has three supports not in a straight line, it will be in equilibrium when the centre of gravity lies on a vertical drawn through any point of the triangle formed by joining these points. An example is shown in Fig. 25, which represents a three-legged table. The centre of gravity being at *g*, the table will be in equilibrium so long as the vertical through that point pierces the triangle formed by uniting the feet of the table.

When it rests on two points? Example. When on three points? Example

If a body has four or more supports, the condition of equilibrium will be analogous to that just explained. In this case, if the outer points of support be joined by lines, they will form a polygon, called the *Polygon of Support*, and the body will be in equilibrium when its centre of gravity is on a vertical drawn through any point of this polygon.

Different kinds of Equilibrium.

42. When bodies are acted upon only by the force of gravity, and have one or more points of support, three kinds of equilibrium may exist: *Stable, Unstable, and Neutral Equilibrium.*

1. *Stable Equilibrium.*—A body is in *stable equilibrium*, when, on being slightly disturbed from its state of rest, it tends of itself to return to that state.

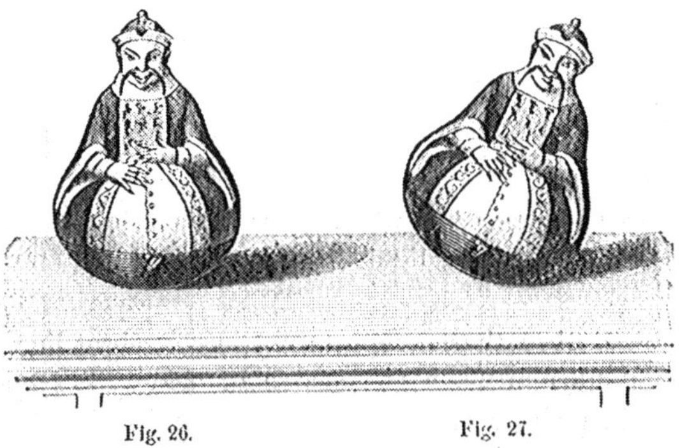

Fig. 26. Fig. 27.

This will be the case when the centre of gravity is lower in its position of rest than it is in any of the neighboring positions, for in this case the weight of the body acting at

When on four or more points? What is the Polygon of Support? (**42.**) What are the three cases of equilibrium? What is Stable Equilibrium? Illustrate.

the centre of gravity tends to keep it in the lowest position. If slightly disturbed from the lowest position, the weight will act to draw it back, and so establish the equilibrium.

We have an example of stable equilibrium represented in Figs. 26 and 27, which represent images often met with in the toy shops. If the image be inclined to one side, as shown in Fig. 27, it will by its own weight right itself, and take the position shown in Fig. 26. These figures are hollow and light, and are ballasted with lead at their lower part so as to throw the centre of gravity very low. The result is, that when the figure is inclined, the centre of gravity is raised, and the weight acts to restore it. The figure settles in its primitive state of rest only after several oscillations, which are due to the inertia of the body. The explanation of this oscillation is the same as that given for the oscillation of the pendulum.

2. *Unstable Equilibrium.*—A body is in *unstable equilibrium*, when, on being slightly disturbed from its state of rest, it does not tend to return to that state, but continues to depart from it more and more.

This will be the case when the centre of gravity is higher in its position of rest than in any of the neighboring positions. When the body is slightly disturbed, the weight acts not only to prevent its return, but also to cause it to descend still lower.

We have examples of unstable equilibrium shown in Figs. 23 and 24. In Fig. 23, the cane may overturn in any direction, whilst in Fig. 24, the man will overturn about the line joining the bottom of his stilts.

3. *Neutral Equilibrium.*—A body is in *neutral equilibrium*, when, on being slightly disturbed, it has no tendency either to return to its primitive state, or to depart further from it.

This will be the case when the centre of gravity is at the

Example. What is Unstable Equilibrium? Illustrate. *Examples.* What is Neutral Equilibrium? Illustrate.

same height in its position of rest as in the neighboring positions.

We have an example of this kind of equilibrium in a ball resting on a horizontal table.

Fig. 28.

Stability of Bodies.

43. From what has been said in the preceding articles, it follows that bodies will in general be most stable when their bases are largest. For in such cases, even after a considerable inclination, the line of direction of the weight will

---

Example. (**43.**) What bodies are most stable?

pass within the original polygon of support, and the weight will act to return the body to its original state of rest. Hence it is that we find chairs, lamps, candlesticks, and many other familiar utensils constructed with broad bases, to render them more stable.

The leaning tower of Pisa is so much inclined that it appears about to fall; yet it stands, because the vertical through the centre of gravity passes within the base of the tower. Fig. 28 represents a tower at Bologna, which is even more inclined than that at Pisa. This tower was built in the year 1112, and received its inclination from unequal settling of the ground on which it was built. It does not fall, because the vertical through the centre of gravity, *G*, passes within its base.

In the cases considered, the position of the centre of gravity remains the same for the same body. With men and animals the position of the centre of gravity changes with every change of attitude, which requires a proper adjustment of the feet, to maintain a position of stability.

Fig. 29. Fig. 30.

When a man carries a burden, as shown in Fig. 29, he leans forward, that the direction of his own weight with that of his burden

Explain the stability of the towers of Pisa and Bologna. How do men and animals maintain a stable position? Illustrate.

3

may pass between his feet. When a man carries a weight in one hand, as shown in Fig. 30, he throws his body toward the opposite side for the same reason.

In the art of rope-dancing, the great difficulty consists in keeping the centre of gravity exactly over the rope. To attain this result the more easily, a rope-dancer carries a long pole, called a balancing pole, and when he feels himself inclining towards one side, he advances his pole towards the other side, so as to bring the common centre of gravity over the rope, thus preserving his equilibrium. The rope-dancer is in a continual state of unstable equilibrium.

The Balance.

44. A BALANCE is a machine for weighing bodies.

Balances are of continual use in commerce and the arts, in the laboratory, and in physical researches; they are consequently extremely various in their forms and modes of construction. We shall only describe that form which is in most common use in the shops.

It consists of a metallic bar, *AB* (Fig. 31), called the *Beam*, which is simply a lever of the first order. At its middle point is a knife-edged axis, *n*, called the *Fulcrum*. The fulcrum projects from the sides of the beam, and rests on two supports at the top of a firm and inflexible standard. The knife-edged axis, and the supports on which it rests, are both of hardened steel, and nicely polished, in order to make the friction as small as possible. At the extremities of the beam are suspended two plates or basins, called *Scale Pans*, in one of which is placed the body to be weighed, and in the other the weights of iron or brass to counterpoise it. Finally, a needle projecting from the beam, and playing in front of a graduated scale, *a*, serves to show when the beam is exactly horizontal.

Explain the principle of rope-dancing. (44.) What is a balance? Explain the details of the common Balance. The Beam. The Fulcrum. The Scale Pans. The Scale.

To weigh a body, we place it in one of the scale pans, and then put weights into the other pan until the beam

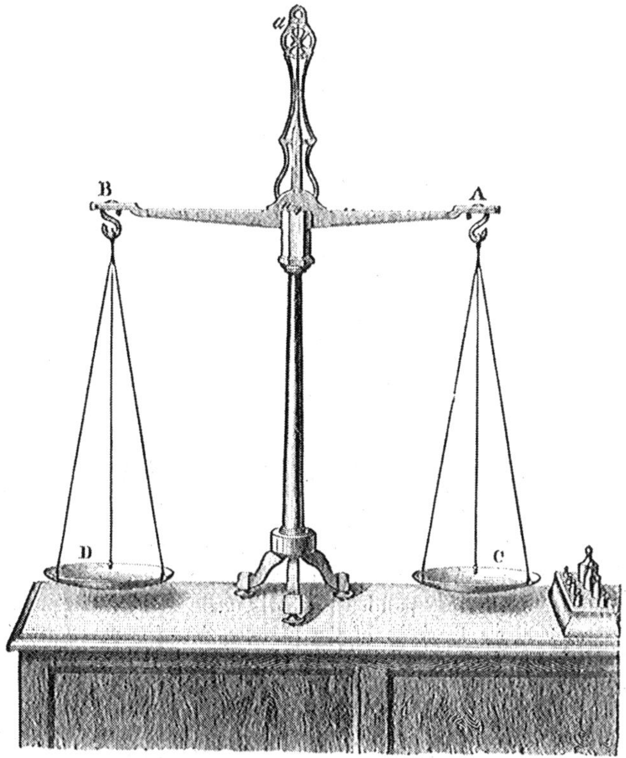

Fig. 31.

becomes horizontal. The weights put in the second pan indicate the weight of the body.

How are bodies weighed?

Requisites for a good Balance.

45. A good balance ought to satisfy the following conditions:

1. The lever arms, *An* and *Bn*, should be exactly equal.

We have seen in discussing the lever, that its arms must be equal, in order that there may be an equilibrium between the power and resistance, when these are equal. If the arms are not equal, the weights placed in one scale pan will not indicate the exact weight of the body placed in the other.

2. The balance should be *sensitive;* that is, it should turn on a very small difference of weights in the two scale pans.

This requires the fulcrum and its supports to be very hard and smooth, so as to produce little friction. By making the needle long, a slight variation from the horizontal will be more readily perceived.

3. The centre of gravity of the beam and scale pans should be slightly below the edge of the fulcrum.

If it were in the edge of the fulcrum, the beam would not come to a horizontal position when the scales were equally loaded, but would remain in any position where it might chance to be placed. If it were above the edge of the fulcrum, the beam would remain horizontal if placed so, but if slightly deflected, it would tend to overturn by the action of the weight of the beam.

The nearer the centre of gravity comes to the edge of the fulcrum, the more accurate it will be; but at the same time, it would turn more slowly, and might finally come to turn too slowly to be of use for weighing.

It is to be observed that when the scale pans are heavily loaded, an increased weight is thrown on the fulcrum, which

(**45.**) Explain the requisites of a good balance. 1. Lever arms. *Illustrate.* 2. Sensitiveness. *Illustrate.* 3. Position of centre of gravity. *Illustrate.*

causes an increase of friction, and consequently a diminution of sensitiveness.

Methods of Testing a Balance.

46. To see whether the arms are of equal length, let a body be placed in one scale pan, and counterbalanced by weights put in the other; then change places with the body and the weights. If the beam remains horizontal after this change, the arms are of equal length, otherwise the balance is *false*.

To test the sensitiveness, load the balance and bring the beam to a horizontal position, then deflect it slightly by a small force and see whether it returns slowly to its former position. It ought to come to a state of rest by a succession of oscillations.

Method of weighing correctly with a false Balance.

47. To weigh a body with a false balance, place it in one scale pan and counterbalance it by any heavy matter, as shot or sand, placed in the other pan. Then take out the body and replace it by weights which will exactly restore the equilibrium of the balance. The weights will be exactly equal to the weight of the body. The reason for this method is apparent.

Laws of falling bodies.

48. When bodies starting from a state of rest fall freely in vacuum, that is, without experiencing any resistance, they conform to the following laws:

1. *All bodies fall equally fast.*

(**46.**) How is a balance to be tested? (**47.**) How may a body be weighed correctly by a false balance? (**48.**) What is the first law of falling bodies?

When resisted by the air, bodies whose bulk is very large in proportion to their weight, fall more slowly than those whose bulk is small; thus, a soap-bubble falls more slowly than a bullet.

2. *The velocities acquired during the fall are proportional to the times occupied in falling.*

A body acquires a velocity of $32\frac{1}{6}$ feet in one second; it will therefore acquire a velocity of $64\frac{1}{3}$ feet in two seconds, a velocity of $96\frac{1}{2}$ feet in three seconds, and so on.

3. *The spaces passed over are proportional to the squares of the times occupied in falling.*

A body falls from rest through $16\frac{1}{12}$ feet in one second; it will therefore fall $4\times16\frac{1}{12}$, or $64\frac{1}{3}$, in two seconds, $9\times16\frac{1}{12}$, or $144\frac{3}{4}$ feet, in three seconds, $16\times16\frac{1}{12}$, or $257\frac{1}{3}$ feet in four seconds, and so on.

Fig. 32.

The first law is verified by the following experiment. A glass tube, six feet long (Fig. 32), is closed at one end, and at the other it has a stop-cock, by which it can be closed or opened at pleasure. A small leaden ball and a feather are introduced within the tube. So long as the tube is full of air, if it be suddenly inverted, it will be observed that the ball reaches the bottom sooner than the feather. If now the air be exhausted by means of an air-

Effect of atmospheric resistance. What is the second law? *Illustrate.* Third law? *Illustrate.* How is the first law verified?

pump, and the tube suddenly inverted, both the ball and the feather will be seen to fall through the length of the tube in the same time. This experiment, besides verifying the law, shows also that the air offers a resistance, which is greater for light than for heavy bodies. This resistance is proportional to the surface offered to the direction of the fall.

The second law is a consequence of inertia and the continued action of gravity. The velocity generated in the first second is to be added to that generated in the next second, to obtain the velocity generated in two seconds. This must be twice that generated in the first second. This again must be added to that generated in the third second, to obtain that generated in three seconds. This then must be three times that generated in the first second, and so on.

The explanation of the third law will be better understood after having considered the nature of the inclined plane, which is discussed in the succeeding articles.

The Inclined Plane.

49. An INCLINED PLANE is a plane which is inclined to a horizontal plane; thus, *AB*, Fig. 33, is an inclined plane.

When a body rests on a horizontal plane, as for example on a table, the action of gravity tending to draw it down is completely counteracted by the resistance of the plane, and it remains at rest. It is not so, however, when a body is placed upon an inclined plane. In this case, the action of gravity may be resolved into two components, one perpendicular to the plane, and the other parallel to it. The action of the first component is counteracted by the resistance of the plane, whilst the second component causes

What other principle does the experiment show? Explain the reason of the second law. (**49.**) What is an Inclined Plane? Explain its principle.

the body to move down the plane. Now, this last force is only a fraction of the weight of the body, as a fourth, a fifth, or a sixth, according to the inclination, but it obeys the same laws that the entire force would, in causing a body to fall.

Verification of the third Law of falling Bodies.

50. To verify the third law of falling bodies, we construct a plane with a slight inclination and divide it into 100 equal parts, as shown in Fig. 33. We then ascertain by successive trials at what division of the scale a leaden

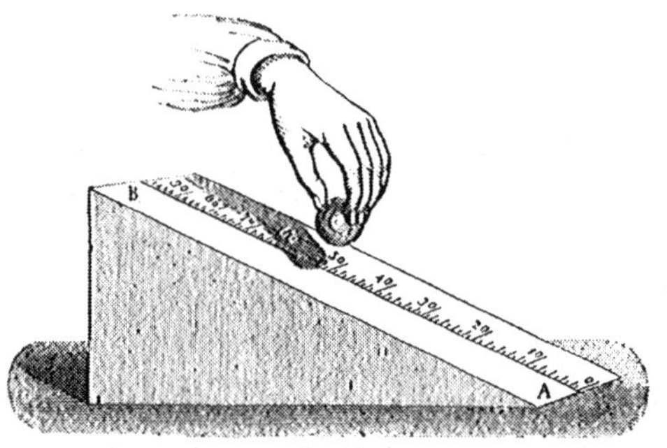

Fig. 33.

ball must be placed to roll to the bottom *A*, in one second; suppose at the sixth division. If now the ball be placed at the twenty-fourth division, it will roll to the bottom in two seconds; if placed at the fifty-fourth division, it will roll down in three seconds; if placed at the ninety-sixth division, it will roll down in four seconds, and so on.

Hence, we conclude that, *the spaces passed over are proportional to the squares of the times.*

(**50.**) How is the third law of falling bodies verified?

Applications of the Inclined Plane.

51. When a body is placed upon an inclined plane, that component of its weight which acts to move it down the plane, becomes smaller as its inclination diminishes. Hence, the force required to draw a body up an inclined plane, will become smaller as the inclination diminishes. This principle is often utilized in the Arts; thus, to raise a heavy body to a height, we construct an inclined plane, up which it may be easily drawn.

It is in accordance with this principle that roads are constructed to ascend high hills and mountains, as shown in Fig. 34. Such a road consists of a succession of planes

Fig. 34.

lying in different directions, which may be equally or unequally inclined to the horizon.

(**51.**) What is the use of the inclined plane in the Arts? Explain its application to roads.

It is according to the principle of the inclined plane that water flows along rivers and canals. The steeper the inclined planes which form their beds, the more rapid their currents.

In mechanics, two inclined planes, wound about a cylinder, constitute the screw; hence the principle of the screw is but a modification of that of the inclined plane. The wedge is made up of two inclined planes, placed back to back; hence its principle is also but a modification of that of the inclined plane.

The Pendulum.

52. A PENDULUM is a heavy body suspended from a horizontal axis about which it is free to vibrate. Thus, the ball *m*, suspended from *C*, by a string, Figs. 35 and 36, is a pendulum.

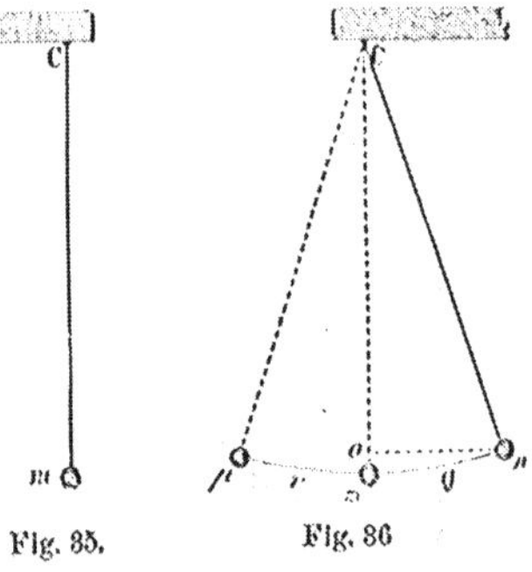

Fig. 35. Fig. 36

When the centre of the ball, *m*, is exactly below the point of suspension *C*, Fig. 35, it is in equilibrium, for in that position the action of gravity is resisted by the tension of the string. If, however, the ball be drawn aside to *n*, Fig. 36, it is no longer in equilibrium, for in that position the force of gravity acts to draw it back to *m*, at which point it will arrive with the same velocity as though it had fallen through the vertical height *om*. In consequence of its inertia and acquired velocity, the ball does not stop at *m*, but moves on towards *p*. In descending from *n* to *m*, the force of gravity acts as an accelerating

Explain the flow of rivers. What is the screw? What is the wedge? On what principle do they act? (**52.**) What is a Pendulum? *What causes the pendulum to vibrate? Explain the action in detail.*

force, but in ascending from m to p, it acts as a retarding force, hence the ball moves slower and slower till it reaches p. The distance mp would be rigorously equal to mn, were it not for the resistance of the air.

The ball, having reached p, is in the same state as it was at n; the weight again acts to draw it back to m, whence, by virtue of its inertia and velocity, it again rises to n, and so on indefinitely.

This backward and forward motion is called *Oscillatory Motion*. A single excursion from n to p, or from p to n, is called a *Simple Oscillation*, or *Vibration*. An excursion from n to p, and back again to n, is called a *Double Oscillation*. The angle, pCn, is called the angle of the *Amplitude* of the oscillation.

In consequence of the resistance of the air, the amplitude is continually diminishing, and the ball eventually comes to rest, though often not till after the lapse of some hours.

Simple and Compound Pendulums.

53. A Simple Pendulum is such a pendulum as would be formed by suspending a single material point, by a string destitute of weight.

Such a pendulum may exist in theory, and is thus useful in arriving at the laws of oscillation, but in practice it can only be approximated to by making the ball very small and the string very fine.

A Compound Pendulum is any heavy body which is free to oscillate about a horizontal axis.

It may be of any form, but in general it consists of a stem T, Fig. 38, which is either of wood or metal. The stem terminates above in a thin and flexible plate, a, usually of steel; it terminates below in a disk of metal L, called the *bob*, which disk is of a lenticular shape, that the resistance of the air to its motion may be as little as possible.

What is Oscillatory Motion? What is an Oscillation or Vibration? What is its Amplitude? What effect has the air on vibration? (**53.**) What is a Simple Pendulum? *Is it real or ideal?* What is a Compound Pendulum? *Explain its construction.*

Laws of Oscillation of the Pendulum.

54. The oscillations of the pendulum take place in accordance with the following laws:

1. *For pendulums of unequal lengths, the times of oscillation are proportional to the square roots of their lengths.*

2. *For the same pendulum, the time of oscillation is independent of the amplitude, provided the amplitude be small.*

3. *For pendulums of the same length, the time of oscillation is independent of the nature of the material.*

Pendulums of wood, iron, copper, glass, all being of the same length, will all oscillate in the same time.

4. *For the same pendulum at different places, the times of oscillation are inversely as the square roots of the force of gravity at those places.*

These laws are deduced from a course of mathematical reasoning on the theoretical simple pendulum, but they may be verified experimentally by employing a very small ball of platinum, or other heavy metal, and suspending it with a very fine silk thread.

To verify the first law with such a pendulum, we begin by making it vibrate, and then counting the number of vibrations in one minute. Suppose, for example, that it makes seventy-two per minute. Now make the string four times as long as before, and it will be found that the pendulum makes only thirty-six oscillations per minute. If the string is made nine times as long as in the first instance, it will be found that the pendulum makes only twenty-four oscillations per minute, and so on. In the second case the time of oscillation is twice as great, and in the third case it is three times as great as in the first case. Now, because two, three, &c., are the square roots of four, nine, &c., it follows that the law is verified.

To verify the second law, let the same pendulum oscillate, at first

---

(**54.**) What is the first law of vibration? The second law? The third law? *Illustrate.* The fourth law? *How are these laws deduced? How is the first law verified? How is the second law verified?*

through an arc, *pn*, and then through any other arc, *rg*; it will be found that the number of oscillations per minute is the same in each case. Hence the law is verified. It is to be observed that the law does not hold true unless the arcs, *pn* and *rg*, are very small, that is, not more than three or four degrees.

The property of pendulums, that their times of oscillation are independent of the amplitude of vibration, is designated by the name *isochronism*, from two Greek words signifying *equal times*; oscillations performed in equal times are called *isochronal*.

GALILEO first discovered the fact that small oscillations of a pendulum were isochronal, towards the end of the sixteenth century. It is stated that he was led to the discovery by noticing the oscillations of a chandelier suspended from the ceiling of the Cathedral of Pisa.

Applications of the Pendulum.

55. On account of the isochronism of its vibrations, the pendulum has been applied to regulate the motion of clocks. It was first used for this purpose in 1657, by HUYGHENS, a Dutch philosopher. The motive power of a clock is sometimes a weight acting by a cord wound around a drum, and sometimes a coiled spring similar to a watch spring. These motors act to set a train of wheel-work in motion, which in turn imparts motion to the hands that move round the dial to point out the hour. It is to impart uniformity of motion to this train of wheel-work that the pendulum is used.

Fig. 38 shows the mechanism by means of which the pendulum acts as a regulator. A toothed wheel, *R*, called a scape wheel, is connected with the train driven by the motor, and this scape wheel is checked by an anchor, *mn*, which is attached to the pendulum and vibrates with it. The anchor has two projecting points, *m* and *n*, called pallets, which engage alternately with the teeth of the scape wheel, in such a manner that only one tooth can pass at each swing

Limitation. What is isochronism? When are vibrations isochronal? Who discovered the pendulum, and when? (55.) What is the principal use of the pendulum? What is the motor in a clock? What is the use of the pendulum? *Explain the action of the pendulum as a regulator.*

of the pendulum. The motor turns the scape wheel in the direction of the arrow until one of the teeth comes in contact with the pallet, *m*, which stops the motion of the wheel-work till a swing of the pendulum lifts the pallet, *m*, from between the two teeth, when a single tooth passes and the wheel-work moves on until again arrested by the pallet, *n*, falling between two teeth on the other side. A second swing of the pendulum lifts out the pallet, *n*, suffers another tooth to pass, when the wheel-work is again arrested by the pallet, *m*, and so on indefinitely. The beats of the pendulum being isochronous, the interval of time between the consecutive escape of two teeth is always constant, and thus the motion of the wheel-work is kept uniform. The loss of force which the pendulum continually experiences, is supplied by the motor through the scape wheel and the anchor. This is called the sustaining power of the pendulum.

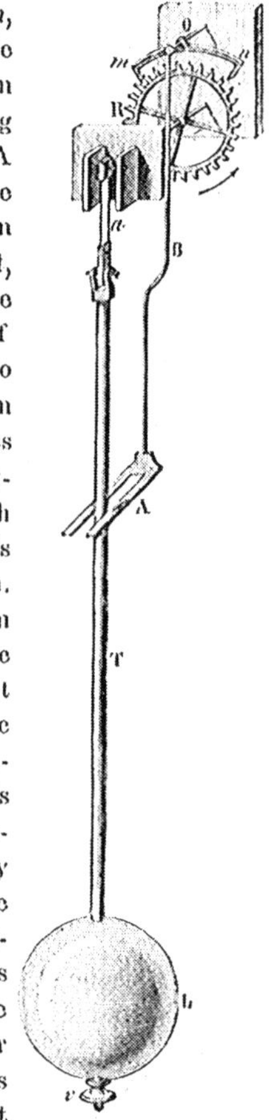

Fig. 38.

Owing to expansion and contraction from variations of temperature, the length of the pendulum varies, and according to the first law, its time of vibration changes. In nice clocks this change is compensated by a combination of metals. In common clocks, it is rectified by lengthening or shortening the pendulum by a nut and screw, shown at *v*, by means of which the lenticular bob may be moved up and down. In summer the pendulum elongates and the clock loses time, or runs too slow; this is rectified by screwing up the nut and shortening the pendulum. In winter the pendulum contracts and the clock gains time; this is rectified by unscrewing the nut and lengthening the pendulum.

What effect have variations of temperature on the pendulum? How are these effects compensated in nice clocks? How in common clocks? Why do clocks lose time in summer and gain time in winter?

In accordance with the principle enunciated in the fourth law, the pendulum has been used to determine the intensity of gravity at different points on the earth's surface. In this way it has been shown that the velocity acquired by a body falling in vacuum for one second, is $32\frac{1}{6}$ feet, in the latitude of the city of New York. It has been found by careful experiment that the length of a pendulum vibrating seconds in New York, is a little over 39 inches.

The length of the seconds pendulum at any place being constant, it has been taken as the basis of the English system of weights and measures, and from the English we have taken our own system.

The pendulum has been successfully employed by M. FOUCAULT, a French physicist of our own day, to demonstrate the daily rotation of our globe. The details of his experiment are too abstruse to be given in this place.

The Metronome.

56. The METRONOME is a sort of pendulum employed by musicians and others to mark equal intervals of time. It is shown in Fig. 39. It consists of a pen-

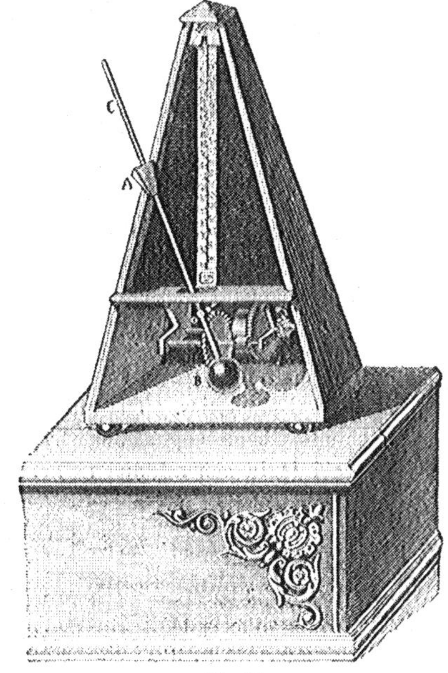

Fig. 39.

What principle enables us to measure the force of gravity? How far does a body fall in one second? What is the length of a seconds pendulum in New York? Application to weights and measures? What application did FOUCAULT make of the pendulum? (**56.**) What is a Metronome?

dulum *CB*, suspended at *O*. A weight, *A*, slides along the rod *C*, and may be set so as to make the vibrations as slow or as rapid as may be desired. The instrument is set by means of a scale, marked on the rod, so that any number of oscillations may be made in a minute. The pendulum is sustained by a coiled spring which sets in motion a train of wheels, somewhat in the manner of a clock. In the drawing the weight is set at 92, which shows that it is to make 92 oscillations per minute.

IV.—PRINCIPLES DEPENDENT ON MOLECULAR ACTION.

Molecular Forces.

57. Besides the forces which act upon bodies from without and at sensible distances, there is another class of forces continually exerted between the molecules of bodies, and acting only at insensible distances. These forces are called *Molecular Forces*, and are both *attractive* and *repellent.*

The molecules of bodies are held in equilibrium by these forces, and it is to them that are to be attributed many of the most important physical properties. The ultimate particles of bodies do not touch each other, being kept asunder by a force of repulsion, which we have said is in general due to heat; they are prevented from receding from each other too far by a force of attraction, and it is only when these forces just balance each other throughout the body, that it is in equilibrium.

When a body is compressed, the forces of repulsion are called into play, and, acting like coiled springs, they tend to restore the body to its primitive form. In like manner, when a body is elongated, or stretched, the forces of attraction are called into action and tend to restore the body to its primitive form.

Describe it. (**57.**) What are Molecular Forces? How divided? *How are molecules held in place? To what is the repellent force due? Explain the effects of compressing and stretching bodies.*

Cohesion.

58. COHESION is the force of attraction which holds the molecules of the same body together, as, for example, in a mass of iron, or of wood.

Cohesion differs from chemical affinity, which determines the molecules by uniting dissimilar atoms according to fixed laws. Chemical affinity unites atoms of carbon, oxygen, and hydrogen, to form molecules of sugar; but it is cohesion that unites the molecules of sugar into a solid body.

The strength of bodies depends upon *cohesion*. When a body offers a strong resistance to forces tending to tear it asunder, it is said to be tenacious; for example, iron or steel wires, and the like, are highly tenacious.

Adhesion.

59. ADHESION is the force of attraction which holds the molecules of dissimilar bodies together. Thus, it is adhesion which causes paint and glue to adhere to wood.

If two polished bodies are brought into contact, and pressed together, they will adhere with considerable force. If two plates of glass be ground so as to fit closely, and a little oil be interposed, it is very difficult to separate them. If two hemispheres of lead be pressed together, after having their plane surfaces well polished, they will adhere very strongly.

It is adhesion which renders it difficult to raise a wooden board from the surface of the water on which it floats. It is also adhesion between the particles of wood and water, that causes water to spread over a piece of wood upon which it is poured.

Solution is due to adhesion. Thus, when sugar dissolves in water, it is because the adhesion between the molecules of sugar

---

(**58.**) What is Cohesion? Example. *Difference between cohesion and chemical affinity? Illustrate.* When is a body tenacious? (**59.**) What is Adhesion? Example. *Explain adhesion of metallic surfaces. Of board to water. Explain the phenomenon of solution.*

and water is stronger than the cohesion between the molecules of sugar. If a liquid tends to spread itself over a solid body, it is said to wet it, as water upon glass. If it gathers in globules, it does not wet it, as quicksilver upon glass.

Capillary Forces.

60. Capillary Forces are molecular forces, exerted between the particles of a solid and those of a liquid. They are called capillary, because their effect is mostly observed in capillary tubes, that is, tubes of the diameter of a hair.

The following are some of the phenomena of capillarity:

1. When a body is plunged into a liquid which is capable of wetting it, as when a glass rod is plunged into water, it is observed that the liquid is slightly elevated about the body, taking a concave form, as shown in Fig. 40.

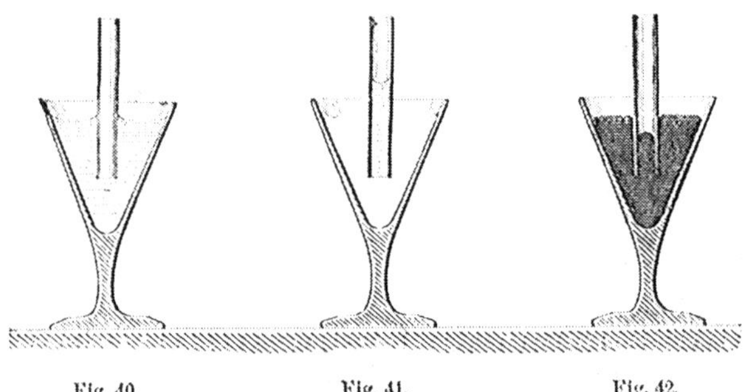

Fig. 40. Fig. 41. Fig. 42.

2. If a hollow tube is used instead of a rod, the liquid will also rise in the tube, as shown in Fig. 41. The smaller the bore of the tube, the higher will the liquid rise, and the more concave will be its upper surface.

(60.) What are Capillary Forces? Why so called? Explain the phenomenon observed when a glass rod is plunged in water. When a tube is plunged into water.

3. When a tube is plunged into a liquid which is not capable of wetting it, as when glass is plunged into quicksilver, the liquid is depressed both on the outside and on the inside, taking a convex surface, as shown in Fig. 42. The smaller the tube, the greater will be the depression, and the more convex will be the upper surface.

These capillary phenomena are due to the resultant action of the cohesion of the liquid and the adhesion of the solid and liquid. When the former predominates, the liquid is depressed in the tube. When the latter predominates, the liquid is raised in the tube.

Applications of Capillarity.

61. It is in consequence of capillary action that oil is raised through the wicks of lamps, to supply the flame with combustible matter. The fibres of the wicks leave between them a species of capillary tubes, through which the oil rises.

If a piece of sugar have its lower end dipped in water, the water will rise through the capillary interstices of the sugar and fill them. This drives out the air and renders the sugar more soluble than when plunged dry into water, in which case the contained air resists the absorption of water, and retards solution.

If a bar of lead be bent into the form of a siphon, and the short arm be dipped into a vessel of mercury, the mercury will rise into the lead by capillary action, and flowing over the edge of the vessel will descend along the longer branch and escape from the lower extremity. In this way the vessel may be slowly emptied of the quicksilver.

Many fluids may be drawn over the edges of the containing vessels by a siphon of candle-wicking or other capillary substance.

When a glass tube is plunged into mercury. Causes of the phenomena. (61.) Why does oil rise in a wick? Water in sugar? Explain leaden siphon. Explain siphon of wicking.

Absorption.

62. ABSORPTION is the penetration into a porous body, of any foreign body, whether solid, liquid, or gaseous.

Carbon, in the form of charcoal, has a great capacity for absorbing gases. If a burning coal be introduced into a bell-glass, filled with carbonic acid, collected over mercury, the volume of the gas is diminished by being absorbed by the coal. It is found that the charcoal absorbs in this way thirty-five times its own volume of the gas. Charcoal also absorbs other gases in even still greater quantities.

Spongy platinum absorbs hydrogen so rapidly as to heat the platinum red-hot.

In vegetables and animals we have many examples of absorption. The roots of plants absorb from the earth the material necessary to the growth of the stem and branches.

In the animal world, absorption plays an important part in the process of nutrition and growth. Animal tissues also absorb solid substances. For example, workmen engaged in handling lead absorb through the skin and lungs more or less of this substance, which often gives rise to very serious diseases.

Imbibition.

63. IMBIBITION is the absorption of a liquid by a solid body.

Imbibition is an effect of capillarity, for the interstices between the molecules, by communicating with each other, form a mass of capillary tubes, into which the liquid penetrates by virtue of the capillary forces. Such is the cause of wood and earth absorbing water and other liquids. If a damp substance be placed in a dry and porous vessel, it

---

(**62.**) What is Absorption? Examples. Carbon. Spongy platinum. Vegetables. Animals. (**63.**) What is Imbibition? What is the cause of imbibition? Examples.

will grow drier, whereas, if placed in a vessel which has no attraction for water, it will remain moist.

When vegetable and animal substances absorb water, they generally augment in volume. This fact explains many phenomena of daily observation.

If a large sheet of paper be moistened, it increases in size, and again contracts when dried. This property is employed by draughtsmen to stretch paper on boards. The paper is moistened, and after being allowed to expand, its edges are glued to a drawing-board; on drying it is stretched, forming a smooth surface for drawing upon. The same property causes the paper to peel from the walls of a room when exposed to moisture.

When a workman would bend a piece of wood, he dries one side and moistens the other. The side which is dried contracts, and the opposite side expands, so that the piece is curved. It is the absorption of moisture that causes the wood-work of houses, furniture, &c., to swell and shrink with atmospheric changes, and which necessitates their being painted and varnished. Paints and varnishes, by filling the pores, prevent absorption.

If two different liquids be separated by a membranous partition, a current will be set up from each liquid to the other through the membrane, and after a time it will be found that there is a mixture of both liquids on each side of the partition. These currents are generally unequal, so that there is an actual gain of substance on one side and a corresponding loss on the other. The current that acts to produce an increase on one side is called *endosmose*, and the opposite current is called *exosmose*. Thus, if a bladder filled with strong syrup be tied to the end of a glass tube, and the whole plunged into a vessel of water, the syrup soon becomes diluted by the flowing in of water, and the mixture rises in the tube; at the same time a portion of the syrup flows out and mixes with the water. The flowing in of the water is *endosmose*, and the flowing out of the syrup is *exosmose*. Similar results are obtained by using other liquids. The phenomena of endosmose and exosmose enable us to explain many interesting facts in animal and vegetable physiology.

What is the effect of imbibition? On paper? Application. Effect on wood? Application. What are endosmose and exosmose? Illustrate.

V.—PROPERTIES OF SOLIDS DEPENDENT ON MOLECULAR ACTION.

Tenacity.

64. TENACITY is the resistance which a body offers to rupture when subjected to a force of traction; that is, a force which tends to tear the particles asunder.

The tenacity of a body may be determined in pounds. For this purpose it is wrought into a cylindrical form, having a given cross-section; its upper end is then made fast, and a scale-pan is attached to the lower end; weights are then placed in the pan until rupture takes place. These weights measure the tenacity of the body.

Metals are the most tenacious of bodies, but they differ greatly from each other in this respect. The following table exhibits the weights required to break wires of $\frac{656}{10000}$ of an inch in diameter, formed of the metals indicated:

Iron	549 lb.
Copper	302 "
Platinum	274 "
Silver	187 "
Gold	150 "
Lead	27 "

It has been shown by theory and confirmed by experiment, that of two cylinders of equal length and containing the same amount of material, one being solid and the other hollow, the latter is the stronger.

This latter principle is also true of cylinders required to support weights; the hollow cylinder is better adapted to resist a crushing force than the solid one of the same weight, and hence it is that columns and pillars for the support of buildings are made hollow. This principle also indicates that the bones and quills of birds, the stems of grasses and other plants, being hollow, are best adapted to secure a combination of lightness and strength.

---

(64) What is Tenacity? *How is it measured? What bodies are most tenacious? Examples. What is the form of greatest strength? Application to grasses, quills, bones, &c.*

Hardness.

65. Hardness is the resistance which a body offers to being scratched or worn by another. Thus, the diamond scratches all other bodies, and is therefore harder than any of them.

After the diamond come the sapphire, the ruby, rock-crystal, &c., each of which is scratched by the preceding one, but scratches the succeeding one.

Hardness must not be confounded with resistance to shocks or compression. Glass, diamond, and rock-crystal are much harder than iron, brass, and the like, and yet they are less capable of resisting shocks and forces of compression; they are more *brittle.*

An alloy or mixture of metals is generally harder than the separate metals of which it is composed. Thus, gold and silver are soft metals, and, in order to make them hard enough for coins and jewelry, they are alloyed with a small portion of copper. In order to render block-tin hard enough for the manufacture of domestic utensils, it is alloyed with a small quantity of lead.

The property of hardness is utilized in the arts. To polish bodies, powders of emery, tripoli, &c., are used, which are powders of very hard minerals. Diamond being the hardest of all bodies, it can be polished only by means of its own powder. Diamond-dust is the most efficient of the polishing substances.

Ductility.

66. Ductility is the property of being drawn out into wires by forces of extension.

Wax, clay, and the like, are so tenacious, that they can easily be flattened by forces of compression, and readily wrought between the fingers. Such bodies are plastic. Glass, resins, and the like become tenacious only when heated. Glass at high temperatures is

---

(65.) What is Hardness? What body is hardest? What bodies come next? *What are brittle bodies? What is the effect of alloying bodies? Explain the operation of polishing? How is the diamond polished? What is the best polishing substance?* (66.) What is Ductility? *Give examples of plastic bodies?*

so highly ductile, that it may be spun into fine threads and woven into fabrics. Many of the metals, as iron, gold, silver, and copper, are ductile at ordinary temperatures, and are capable of being drawn out into fine wires, by means of wire-drawing machines.

The following metals are arranged in the order of their ductility: *platinum, silver, iron, copper, gold, zinc, tin, lead.*

Malleability.

67. MALLEABILITY is the property of being flattened or rolled out into sheets, by forces of compression.

This property often augments with the temperature; every one knows that iron is more easily forged when hot than when cold. Gold is highly malleable at ordinary temperatures. Gold is reduced to thin sheets by being rolled out into plates by a machine; these plates are cut up into small squares, and again extended by hammering until they become extremely thin. They are then cut up again into squares, and hammered between membranes, called gold-beater's skins. By this process gold may be wrought into leaves so thin, that it would take 282,000, placed one upon another, to make an inch in thickness. These leaves are employed in gilding metals, woods, paper, and the like. Silver and copper are wrought in the same manner as gold.

The following metals are amongst the most malleable under the hammer: *gold, silver, platinum, iron, tin, zinc, copper, lead.*

When metals are alloyed, they are generally harder and less malleable, as well as less ductile.

Is gold ductile? When? Give examples of ductile metals. (**67.**) What is Malleability? *Effect of temperature? How is gold formed into sheets? What is the order of malleability of metals? Effect of alloying.*

CHAPTER II.

MECHANICS OF LIQUIDS.

I.—GENERAL PRINCIPLES.

Definition of Hydrostatics and Hydrodynamics.

68. THE Mechanics of Liquids is divided into two branches: HYDROSTATICS, which treats of the laws of equilibrium of liquids, and HYDRODYNAMICS, which treats of the laws of motion of liquids.

Properties of Liquids.

69. The following properties are common to all liquids:

1. The molecules of liquids are extremely movable, yielding to the slightest force.

There is very little cohesion between the molecules of liquids, whence their readiness to slide amongst each other. It is to this principle that they owe their fluidity.

2. Liquids are only slightly compressible.

Liquids are so slightly compressible, that for a long time they were regarded as absolutely incompressible. In 1823, ŒRSTED demonstrated, by an apparatus which he contrived, that liquids are slightly compressible. He showed that for a pressure of one atmosphere, that is, of 15 lbs. on each square inch of surface, water is compressed the $\frac{40}{1000000}$th of its original volume. Slight as is

(**68.**) Define Hydrostatics. Hydrodynamics. (**69.**) What is the first property of Liquids. *Illustrate.* Second property? *Illustrate.*

the compressibility of water, it is nevertheless ten times as compressible as mercury.

3. Liquids are porous, elastic, and impenetrable, like other bodies.

That liquids are porous, has already been shown (Art. 9). That they are elastic, is shown by their recovering their volume after the compressing force is removed. It is also shown by the fact that they transmit sound. Their impenetrability is shown by plunging a solid body into a vessel filled with a liquid. If there is no imbibition, a volume of water will flow over the vessel just equal to that of the solid introduced.

Upon these three properties of liquids depends their property of transmitting pressures in all directions.

Transmission of Pressures.—Principle of Pascal.

70. Let a bottle be filled with water and corked, as represented in Fig. 43. If the cork be pressed inwards, the pressure will be transmitted to the molecules in contact with it; these molecules will in their turn press upon the neighboring ones, and so on until the pressure is finally transmitted to every point of the interior surface of the bottle.

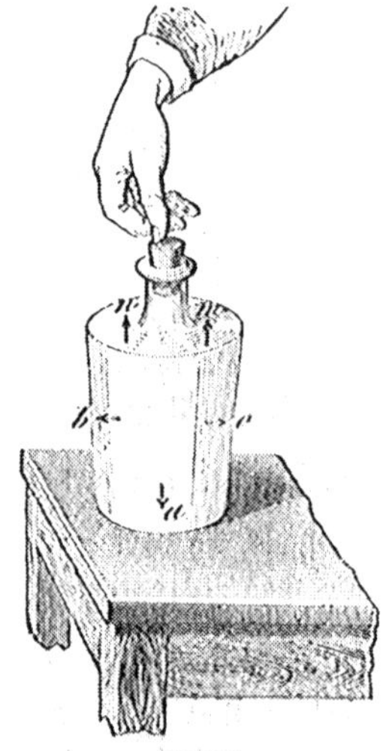
Fig. 43.

It is shown by experiment that the pressure thus transmitted is equal to that applied to the cork; that is, the pressure upon each square inch of

Third property? *Illustrate.* (**70.**) What is the Principle of Pascal?

the interior surface of the vessel is equal to that upon a square inch of the cork. The pressure is everywhere perpendicular to the surface, as shown by the arrow-heads.

This principle is called the *Principle of Pascal*, because it was first demonstrated by BLAISE PASCAL in the seventeenth century. Upon it depends the whole theory of Hydrostatics.

Pressure due to the Weight of Liquids.

71. If a cylindrical vessel is filled with a heavy liquid, its weight produces a pressure upon the walls of the vessel. If we suppose the liquid divided into horizontal layers of equal thickness, it is plain that the second layer from the top supports a pressure equal to the weight of the first, the third layer supports a pressure equal to the weight of the second and first, and so on to the bottom. Hence, *the pressure upon any layer is proportional to its depth below the upper surface, and is equal to the weight of the column of fluid above it.*

In consequence of the principle of PASCAL, this pressure is transmitted laterally, and acts against the sides of the vessel with an equal intensity. Hence, *every part of the surface is pressed with a force equal to the weight of a column of liquid whose base is the surface pressed, and whose height is equal to the distance from that surface to the upper level of the fluid.*

The same principle holds, whatever may be the form of the vessel.

Why so called? How illustrated? (**71.**) What is the measure of the pressure on any horizontal layer of a liquid? How shown? How is it transmitted? What pressure is exerted on the surface of a containing vessel?

Lateral Pressures.—Reaction Wheel.

72. The fact that liquids exert lateral pressures upon the walls of vessels, is demonstrated by means of the reaction wheel. This wheel is shown in Fig. 44; it consists of a vertical cylindrical tube *C*, turning freely in a ring, *n*, near its upper extremity, and resting upon a pivot at its lower extremity. Just above the pivot, the tube terminates in a cubical box, from the faces of which project four tubes, having their ends curved, as shown in the figure. Water is supplied from a cistern through the funnel *D*. When the water is admitted, it flows down the tube *C*, and escaping through the curved tubes at the bottom, the wheel is turned in the direction indicated by the arrow-head.

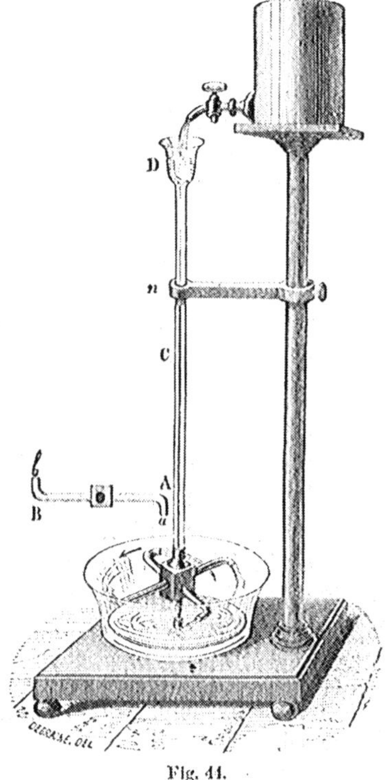

Fig. 44.

The reason of this will be plain from a consideration of the small figure *ab*, which is a plan of two of the tubes. The weight of the water causes a pressure upon *A*, which, were

(**72.**) How is the lateral pressure demonstrated? Describe the reaction wheel. *Explain its action.*

a closed, would be exactly counterbalanced by the pressure upon it; but *a* being open, the pressure upon *A* is not counterbalanced, but acts from *a* towards *A*, producing rotary motion. The pressures in all of the tubes conspire to produce rotation in the same direction.

Pressure upwards.

73. That liquids exert a pressure upwards is demonstrated by means of the apparatus shown in Fig. 45. It consists of a tube of glass, with a movable disk, *a*, ground so as to fit the bottom of the tube. The disk being held closely against the tube by a string, *b*, the whole is plunged into a vessel of water. In this state, the disk, though heavier than water, does not fall to the bottom, showing that it is held in place by an upward pressure. If water now be poured into the tube in a gentle stream, the disk will adhere till the latter is filled to the level of the fluid on the outside. This shows that the upward pressure is equal to the weight of a column of water whose base is that of the tube, and whose altitude is its distance below the upper surface of the fluid.

Fig. 45.

The upward pressure of fluids is called their *Buoyant Effort.* It is in consequence of their buoyant effort that fluids sustain lighter bodies on their surfaces. The same principle causes fluids to buoy up bodies of all kinds, diminishing the weight of heavy ones, and causing light ones *to float.*

(**73.**) How is upward pressure demonstrated? *What is the Buoyant Effort? Its effect on bodies?*

Pressure on the Bottom of a Vessel independent of its Shape.

74. The pressure on the bottom of a vessel, arising from the weight of a liquid, is entirely independent of the shape of the vessel, as well as of the quantity of liquid which it contains. It depends only on the size of the surface pressed, and its distance below the upper surface of the liquid.

This principle may be demonstrated by means of an apparatus, shown in Fig. 46. The apparatus consists of a tube, *o*, firmly attached to the cover of a glass vessel, *P*. By means of a screw joint, different shaped vessels, *A*, *B*, *C*, may be attached to the upper end

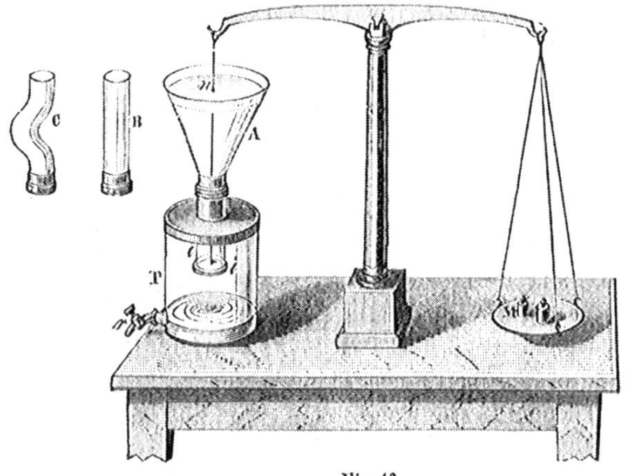

Fig. 46.

of the tube. A disk, *i*, of ground glass is held in contact with the lower end of the tube by a string, which is secured at its upper extremity to an arm of a balance.

The vessel, *A*, is screwed on, and filled with water until the downward pressure exactly counterpoises a given weight in the

(74.) Does the pressure on the bottom of a vessel depend upon the shape of the vessel? *How shown? Explain the experiment in detail.*

scale-pan, *M*, when the upper surface of the water is marked by a sliding bead, *n*. The other vessels, *B* and *C*, are successively screwed on, and filled with water up to the level, *n*; if any more water is poured into either, the downward pressure overcomes the weight, *M*, and the water escapes into the vessel, *P*.

This principle of pressure on the bottom of vessels is sometimes called the *Hydrostatic Paradox*. It is so called, because the same pressure may be obtained by using very different quantities of the same liquid.

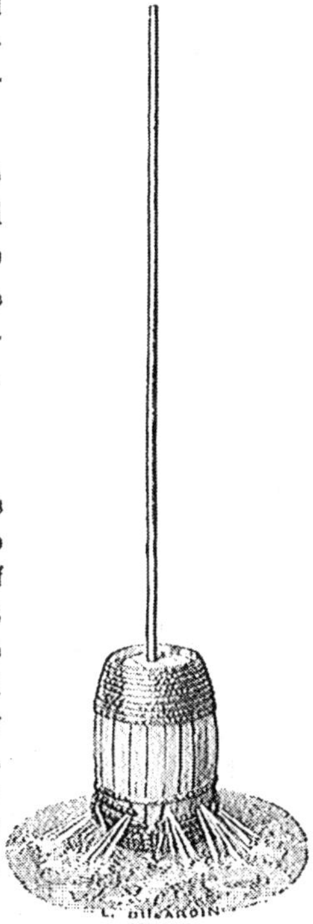

Pascal's Experiment.

75. The following experiment was made by PASCAL, in 1647. He fitted into the upper head of a strong cask a tube of small diameter and about thirty-four feet in length, as shown in Fig. 47. The cask being filled with water, he succeeded in bursting it by pouring a comparatively small quantity of water into the tube. In this case the pressure exerted laterally was the same as though the tube had been throughout of the same diameter as the cask, or even greater.

Fig. 47.

Hydraulic Press.

76. The principle of equal pressures has been applied in the construction of a press, by means of which a single man may exert an enormous power. This press is shown in perspective in Fig. 48, and in section in Fig. 49, the letters in both figures corresponding to the same parts.

What is this principle of pressure called? Why? (**75.**) *Explain Pascal's experiment.* (**76.**) What is the principle of the Hydraulic Press?

The press consists of two cylinders, *A* and *B*, of unequal diameters. In the cylinder, *B*, is a solid piston, *C*, which rises as the water is forced into *B*, and thus forces up a platform, *K*. The cylinder, *A*, forms the barrel of a pump by means of which water is raised from a reservoir, *P*, and forced into the cylinder, *B*. This pump is worked by a lever, *O*, attached to a solid piston, *a*. When the piston, *a*, is raised, a vacuum is formed behind it, which is filled by water from

Fig. 48.

the reservoir, *P*, which enters by opening the valve, *S*. When the piston is depressed, the valve, *S*, closes, the valve, *m*, is opened, and a portion of the water is forced through the pipe, *d*, into the cylinder, *B*. By continuing to work the piston, *a*, up and down, additional quantities of water are forced into the large cylinder.

Describe the press in detail. Explain its action.

In consequence of the principle of *equal pressures*, the force applied to the piston, *a*, is transmitted through the tube, *d*, and is finally exerted upwards against the piston, *C*, its effect being multiplied by the number of times that the section of the piston, *C*, is greater than that of the piston, *a*. For example, if the section of *C* is 150 times as great as that of *a*, every pound of pressure on the latter will produce 150 lbs. of pressure on the former. This effect is further multiplied by means of the lever, *O*. The pressure exerted upon *C*, forces up the platform, *K*, with an energy that may be utilized in compressing any substance placed between it and the top of the press, *MN*. This upward pressure may also be used for raising heavy weights.

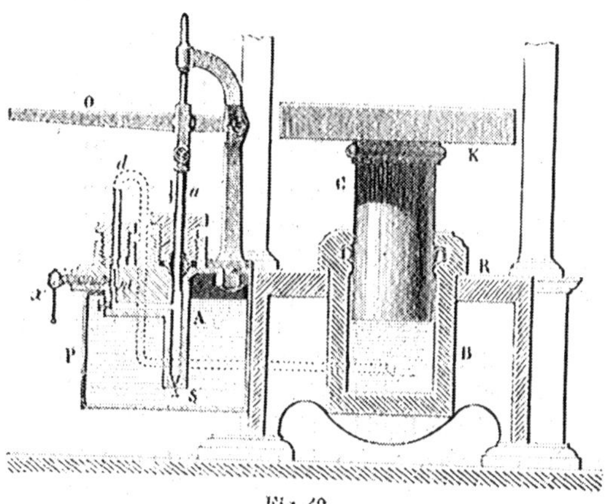

Fig. 49.

By varying the relative dimensions of the parts of the machine, an immense power may be exerted. In the arts, presses of this kind are constructed capable of exerting a force of more than a hundred thousand pounds.

The hydraulic press is used in compressing seeds to obtain oils, in packing hay, cotton, and other goods for shipment, in pressing books for the binder, and in a great variety of other operations. The immense tubular bridge over the Menai Straits was raised from the level of the water to the top of the piers by means of presses of this

Illustrate its power by an example. What are its uses?

description. The hydraulic press was also used in launching the Great Eastern, the heaviest movable structure ever constructed by man.

II.—EQUILIBRIUM OF LIQUIDS.

Conditions of Equilibrium.

77. A solid body is in equilibrium when its centre of gravity is supported, because the particles of the body are held together by cohesion. In liquids the particles do not cohere, and unless restrained they would flow away and spread out indefinitely. A liquid can be in equilibrium only when restrained by a vessel, or something equivalent. Furthermore, each particle must be equally pressed in all directions, which requires that the free surface should be level, that is, everywhere perpendicular to the force of gravity.

In saying that the free surface must be level, we suppose that the liquid is acted upon only by the force of gravity, which is the ordinary case. If, however, it is acted upon by other forces, the free surface must, at every point, be perpendicular to the resultant of all the forces acting at that point; for if it were not so, this resultant might be resolved into two components, one perpendicular to the surface, and the other parallel to it. The former would be resisted by the reaction of the liquid, and the latter, being uncompensated, would produce motion, which is contrary to the hypothesis of equilibrium.

Level Surface.

78. The surface of a liquid is LEVEL when it is everywhere perpendicular to the direction of gravity. Small level surfaces coincide sensibly with horizontal planes. Large level surfaces are curved so as to conform to the general form of the earth's surface. That the surface of the ocean is curved is shown by the phenomena presented by a

---

(**77.**) Explain the difference between equilibrium of solids and liquids. When is a liquid in equilibrium? *How is the upper surface when other forces than gravity act? Why?* (**78.**) What is a level surface? Nature of a small level surface? Of a larger one? Illustrate.

ship viewed from the shore, as exhibited in Fig. 50. As the vessel recedes, we first lose sight of her hull, then her lower sails disappear, then her higher sails, until at last the entire vessel is lost to view.

Fig. 50.

In defining a level surface, we said that it is everywhere perpendicular to the direction of gravity; more strictly speaking, it is perpendicular to the resultant of gravity and the centrifugal force due to the earth's rotation on its axis. Were it not for the centrifugal force, the surface of the ocean would be perfectly spherical, but in consequence of that force, it is ellipsoidal; that is, the oceans are elevated about the equator and depressed about the poles.

The general level of the ocean is called the *true level;* a horizontal plane at any point is called the *apparent level.*

Equilibrium of Liquids in Communicating Vessels.

79. When a liquid is contained in vessels which communicate with each other, it will be in equilibrium if its

---

Explain the effect of the centrifugal force on the form of a level surface. What is a true level? An apparent level? (**79.**) What are the conditions of equilibrium in communicating vessels?

upper surface in all of the vessels is in the same horizontal plane.

This principle is demonstrated by means of the apparatus represented in Fig. 51. This apparatus consists of a system of glass vessels of different shapes and capacities, all of which communicate by a tube, *ac*. If any amount of water or other liquid be poured

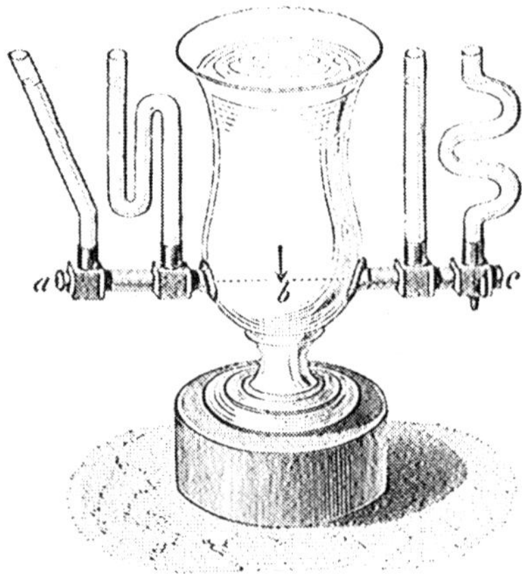

Fig. 51.

into one of the branches and allowed to come to rest, it will be seen that its upper surface in all of the vessels is in the same horizontal plane. The reason of this is, obviously, a necessary consequence of the principle of equal pressures.

Case of Vessels containing Liquids of different Densities.

80. When liquids of different densities are contained in communicating vessels, they will be in equilibrium when the heights of the columns are inversely as their densities.

This principle is demonstrated by means of an apparatus shown in Fig. 52. The apparatus consists of two glass tubes, *A* and *B*, open

How is this demonstrated? (**80.**) What are the conditions of equilibrium in the case of liquids of different densities? *How is this demonstrated?*

at top, and communicating at bottom by a smaller tube. If a quantity of mercury be poured into one of the tubes, it will come to a level in both tubes, according to the principle explained in the preceding article. If a quantity of water be poured into the tube *A*, the level of the mercury in that tube will be depressed, whilst it will be elevated in the tube *B*. The difference of level, *dc*, can be determined by the graduated scales on the tubes. It will be found by measurement, that the column of water, *ab*, is 13.6 times as high as the column of mercury, *dc*, which it supports. It will be shown hereafter, that mercury is 13.6 times as dense as water; hence the principle is proved. Other liquids may be employed with similar results.

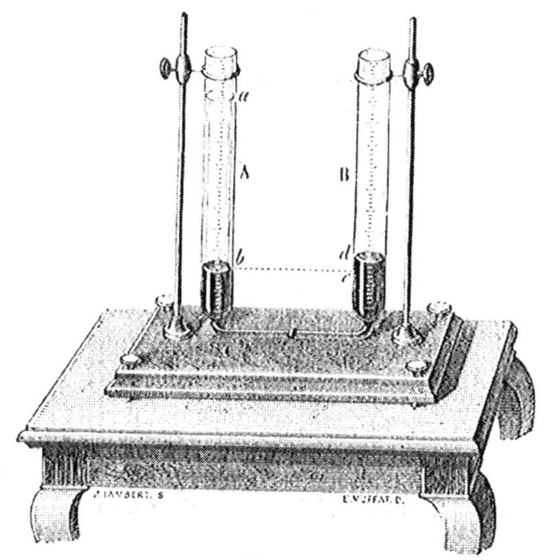

Fig. 52.

Equilibrium of Heterogeneous Liquids.

81. If liquids of different densities, but which do not mix, be poured into a vessel, they will arrange themselves

(**81.**) What are the conditions of equilibrium of heterogeneous liquids?

in the order of their densities, the heaviest being at the bottom, and the upper surface of each will be horizontal.

This is shown by a vial, Fig. 53, containing liquids of different densities, as mercury, water, and oil. If the vial be shaken, the liquids appear to mix, but if allowed to stand, they arrange themselves in horizontal layers, the densest liquid at the bottom.

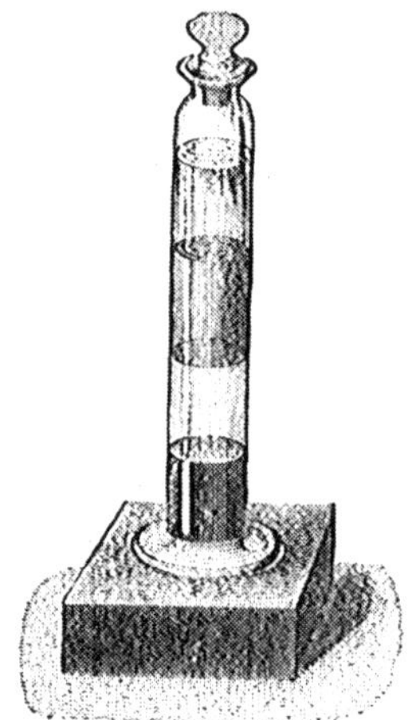

Fig 53.

The vial in the figure is represented as containing four liquids. It was formerly called the *vial of four elements*.

It is in accordance with this principle that cream rises on milk, and oil on water. The principle is often employed to separate liquids of different density by the process of decanting.

III.—APPLICATIONS OF THE PRINCIPLE OF EQUILIBRIUM.

The Water Level.

82. A WATER LEVEL is an instrument employed for determining the difference of level between two points. It consists of a horizontal tube of tin, 2½ or 3 feet in length, into the extremities of which two glass tubes are inserted perpendicular to it. The whole rests upon a three-legged support, called a *tripod*, as shown in Fig. 54. A quantity of water tinged with carmine or other coloring matter is introduced into one of the glass tubes, which, flowing through the horizontal tube, rises to the same level in the other. A visual ray directed along the surfaces of the

How shown? **(82.)** What is a Water Level? Describe it and its use.

water in the two glass tubes will be a horizontal line, or a *line of apparent level.* The use of the instrument is evident from the figure.

Fig. 54.

The Spirit Level.

83. The SPIRIT LEVEL consists of a tube of glass nearly filled with alcohol, and closed at its two extremities. The tube is slightly curved, and when placed horizontally, the bubble of air which it contains rises to the middle of the upper side of the tube. If either end be depressed, the

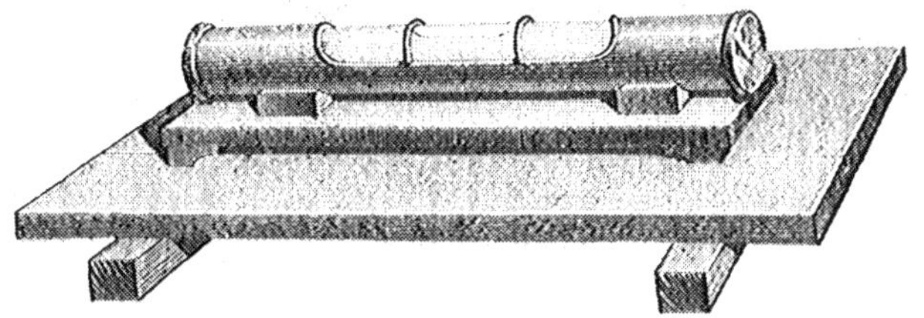

Fig. 55.

(**83.**) Describe a Spirit Level. How mounted?

bubble runs towards the other end. When used it is ordinarily mounted in a wooden case.

This form of *level* is much used by masons, carpenters, and other artisans. To ascertain whether a surface is level, the instrument is laid upon it, and the position of the bubble noticed. If the bubble is in the middle of the tube, the surface is level.

This form of level is also attached to many kinds of surveying and astronomical instruments.

Fig. 56.

Springs.—Fountains.—Rivers.

84. It is the principle of equal pressures that causes water to rise in springs and fountains. The water which feeds them is contained in natural or artificial reservoirs higher than the spring or fountain. These reservoirs communicate with the springs or fountains by natural or artificial channels, and the pressure of the water in them

What are its uses? Applications. (**84.**) What is a Spring? Fountain?

causes that in the spring or fountain to boil up, or sometimes to shoot up in a jet.

Fig. 56 represents a fountain called a *jet d'eau*. The reservoir is on the hill to the left, and the water reaches the bottom of the basin by a pipe represented by dotted lines.

The water of the jet tends to rise to the level of that in the reservoir, and would do so were it not for the resistance of the air, the friction of the water against the pipe, and the resistance offered by the falling particles, all of which combine to render the jet lower than the fountain-head.

The same principle determines the flow of streams from the higher to the lower grounds. The water of lakes, seas, and oceans is continually evaporating to form vapors and clouds. These are condensed in the form of rain, and the particles of water, urged by their own weight, seek a lower level. The rivulets gather to form brooks, and these unite to form rivers, by which the water is once more returned to the oceans and lakes. All of the water does not flow back to the ocean along the surface, but a portion percolates through the porous soils and accumulates in cavities to feed our springs and wells.

Artesian Wells.

85. ARTESIAN WELLS are deep wells, formed by boring through rocks and strata of various kinds of earth to reach a supply of water. These wells are named from the province of Artois, in France, where they were first used.

Fig. 57 illustrates the principle of these wells. *H* is the natural surface of the earth. *AB* and *CD* are curved strata of clay or rock which do not allow of the percolation of water. *KK* is an intermediate stratum of sand or gravel, which permits water to penetrate it. When a hole, *I*, is bored down to strike the water-bearing stratum, *KK*, the pressure of the water in the stratum forces it up in a jet. The well of Grenelle, in Paris, is nearly 1800 feet

---

Explain the jet d'eau? What causes the flow of streams? How are they fed? (**85.**) What are Artesian Wells? *Explain their action? How deep is that at Paris?*

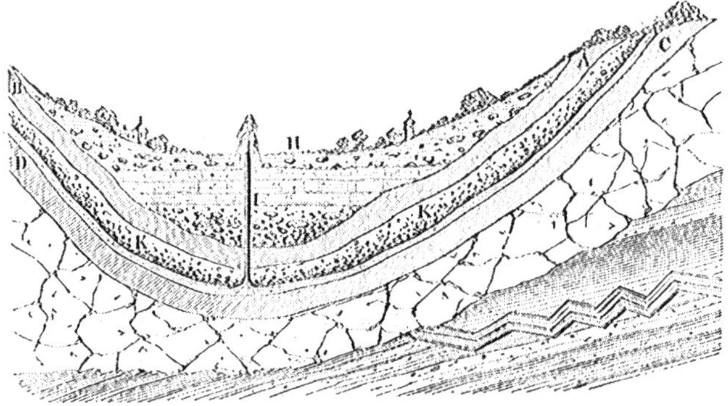

Fig. 57.

deep, and is fed by water coming from the hills of Champagne, which are much higher than Paris. The supply of water from this well is immense.

Many Artesian wells have been sunk in our own country.

IV.—PRESSURE ON SUBMERGED BODIES.

Principle of Archimedes.

86. IF a body is submerged in a fluid, it will be pressed in all directions, but not equally.

To illustrate, suppose a cube immersed in water, as shown in Fig. 58. The lateral faces, *a* and *b*, will be equally pressed and in opposite directions. The same will be true for the other lateral faces. Hence, the horizontal pressures will exactly neutralize each other. The upper and lower faces, *c* and *d*, will be unequally pressed, and in opposite directions. The face, *c*, will

Fig. 58.

(**86.**) Are submerged bodies pressed equally in all directions? *Illustrate in detail.*

be pressed upwards by a force equal to the weight of a column of the liquid whose cross-section is that of the cube, and whose height is the distance of *c* from the surface of the fluid. The face, *d*, will be pressed downwards by the weight of a column of the liquid, having the same cross-section as the cube, and a height equal to the distance of *d* from the surface of the liquid; the resultant of these two pressures is an upward force, equivalent to the weight of a volume of the liquid equal to that of the cube. This upward pressure is the *buoyant effort* of the fluid.

The principle just explained is called the *Principle of Archimedes.* It may be expressed by saying that, *a submerged body loses a portion of its weight equal to that of the displaced fluid.*

Hydrostatic Balance.

87. A HYDROSTATIC BALANCE is a balance having a hook attached to the lower face of each scale pan, and so constructed that the beam may be raised or lowered at pleasure.

Fig. 59 represents a hydrostatic balance. The cylinder, *c*, is solid, and fitted to slide up and down in the hollow cylinder, *d*. The cylinder, *c*, may be confined in any position by means of a clamp screw, *n*.

Cylinder and Bucket Experiment.

88. The principle of ARCHIMEDES' may be illustrated by what is called the *Cylinder and Bucket Experiment*, as shown in Fig. 59. A hollow cylinder or bucket, *b*, of brass, is attached to the hook of one of the scale pans, and from it is suspended a solid cylinder of brass, just large enough to fill the bucket, and the two are balanced by weights placed in the opposite scale pan. A glass vessel

Enunciate the principle of ARCHIMEDES. (**87.**) What is a Hydrostatic Balance? *Explain its construction.* (**88.**) Explain the Cylinder and Bucket Experiment.

having been placed beneath the cylinder, water is gradually poured into it, until the cylinder is immersed. The opposite scale pan will descend, showing that the cylinder is

Fig. 59.

buoyed up by some force. If we now fill the bucket, *b*, with water, the equilibrium will be restored, and the beam will come to a level. Because the water poured into the bucket is equal to that displaced by the cylinder, we infer that the buoyant effort is exactly equal to the weight of the displaced fluid.

The principle of ARCHIMEDES is so called, because it was first discovered by the illustrious philosopher of that name. He was led to the discovery in an attempt to detect a fraud, perpetrated upon

Why is the principle of ARCHIMEDES *so called?*

HIERO of Syracuse, by a goldsmith, who had been employed to make a golden crown. The artisan mixed a portion of silver with the gold that was given him for making the crown, but by means of the principle above explained, ARCHIMEDES was able to determine the exact amount of each material employed.

Floating Bodies.—Principles of Flotation.

89. When a body is plunged into a liquid, it is urged downward by its proper weight, and upward by the buoyant effort of the liquid, and, according to the relative intensities of these two forces, three cases may arise:

1. If the density of the immersed body is the same as that of the liquid, its weight will be equal to the buoyant effort of the liquid, and it will remain in equilibrium wherever it may be placed. This is practically the case with fishes. They maintain themselves in any position in which they may happen to be, without effort.

2. If the density of the body is greater than that of the liquid, its weight will be greater than the buoyant effort, and the body will sink to the bottom. This is what happens when a stone or piece of iron is thrown into water.

3. If the density of the body is less than that of the liquid, its weight will be less than the buoyant effort, and the body will rise to the surface. The body will continue to rise until the weight of the displaced liquid equals that of the body, when it will come to rest. It is then said *to float*. Thus, a piece of wood floats upon water, and in like manner a piece of iron floats upon mercury.

When a floating body comes to rest on a liquid, the plane of the upper surface of the liquid is called the *Plane of Flotation*.

Explain the method of its discovery. (89.) When a body is plunged into a liquid, what three cases may arise? Explain the first case. The second case. The third case. What is the Plane of Flotation?

It sometimes happens that a body which is more dense than a liquid floats upon it. Thus, a porcelain saucer floats upon water. This arises from its form being such, that it displaces its own weight of water, when only partially immersed. For the same reason iron ships float freely on the ocean.

Illustration of the Principles of Flotation.

90. The principles of flotation may be illustrated by an instrument shown in Fig. 60, which under various forms is sold in the shops as a child's toy.

In the form shown, it consists of a high and narrow glass vessel, surmounted by a brass cylinder, *A*, in which is an air-tight piston that may be raised or depressed by the hand. The vessel is partially filled with water, and contains a light body, as a fish, hollow and of porcelain or glass. The fish is attached to a sphere of glass, *m*, filled with air, and with a small hole, *o*, at its lower side, through which water can flow in or out, as the pressure is increased or diminished.

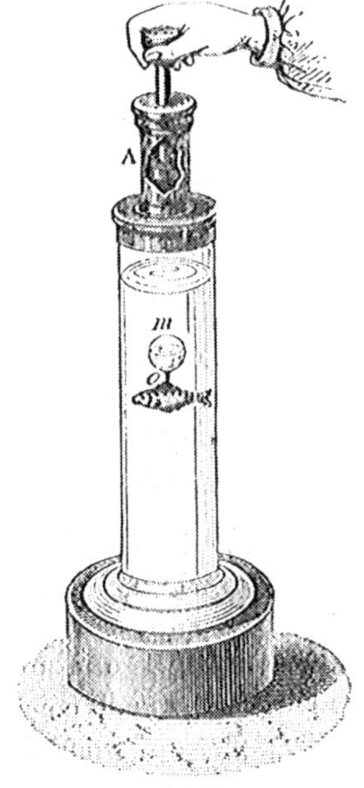

Fig. 60

Under ordinary circumstances the sphere, *m*, with its attached fish, floats at the surface of the water. If the piston is depressed, the air beneath it is compressed, and acting upon the water forces a portion of it into the globe. The apparatus then becomes more dense than the water, and sinks. By relieving the pressure, the air in the globe expands and drives the water out, when it again floats on the surface. The experiment may be repeated at pleasure.

Explain the case of a dense body floating on a liquid. (**90.**) *What instrument illustrates the laws of flotation? Explain its use and action.*

Swimming Bladder of Fishes.

91. In many fishes there is a bladder filled with air, situated directly under the backbone. This is called the *Swimming Bladder.*

When the fish wishes to descend, it compresses this bladder by a muscular effort, and then, as the quantity of water displaced is less than before, the weight of the fish prevails over the buoyant effort, and the fish sinks. On relaxing the effort, the bladder expands, the buoyant effort of the water prevails over the weight of the fish, and it rises.

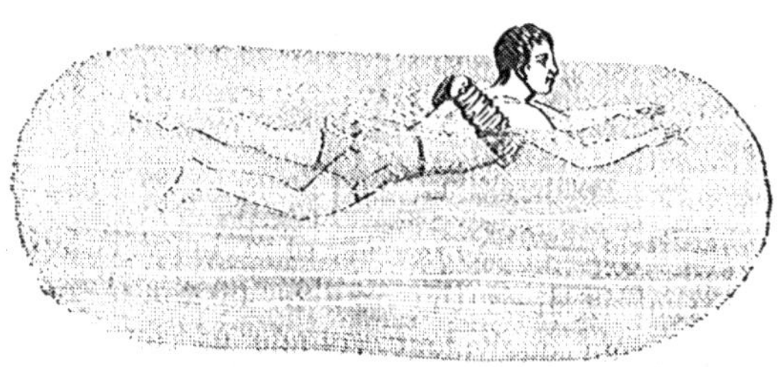

Fig. 61.

Swimming.

92. The human body is lighter than water, especially than the salt water of the ocean, and tends naturally to float when immersed. The only reason why men do not swim naturally, is the difficulty of keeping the head out of water, so as to be able to breathe. The head is the heaviest part of the body, and tends continually to sink into the water.

Many quadrupeds swim naturally, because the head is small in proportion to the body, and is so placed upon the trunk, that it is easy to keep it above the surface.

The safest position for a person in the water, who does not know

---

(91.) What is the Swimming Bladder of a fish? *Explain its action.* **(92.)** *Explain the phenomenon of swimming. Why do some quadrupeds swim naturally? What is the safest position in the water?*

how to swim, is upon the back. The tendency to raise the arms out of the water should be resisted, as this diminishes the buoyant effort of the fluid without diminishing the weight.

In learning to swim, it is often the custom to place bladders filled with air, or blocks of cork, under the arms, as shown in Fig. 61. These act to increase the buoyant effort of the fluid, without sensibly increasing the weight. It is on this principle that life-preservers are constructed.

Many kinds of birds, as ducks, geese, swans, and the like, swim naturally and without effort. They owe this faculty to a thick layer of down and feathers which are very light and impermeable by water. They, therefore, displace a large volume of water in proportion to their weight, giving rise to a strong buoyant effort.

V.—SPECIFIC GRAVITY OF BODIES.

Definition of Specific Gravity.

93. The SPECIFIC GRAVITY of a body is its *relative weight;* that is, it is the number of times the body is heavier than an equivalent volume of some other body taken as a standard.

It is a matter of daily observation, that some substances are heavier than others under the same volume. Thus, gold is heavier than silver, lead than iron, stones than wood, and so on. In order to compare the relative weights of different bodies, all are referred to a common standard.

Distilled water is generally adopted as a standard, and because water varies in density at different temperatures, it is usual to take it at the temperature of 39°.2 Fahrenheit, water being most dense at that temperature.

In order to find the specific gravity of any body, all that we have to do is, to find how many times heavier any

What is the principle of the life-preserver? Why do some birds swim naturally? (**93.**) What is Specific Gravity? *Illustrate. What is taken as a standard? At what temperature? Why?* What is the process of finding the specific gravity of a body?

given volume of the body is, than an equivalent volume of distilled water at 39°.2 F. This is the method of fixing the specific gravity of solids and liquids; we shall see hereafter how it is possible to fix the specific gravity of gases and vapors.

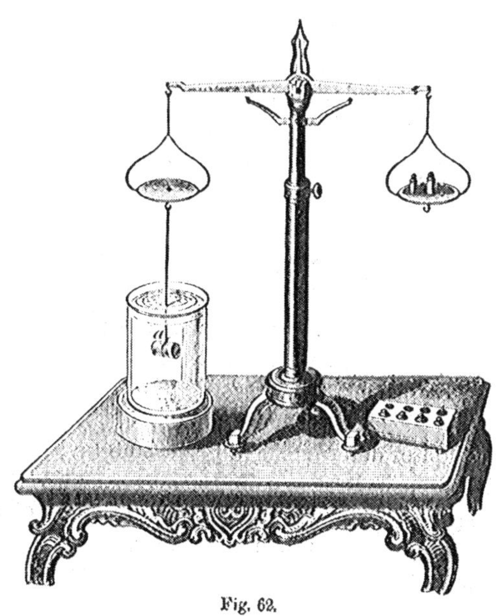

Fig. 62.

Specific Gravity of Solids.

94. The following are some of the methods of determining the specific gravities of solids:

1. *By the Hydrostatic Balance.*—Place the body in one of the scale pans and balance it by known weights in the other pan. These will give the weight of the body in air.

(**94.**) Explain, in detail, the method of finding the specific gravity of a solid by means of the hydrostatic balance.

Next suspend the body in a vessel of distilled water by means of a thread or wire attached to one of the scale pans, as shown in Fig. 62, and balance it by weights placed in the other pan. On account of the buoyant effort of the water, the weight of the body in water will be less than that in air. Subtract the weight of the body in water from that in air, and the difference will be the weight of the displaced water, that is, the weight of a volume of water equal to that of the body. Having found the weight of the body in air, and the weight of an equivalent volume of water, divide the former by the latter, and the result will be the specific gravity required.

2. *By Nicholson's Hydrometer.*—NICHOLSON'S HYDROMETER consists of a hollow cylinder of glass, as shown in Fig. 63, weighted at the bottom by a heavy body, *d*, to make it float erect, and terminating above by a thin stem, *c*, which supports a scale pan, *a*. The instrument is so constructed that when a given weight, say 500 grains, is placed in the pan, it will sink in distilled water to a notch, *c*, on the stem.

The method of determining the specific gravity by means of this instrument is shown in Figs. 64 and 65. Suppose it were requred to determine the specific gravity of a small bar of iron weighing less than 500 grains.

The bar is placed in the pan and weights added till it sinks to the notch in the stem as shown in Fig. 64. These weights, subtracted from 500 grains, give the weight of the bar in air. Next place the bar in the cup, *d*, as shown in Fig. 65, and add weights enough to make the instrument sink again to the notch in the stem. The last weights will denote the buoyant effort of the fluid, or the weight of the water displaced by the bar. Divide the weight of the bar in air by the weight of the displaced water, and the result will be the specific gravity sought.

What is Nicholson's Hydrometer? How used for determining the specific gravity of a solid?

3. *By a flask.*—This method is used when a body exists in a state of powder, or in fine particles like sand. A small flask, whose exact weight is known, is first filled with the powder and the whole carefully weighed. The entire weight, diminished by that of the flask, is the weight of the body.

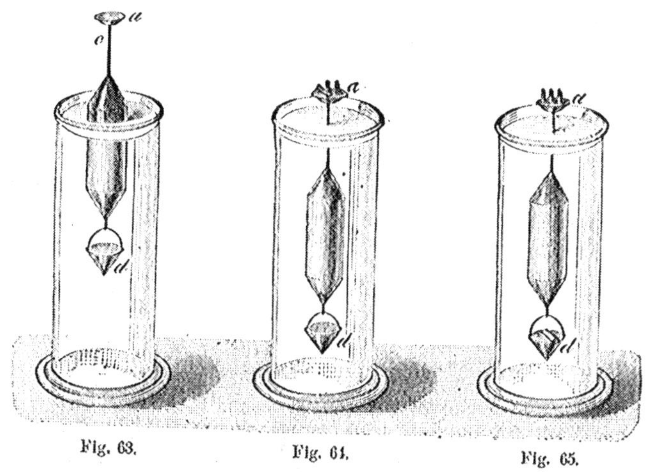

Fig. 63. Fig. 64. Fig. 65.

The flask is then filled with water and weighed. This weight, diminished by that of the flask, is the weight of an equivalent volume of water. Divide the weight of the body by that of its equivalent volume of water, and the result will be the specific gravity required.

Specific Gravity of Liquids.

95. The following are some of the principal methods of determining the specific gravities of liquids:

1. *By the Hydrostatic Balance.*—Select a heavy body which is not soluble either in water or in the liquid whose

Explain the method by means of a flask. (**95.**) How is the specific gravity of a liquid found by means of the balance?

specific gravity is to be determined, as, for example, a ball of platinum. Weigh this body first in air, then in water, and finally in the liquid in question. Subtract the second and third weights from the first separately; the results obtained will be respectively the weights of a volume of water, and of the liquid, equal to that of the platinum ball. Divide the latter by the former, and the quotient will be the specific gravity required.

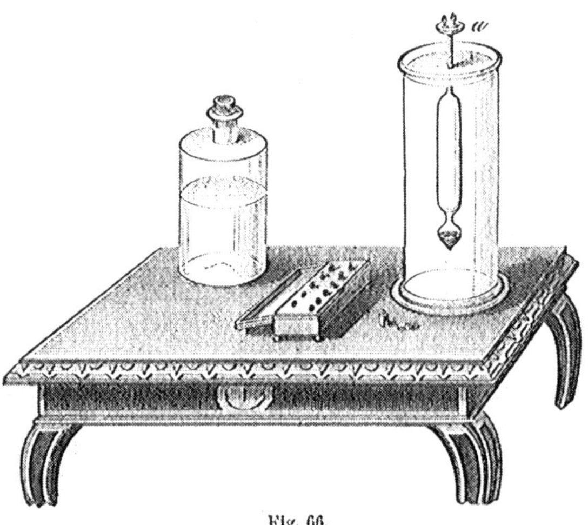

Fig. 66.

2. *By Fahrenheit's Hydrometer.*—FAHRENHEIT'S HYDROMETER consists of a glass cylinder ballasted at the bottom by a small globe filled with mercury, and provided at top with a stem and scale pan as shown in Fig. 66. Its weight is carefully determined.

To use the hydrometer, it is first plunged into distilled water, and weights placed in the scale pan till it sinks to the

---

Describe Fahrenheit's Hydrometer. How is it used to find the specific gravity of a liquid?

notch filed on the stem. These weights, increased by that of the instrument, will give the weight of the displaced water. The instrument is next plunged into the liquid in question, and weights are placed in the pan till the instrument again sinks to the notch. These weights, added to that of the instrument, give the weight of the displaced liquid. Now the volumes displaced are the same in both cases, each being that of the submerged instrument; hence, if we divide the weight of the displaced liquid by that of the displaced water, the quotient will be the specific gravity required.

3. *By the flask.*—A flask is constructed so as to hold a given weight of distilled water, say 1000 grains. This flask is first weighed when empty, and then when filled with the liquid in question. The difference of these results is the weight of the liquid, and this, divided by 1000 grains, will be the specific gravity required.

The specific gravities of some of the most important substances are given in the following table:

TABLE,

SHOWING THE SPECIFIC GRAVITIES OF SOLIDS AND LIQUIDS.

Platinum (rolled)	22.07	Mercury	13.60
Gold (stamps)	19.36	Sulphuric Acid	1.84
Lead (cast)	11.35	Milk	1.03
Silver (cast)	10.47	Sea Water	1.03
Iron (bar)	7.79	Distilled Water	1.00
Zinc (cast)	6.86	Bordeaux Wine	0.99
Diamond	3.53	Olive Oil	0.91
White Marble	2.84	Spirits of Turpentine	0.87
Glass (flint)	3.33	Absolute Alcohol	0.79
Ivory	1.92	Ordinary Ether	0.72

It will be seen that platinum is the heaviest solid, and that mercury is the heaviest liquid.

How is the specific gravity of a liquid determined by means of a flask? *Which is the heaviest solid? Liquid?*

A knowledge of the specific gravities of bodies is of frequent application. In mineralogy it aids in determining mineral species. The jeweller determines by its aid the precious stones. It enables us to find the weight of a body when we know its volume. Thus, a cubic foot of iron weighs 11.35 times as much as a cubic foot of water; but a cubic foot of water weighs 1000 ounces, hence a cubic foot of iron weighs 11,350 ounces, or about 709 lbs.

Beaumé's Areometer.

96. BEAUMÉ'S AREOMETER consists of a bulb of glass, ballasted at bottom by a second bulb containing mercury, and terminating at top in a cylinder of uniform diameter, as shown in Fig. 67.

When plunged into liquids, it sinks till the weight of the displaced fluid equals that of the areometer. In light fluids it therefore sinks deeper than in heavy ones.

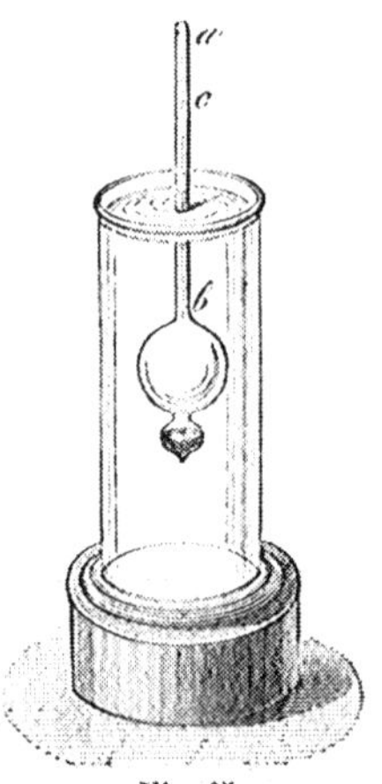

Fig. 67.

The plan of graduating BEAUMÉ'S areometer is as follows. It is ballasted so that in distilled water it will sink to the point *a*, on the stem, which is marked 0. A mixture of salt and pure water is then formed, in the proportion of 15 of the former to 85 of the latter, into which the instrument is plunged. The upper surface then cuts the stem at some point, *c*, which is marked 15. The intermediate space between *a* and *c*, is divided into 15 equal parts, and the division is continued downwards on the stem. The divisions and numbers are on a slip of paper in the interior of the stem.

What are some of the applications of the specific gravity of bodies? (**96.**) Describe BEAUMÉ'S Areometer? How is it graduated?

The use of the instrument thus graduated is to ascertain the amount of salt in any solution of salt in water. It is plunged into the solution in question, and the number to which it sinks, denotes the degree of saturation of the solution.

Instruments constructed on this principle have been devised for determining the strength of other solutions, whether of acids or salts. Also for determining the strength of saccharine solutions and the like.

The Alcoholometer.

97. The ALCOHOLOMETER is similar in its construction to the areometer just described. It is graduated so as to show the percentage of alcohol in any mixture of alcohol and water.

The instrument is first ballasted so that when plunged in pure water it will float with nearly all of its stem above the water. The line of flotation is marked 0. Mixtures are then formed, containing 1, 2, 3, &c., per cent. of pure alcohol and water, and the instrument is plunged into them in succession. The lines of flotation are marked 1, 2, 3, &c., as in the instrument previously. In this case the numbers run upwards. It is necessary to graduate it throughout by trial, as the divisions are not uniform.

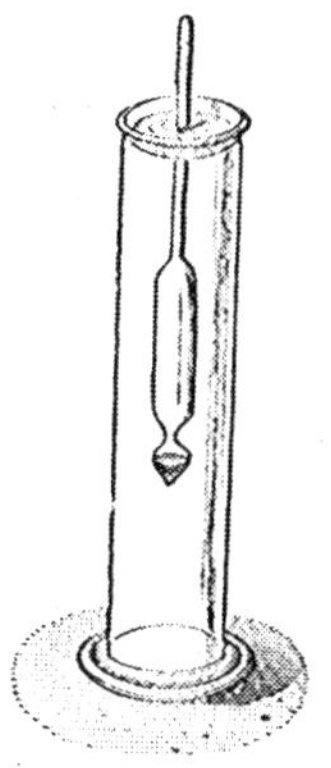

Fig. 68.

To use the instrument, it is plunged into the mixture of alcohol and water to be tested, and the per-centage is read off on the paper scale within the tube, or else the scale is scratched upon the stem with a diamond.

What is its use? How used? What other instruments are constructed on the same principle? (**97.**) Describe the Alcoholometer. How is it graduated? How used?

The Lactometer.

98. The LACTOMETER is entirely analagous in principle to BEAUMÉ'S areometer, and is used to determine the purity of milk. The instrument, and the method of using it, are shown in Fig. 69.

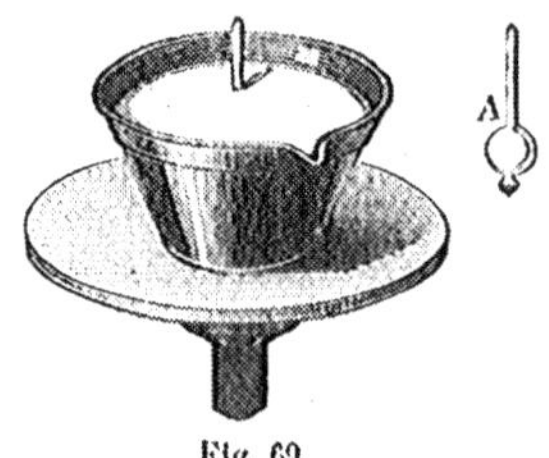

Fig. 69.

It is graduated by trial, using mixtures of milk and water. In the first trial pure water is used, then mixtures containing 10, 20, 30, 40, &c., per cent. of milk. The scale is therefore divided into 10 parts, between pure water and pure milk.

(**98.**) What is a Lactometer? How graduated and used?

CHAPTER III.

MECHANICS OF GASES AND VAPORS.

I.—THE ATMOSPHERE.

General Properties of Gases and Vapors.

99. GASES and VAPORS have been defined to be highly compressible fluids.

The distinction between a gas and a vapor, is not very clear. When a body in a gaseous form can, by moderate pressure, be reduced to a liquid form, it is usually called a vapor. For most of the purposes of Physics the distinction is unimportant.

Besides the property of compressibility, or rather as a consequence of it, gases and vapors continually tend to expand so as to occupy a greater space. The force which they exert in this way, is called their *Tension*, or their *Elastic Force*.

Thirty-four gases are known, thirty of which are compound, and four are simple. The four simple gases are, *oxygen*, *hydrogen*, *nitrogen*, and *chlorine*. Most of the gases are colorless, but some are not so.

Of the thirty-four gases, all but five have been liquefied by pressure, and the application of cold. The five that have thus far resisted are, *oxygen*, *hydrogen*, *nitrogen*, *deutoxyde of nitrogen*, and *carbonic oxyde*.

(99.) What are Gases and Vapors? *What is the difference between them?* What is meant by Tension? *How many known gases are there? Which are simple? Which have not been liquefied?*

Description of the Atmosphere.

100. The air we breathe is a mixture of *oxygen* and *nitrogen*, with a slight quantity of *carbonic acid*, *watery vapor*, and some accidental impurities. The oxygen and nitrogen are mixed in the proportion of 21 to 79.

The oxygen of the air supports life and combustion; without it, neither could long exist. The nitrogen serves to dilute it. Were the air composed entirely of oxygen, bodies would burn with too much rapidity, even many of the metals would be consumed. Animal life, too, would soon be exhausted by overaction in such an atmosphere.

The atmosphere is transparent, without odor, and colorless, except when seen in masses. In masses, it assumes a blue tint, and it is this which causes the sky to take a blue color.

Without an atmosphere, the celestial vault would appear perfectly black; in ascending high mountains, the sky gradually loses its blueness, and approaches a hue of black; this is because the mass of air above the observer rapidly diminishes as we ascend.

The air, by virtue of its elasticity, serves as a medium for the transmission of sound; it also serves as a means of transporting the vapors of oceans and lakes to fall upon the land in the form of rain, snow, and the like.

Expansive Force of Air.

101. Air, like simple gases, always tends to assume a greater volume.

To show this property, take a bladder fitted with a stop-cock, as shown in Fig. 70. Having moistened the bladder to make it more flexible, open the cock, squeeze out most of the air, and then close it.

(**100.**) Describe the composition of the atmosphere. *What is the use of the oxygen? Of the nitrogen?* What is the color of air? *What effect has the air on celestial appearances?* Mention some of the uses of the atmosphere. (**101.**) *How is the expansive force of air shown?*

Place the nearly empty bladder under the receiver of an air-pump, and exhaust the air. As the air becomes rarer in the receiver, the bladder will be seen to expand, showing that the air within it is expansible. In the same way, it may be shown that any gas is expansible.

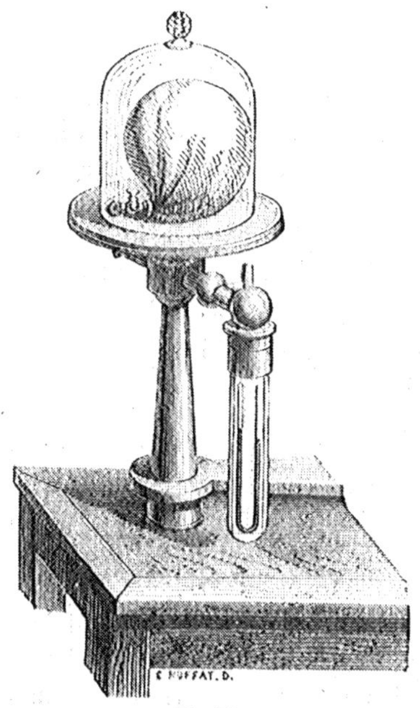

Fig. 70.

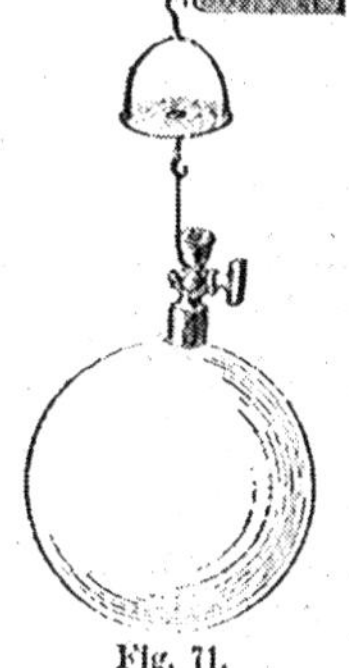

Fig. 71.

Weight of Air.

102. Air, like other bodies, has weight.

To show this, take a hollow globe of glass, fitted with a stop-cock, as shown in Fig. 71. Having attached it to one scale pan of a delicate balance, counterpoise it by weights placed in the other. Then by means of the air-pump exhaust the air from the globe; the opposite scale pan will descend, and some weights will have to be added

(**102.**) *How is it shown that air has weight.*

to the first scale pan to restore the equilibrium The weights added will indicate the weight of the exhausted air.

Composition of the Atmosphere.

103. It has been stated that our atmosphere is composed principally of oxygen and nitrogen, with small quantities of carbonic acid and watery vapor.

The amount of watery vapor depends upon the place, the season, the temperature, and the direction of the wind; under all circumstances it forms but a small per-centage of the entire atmosphere.

The carbonic acid in the atmosphere arises in a great measure from respiration and combustion. A continual supply of this gas is afforded by volcanoes. On the other hand, it is being continually taken up in the process of vegetation. Plants continually absorb it, appropriating the carbon, and giving off the oxygen which it contains. Another cause of diminution in the amount of carbonic acid in the air, is absorption by the water of our streams. Water absorbs large quantities of it, which thus becomes the means of dissolving earthy matters, and eventually of causing calcareous deposits.

It is the result of observation, that the supply and loss are very nearly balanced, so that the per-centage of carbonic acid in the atmosphere remains nearly constant. It amounts to about a thousandth part of the entire atmosphere.

Atmospheric Pressure.

104. The atmosphere, by virtue of its weight, exerts a force of pressure upon the surface of the earth as well as upon every object with which it is in contact. This force is called the *Atmospheric Pressure.*

(**103.**) Upon what circumstances does the watery vapor in the air depend? Whence is carbonic acid supplied? What becomes of the excess of carbonic acid? How do the supply and loss compare? What is the amount in the atmosphere?
104.) What is the Atmospheric Pressure?

This pressure decreases as we ascend into the atmosphere.

If we suppose the atmosphere to be divided into layers parallel to the surface of the earth, it is evident that each layer is pressed down by the weight of all above it. Hence, the higher layers are less compressed than those below them. Being less compressed, they expand, or become *rarefied*. The existence of atmospheric pressure may be shown by a variety of experiments, some of which will be explained below.

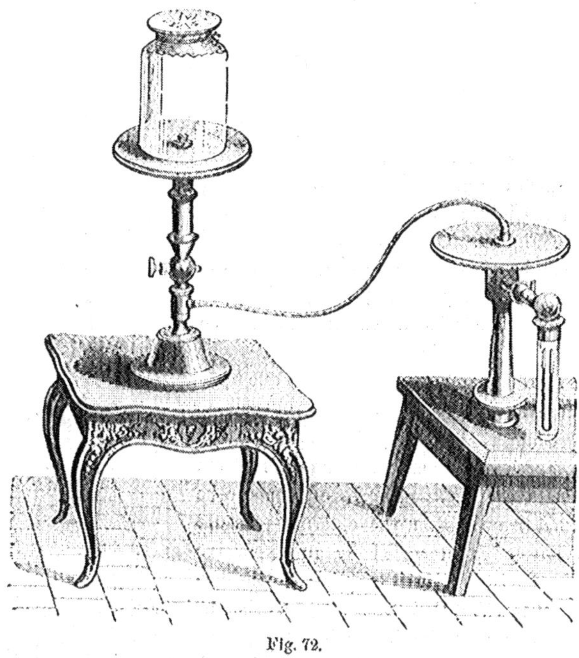

Fig. 72.

Bursting a Membrane.

105. A glass cylinder, open at both ends, has its upper end covered by a stretched membrane, such as is used by gold-beaters,

How does it vary as we ascend? *How shown that the air becomes rarer in ascending?* (**105.**) *Explain the experiment of bursting a membrane.*

and its lower end is ground so as to fit the plate of an air-pump, as shown in Fig. 72.

In its natural condition, the membrane is pressed down by the weight of the atmosphere above it, and this pressure is resisted by the tension of the air within the cylinder. If now the air be exhausted from the cylinder, the membrane will no longer be pressed from within, and will finally burst with a loud report.

The bursting of the membrane shows the pressure of the air. The report arises from the sudden rush of air to fill up the exhausted cylinder.

The Magdeburg Hemispheres.

106. This apparatus, named from the city where it was invented, consists of two hollow hemispheres of brass, which are ground so as to fit each other with an air-tight joint. The hemispheres are shown in Fig. 73. One of them is so prepared that it can be attached to an air-pump, and is provided with a stop-cock, by means of which a communication with the external air can be opened or closed at pleasure.

Fig 73.

The two hemispheres being placed one upon the other, the pressure of the external air is exactly counterbalanced by the tension of that within, and no obstacle prevents them from being drawn apart. If, however, the air be exhausted from within, the external pressure is no longer counteracted by an expansive force from within, and it requires a considerable effort to effect their separation, as shown in Fig. 74. We shall see hereafter that the hemispheres are pressed together by a force equal to 15 lbs., multiplied by the number of square inches in their common cross section.

What causes the bursting? The report? (**106**) *What are the Magdebourg Hemispheres? Describe the experiment, and explain it.*

Fig. 74.

The experiment was devised by OTTO VON GUERICKE, of Magdebourg. He constructed two hemispheres more than two feet in diameter, and after having exhausted the air, it is reported that it required several horses to draw them asunder.

Torricellian Tube.—Measure of the Atmospheric Pressure.

107. The preceding experiments show that the atmosphere exerts a force of pressure; the intensity of that force may be measured by other means.

TORRICELLI, a pupil of GALILEO, showed in 1643, that this pressure amounts to about 15 lbs. on each square inch of surface, at the level of the sea.

What experiment was made by OTTO DE GUÉRICKE? (**107.**) What is the pressure of the atmosphere on a square inch?

In order to repeat TORRICELLI'S experiment, take a glass tube about three feet in length, closed at one end and open at the other. Turning the closed end downwards, let it be filled with mercury. Then holding the finger over the open end, let it be inverted in a vessel of mercury, as shown in Fig. 75. On removing the finger, the mercury sinks in the tube until the column, *AB*, is about 30 inches high, when it comes to a state of equilibrium.

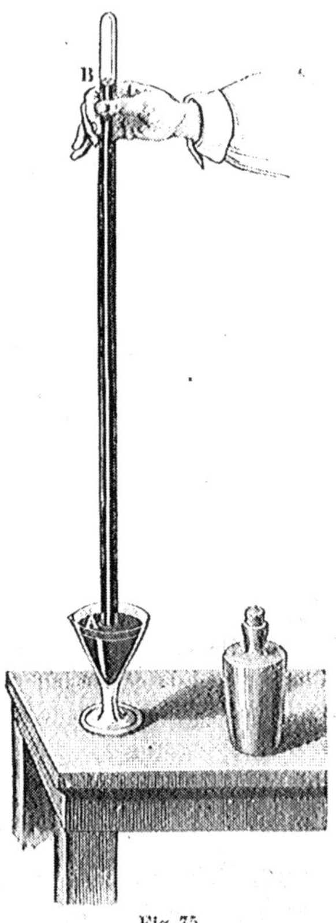

Fig. 75.

In this condition, the mercury is sustained by the pressure of the air upon the surface of the free mercury in the vessel, transmitted according to the law explained in Article 70. At the level of the sea, the height of the column *AB*, is on an average not far from 30 inches, or 2½ feet.

If we suppose the cross-section of the tube to be one square inch, the atmospheric pressure upon that surface must be sufficient to balance the weight of 30 cubic inches of mercury. Now the weight of 30 cubic inches of mercury is a little less than 15 lbs.; hence, we say the measure of the atmospheric pressure is 15 lbs. on each square inch.

A pressure of 15 lbs. on each square inch, is often called *an atmosphere*, and this becomes a unit for expressing the pressures of gases and vapors. Thus, when we say, in any given case, that the pressure of steam in a boiler is four

Describe TORRICELLI'S *experiment. How shown that the pressure is 15 lbs. on an inch?* What unit of pressure is adopted for all gases and vapors? Example.

atmospheres, we mean that it exerts a pressure of 60 lbs. on each square inch of surface.

Pascal's Experiments.

108. As soon as TORRICELLI's experiment was known in France, BLAISE PASCAL undertook to ascertain by experiment whether the mercury was actually retained in the tube by the pressure of the atmosphere, or by some other cause.

He caused a friend to repeat TORRICELLI's experiment upon the top of the mountain of Puy-de-Dome, correctly reasoning, that if the height of the mercurial column is due to atmospheric pressure alone, it ought not to be so great on the mountain top as at the level of the sea. The result of the experiment showed that the height of the column was less on the top of the mountain than at its base.

He next reasoned, that if the tube were filled with any liquid less dense than mercury, the height of the column ought to be proportionally greater. Consequently, he made at Rouen, in 1646, the following experiment. He took a tube, similar to that of TORRICELLI, but nearly 50 feet in length, and after filling it with wine, inverted it in a vessel of the same liquid. PASCAL observed that the column fell, until it was about 35 feet high, when it came to rest. In this case, the column was fourteen times as high as when mercury was used, and as mercury is fourteen times as dense as wine, he concluded that the sole cause of the phenomenon in question was the pressure of the atmosphere.

The Barometer.

109. A BAROMETER is an instrument for measuring the pressure of the air. If, to TORRICELLI's tube, were fitted a

(**108.**) Describe PASCAL's experiments in detail, and his mode of reasoning. What conclusion is derived from PASCAL's experiments? (**109.**) What is a Barometer? What is its principle?

scale for measuring the exact altitude of the mercurial column, it would be a barometer.

Several forms have been given to the barometer, some of which will be described in the following articles.

The Cistern Barometer.

110. Fig. 76 represents a CISTERN BAROMETER, such as is in common use in France and in this country.

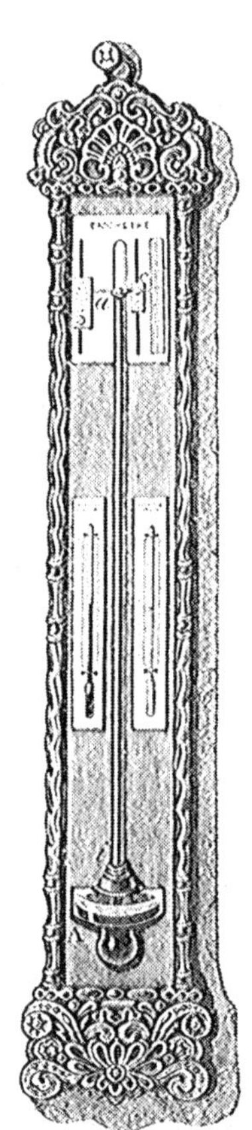

Fig. 76.

It consists of a glass tube, *ai*, about 34 inches long, closed at the top and open at the bottom. This tube has a diameter of about four-tenths of an inch. It is filled with mercury and inverted in a cistern, *A*, which is partially filled with the same liquid, as explained in Article 107. The mercury settles in the tube till the height of the column is about 30 inches at the level of the sea.

The cistern *A*, is 3 or 4 inches in diameter, and it is so adapted to the tube *ai*, as to permit the air to penetrate to the cistern at the joint *i*. Only a part of the cistern is seen in the figure, the remainder being let into the frame which supports the whole instrument. At the top of the

(**110.**) Describe the Cistern barometer. The tube. The cistern.

frame is a scale, *C*, having its 0 point at the level of the mercury in the cistern; or on the opposite side, is a scale on which are marked certain weather indications.

A curved piece of metal embraces the tube and carries an index, which, as the piece is raised or depressed to correspond to the top of the column, points out upon the scale *C*, the height of the column. Two thermometers, one of mercury and one of alcohol, are also attached to the frame, which serve to show the temperature of the instrument and of the mercury which it contains.

The 0 point, or beginning of the scale, is at the surface of the mercury in the cistern. When the pressure of the air increases, a portion of the mercury in the cistern is forced up into the tube, and the 0 point descends; when the pressure diminishes, the reverse takes place. But inasmuch as the surface of the mercury in the cistern is very great in comparison with that in the tube, this rise and fall is, for most purposes, quite unimportant. When great accuracy is required, the bottom of the cistern is made of leather, and can, by means of a screw, be raised or depressed until the surface of the mercury in the cistern just grazes the point of an ivory pin projecting from the top of the cistern. This improvement, devised by FORTIN, is now in general use.

To determine the height of the barometer, the 0 point is first adjusted, then the curved piece is slid up or down till it coincides with the surface of the mercury in the tube, and the height is then read off on the scale *c*. The height of the thermometer should also be noted.

In the instrument described, the scale *c* does not extend throughout the whole length of the instrument, because, in ordinary cases, only a small part of the scale is needed. When a barometer is to be used in high altitudes, the scale is continued downwards as far as necessary.

Describe the scale. The index. The thermometers. *Where is the 0 point of the scale? How is the 0 point regulated in accurate barometers?* How is the height of the barometer determined?

The Siphon Barometer.

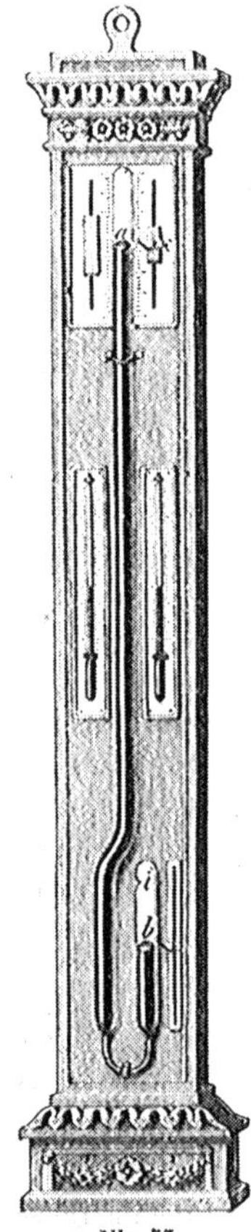

Fig. 77.

111. Fig. 77 represents a SIPHON BAROMETER. It consists of a curved tube, *ab*, having two unequal branches, the shorter one acting as a cistern. In the longer branch, there is a vacuum above the mercury, but the shorter one is supplied with air, which communicates with the external atmosphere through a small opening, *i*. There are two scales, one at the upper part of each branch, and in front of each is a movable index which may be raised or depressed until it comes to the free surface of the mercury in each branch. By means of these scales, the difference of level in the two branches may be measured. This difference is the height of the barometric column.

To prevent violent oscillations when the instrument is moved from place to place, the two branches communicate through a fine, almost capillary, tube. This arrangement also prevents the possibility of a bubble of air penetrating from the shorter to the longer branch, when the instrument is inclined.

Properties of a good Barometer.

112. The space at the top of the tube should be per-

(**111.**) Describe the Siphon barometer. What takes the place of a cistern? How many scales are needed, and how are they arranged? How is the difference of level determined? *How are oscillations obviated?* (**112.**) What are the qualifications of a good barometer?

fectly free from air or moisture, otherwise they would, by their elastic force, prevent the mercurial column from rising to its proper height.

The elastic force of vapor of water is, as will be shown, very considerable, even at ordinary temperatures. To expel both air and moisture, the mercury should be boiled in the tube before the latter is inverted into the cistern.

The mercury should be pure, the bore of the tube should be sufficiently large, and the scale should be accurate. Mercury may be purified by distillation.

Thus far, mercury has been preferred to all other liquids for filling barometers. It is true, other liquids might be used, but in such case, the tube would become unwieldly from its length. In the case of water, a tube of about 35 feet would be required. There is another objection to using water, which arises from its tendency to form vapor even at ordinary temperatures. The formation of vapor at the top of the tube, would, as we have just seen, prove highly injurious to the working of the instrument.

Mean Height of the Barometer.

113. The height of the barometer is constantly fluctuating. The difference between the greatest and least heights observed at Paris, amounts to as much as one-thirteenth part of the greatest. The fluctuations become greater as we approach the poles, and less as we approach the equator.

The mean or average height at any place can be found only from a great number of observations. If we take hourly observations for one day and divide the sum of the heights by 24, the result is called *the mean height for that day*. This does not differ much from the height observed at midnight. If we take the sum of the mean daily heights for a year, and divide by 365, the result is *the mean height*

What liquid is best for filling barometers? Objections to other liquids? (**113.**) Where are the fluctuations of the barometer greatest? Least? Amount at Paris? *How is the mean height for a day determined? For a year?*

for that year. By taking the sum of the mean annual heights for many years and dividing it by the number of years, the result is *the mean height* for that place.

At the level of the sea, the mean height is not far from 30 inches, as already stated.

Causes of Barometrical Fluctuations.

114. The cause of the fluctuations observed in the barometer, is a change in the weight of the column of air above it. Since the weight of the entire atmosphere is constant, if it become heavier at one point on the earth's surface, it must become lighter at some other point; a fact which is confirmed by observations by means of the barometer.

The cause of the change of weight in the column of air over the barometer, is a change of temperature. When the temperature at any place is elevated, the air expands and rises upward until its lateral tension is greater than that of the surrounding air, when it flows away to the neighboring regions. When, on the contrary, the temperature is diminished, the air contracts and an additional quantity flows in from the neighboring regions.

The barometer, then, falls where there is a dilatation, and rises where there is a contraction, of the air.

The barometer serves as a weather-glass. It stands high in fair weather, and low in foul weather. A sudden fall of the barometer indicates an approaching storm, and a sudden rise, in general, indicates approaching fair weather.

The Index Barometer.

115. Fig. 78 represents an ornamental form of an INDEX BAROMETER. The manner in which the index is made to show the

For any place? What is the mean height at the level of the sea? (**114**.) What is the cause of the fluctuations observed? What is the cause of the change of weight in the aerial column? When does the barometer rise? Fall? *Use of the barometer as a weather-glass?* (**115**.) *Explain the Index barometer.*

fluctuations of the barometer, is shown in Fig. 79. The index is attached to an axis which bears a pulley. Passing over this pulley is a fine wire, at one extremity of which is attached an iron weight,

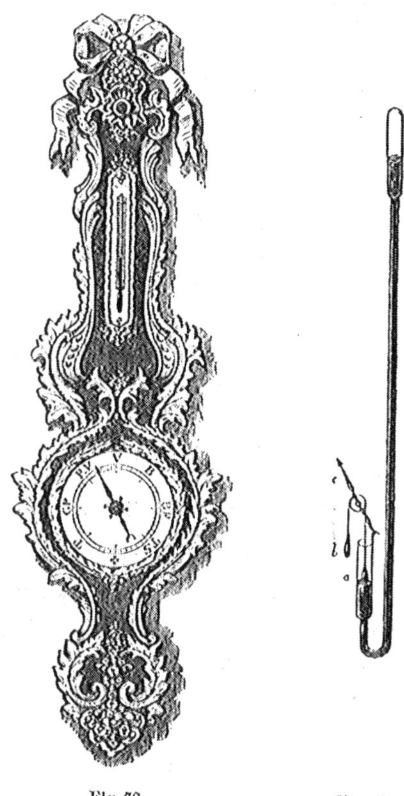

Fig. 78. Fig. 79.

a, which rises when the height of the mercury diminishes, and falls when this height increases. At the second extremity is a counterpoise, *b*, which keeps the wire tense, and causes the wheel to turn as the weights rise and fall.

The index plays in front of a dial-plate, around which are marked certain letters indicating the weather to be expected when the index stands at any one of them. The instrument shown in the figure is of French construction, and the letters are the initials of the French names of the different kinds of weather, as exhibited below. In the annexed table is shown the height of the barometer corresponding to each indication:

TABLE.

HEIGHT OF BAROMETER.	LETTERS.	FRENCH.	ENGLISH.
28.78 inches.	T.	Tempête.	Tempest.
29.13 "	G. P.	Grande pluie.	Heavy rain.
29.48 "	P. V.	Pluie ou vent.	Rain or wind.
29.84 "	V.	Variable.	Variable.
30.19 "	B.	Beau temps.	Fine weather.
30.54 "	B. F.	Beau fixe.	Settled weather.
30.90 "	T. S.	Très-sec.	Drought.

The above table is only given to illustrate the method of employing the instrument. It is evident that different tables would be required at different places. But little reliance is to be placed on barometers of this kind, as weather indicators.

Measure of Mountain Heights by the Barometer.

116. One of the most important applications of the barometer, is to the measurement of the height of any place above the level of the sea.

As we ascend above the level of the sea, the pressure of the air diminishes, and the barometer falls. Formulas have been deduced, by means of which the difference of level between any two places can be found, when we have the heights of the mercurial columns at the two places, together with the temperatures of the air and mercury at these places.

A detailed explanation of the method of making the observations,

---

Its construction and use. (**116.**) *On what principle is the barometer used for measuring heights?*

and deducing the difference of level, does not come within the plan of this work. For information on this subject, the reader is referred to Mechanics, Art. 188.

Height of the Atmosphere.

117. The density of the air at the surface of the earth is about 10,400 times less than that of mercury. Were there no decrease in density as we ascend, its height would be 10,400 times 30 inches, or 26,000 feet; that is, about five miles. But on account of the rapidly decreasing density upwards, the actual height is very much greater. It has been estimated to be not far from forty-five miles in height.

Atmospheric Pressure transmitted in all directions.

118. Gases, as well as liquids, transmit pressures in all directions, from which it results that the pressure of the air is not only felt downwards, but laterally in all directions. This is shown by the Magdebourg hemispheres, which adhere with equal force, whether the force to draw them asunder be exerted vertically, laterally, or in any oblique direction.

The same fact may be illustrated as follows: Let a tumbler be filled with water, and covered with a sheet of paper; then, holding the paper in contact with the water, let the tumbler be inverted. If the hand be withdrawn, the water remains in the tumbler, being held there by the pressure of the atmosphere, directed upwards, as shown in Fig. 80.

The wine-taster, shown in Fig. 81, is constructed on this principle. It consists of a tube open at both ends, the lower opening being quite small. The instrument is introduced into a cask of

---

(**117.**) Were the density the same as at the earth's surface, what would be its height? What is its estimated height? (**118.**) How are pressures transmitted through gases? How is the principle illustrated? *What is the principle of the wine-taster?*

wine through the bung-hole, and when it has become filled to the level of the liquor in the cask, the thumb is placed over the upper end, and the instrument is withdrawn. A portion of the wine is

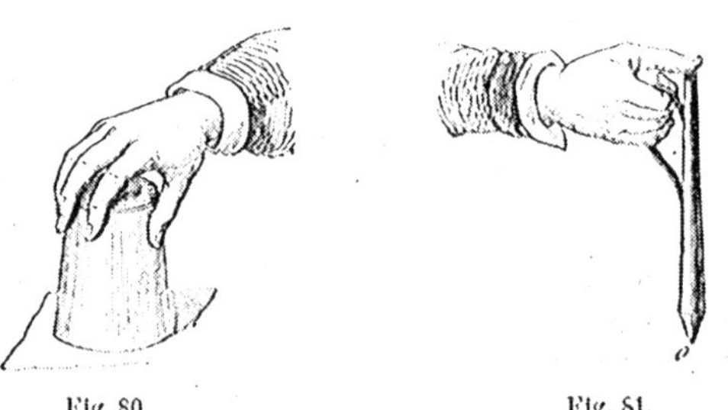

Fig. 80. Fig. 81.

held in the tube, being retained by the atmospheric pressure, and if the tube be placed over a tumbler, and the thumb be raised, the wine will flow out. This is the principle of the dropping tube, employed by druggists and others.

Pressure on the Human Body.

119. The pressure on each square inch of the body is 15 lbs.; hence, on the whole body the pressure is enormous. If we take the surface of the human body equal to 2000 square inches, which is not far from the average in the case of an adult, the pressure amounts to 30,000 pounds, or 15 tons.

If it be asked why the body is not crushed by this enormous pressure, the answer is, because it is uniformly distributed over the whole surface, and is resisted by the elastic force of air, and other gases, distributed through the tissues of the body.

The following experiment shows that the tissues of the human

Describe it and its use. What is the dropping tube? (**119.**) What is the amount of atmospheric pressure on the human body? How is this pressure resisted?

body contain air and gases, whose elasticity resists the atmospheric pressure. Let the hand be pressed closely upon the mouth of a glass cylinder, whose interior communicates with the air-pump, as shown in Fig. 82. No inconvenience will be felt. But if the air be exhausted from the cylinder, the flesh of the hand will be forced into the cylinder by the pressure from without, which is no longer resisted by the pressure of the air. The hand swells, and the blood tends to flow out through the pores.

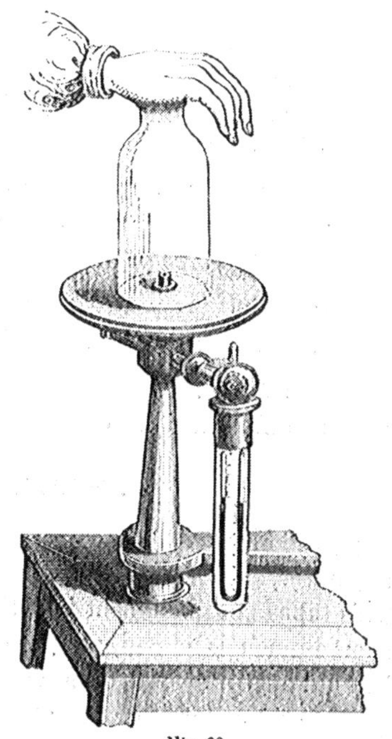
Fig. 82.

The question may be asked, why, when the hand is placed upon a body, it is not retained there by the pressure of the atmosphere. The answer is, there is a thin layer of air between the hand and the body, which exactly counterbalances the effect of the external pressure. Were the air perfectly excluded from between the hand and the body, there would be a strong tendency to adherence between them.

The operation of cupping, in medicine, depends upon the principle just explained.

How shown that the tissues of the body contain gases? Explain experiment. Principle of cupping.

II.—MEASURE OF THE ELASTIC FORCE OF GASES.

Mariotte's Law.

120. When a given mass of any gas or vapor is compressed, so as to occupy a smaller space, its elastic force is increased; on the contrary, when the volume is increased, its elastic force is diminished.

The law of increase and diminution of elastic force was first made known by MARIOTTE; hence it was called by his name. MARIOTTE'S law may be enunciated as follows:

The elastic force of any given amount of gas, whose temperature remains the same, varies inversely as its volume.

As a consequence of this law it follows that,

If the temperature remains constant, the elastic force varies as the density.

Mariotte's Tube.

121. MARIOTTE'S law may be verified by means of an apparatus, shown in Figs. 83 and 84, called *Mariotte's Tube.* This tube is of glass, bent into the shape of a letter J. The short branch is closed, and the long one open at the top. The tube is attached to a wooden frame, provided with suitable scales for measuring the heights of mercury and air in the two branches.

The instrument having been placed vertical, a sufficient quantity of mercury is poured into the long branch to cut off communication between the two branches, as shown in Fig. 83. The level of the mercury in the two branches is the same, and this level is at the 0 point of the two scales. The air in the short branch is of the same density, and has the same tension as that of the external atmosphere.

(**120.**) What is MARIOTTE'S Law? Consequence? (**121.**) Describe Mariotte's Tube.

If an additional quantity of mercury be poured into the longer branch of the tube, it will press upon the air in the shorter branch, and compress it. If the difference of level

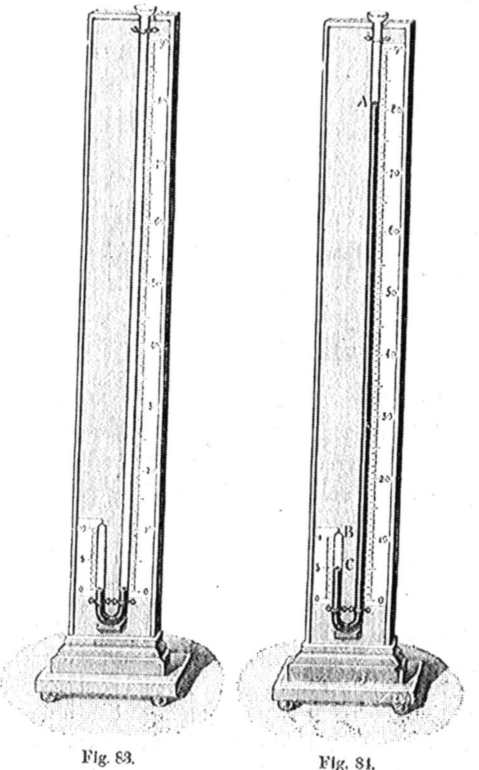

Fig. 83. Fig. 84.

in the two branches be made equal to the height of the barometrical column, as shown in Fig. 84 (where the difference is 76 centimetres, or 29.92 inches), the air will be compressed into *BC*, one half of its original bulk.

How used to verify the law?

In the figure, the air in BC is subjected to the pressure of two atmospheres, one from the actual atmosphere, transmitted through the mercury, and an equal pressure from the weight of the mercury, AC, which is equal to that of an atmosphere.

If the difference of height, AC, be made equal to two, three, four, &c., times that of the barometric column, the air in BC will be reduced to one third, one fourth, one fifth, &c., of its original bulk.

Manometers.

122. A MANOMETER is an apparatus for measuring the elastic force of a gas or vapor.

There are two principal kinds of manometers, the *open* and the *closed* manometer.

The Open Manometer.

123. Fig. 85 represents an OPEN MANOMETER, such as is often used for measuring the pressure of steam in a boiler.

It consists of a narrow tube of glass fixed against a vertical wall, and communicating with a cistern of mercury, C. A pipe leads from the boiler to the cistern, C, and by means of a stop-cock, steam may be admitted to the cistern, or cut off at pleasure.

When the tension of the steam in the boiler is just equal to that of the atmosphere, the mercury stands at the same level in the tube and cistern. When the tension of the steam becomes equal to twice that of the atmosphere, the mercury is forced from the cistern into the pipe, where it rises till the difference of level is 30 inches. This is marked 2 on the tube, and when the mercury is at this division, the tension of the steam is two atmospheres. The divisions 3, 4, 5, &c., are placed at distances of 30 inches, and when the mercury stands at any one of them, the manometer indicates a tension of the corresponding number of atmospheres.

(**122.**) What is a Manometer? How many kinds are employed? What are they?
(**123.**) Describe the Open Manometer. *Explain its action.*

Fig. 85.

In the figure, the tension indicated is 3½ atmospheres.

The Closed Manometer.

124. The CLOSED MANOMETER is shown in Fig. 86, and differs from the one just described, in having its vertical tube closed at the top. It is graduated on the principle enunciated in MARIOTTE'S law.

(**124.**) Describe the Closed Manometer. How is it graduated?

When the pressure in the boiler is one atmosphere, the mercury in the cistern and tube are at the same level, the tension of the steam and the elastic force of the air just balancing each other. When the pressure becomes two, three, four, &c., atmospheres, the air in the closed tube will occupy one half, one third, one fourth, &c., the space it did before, allowance being made for the weight of the mercury which is forced up into the tube. The instrument having been graduated, its use is evident. When it is desired to ascertain the tension of the steam in the boiler, the cock is turned, and the height to which the mercury ascends in the tube, indicates the tension in atmospheres. Any number of subdivisions may be made in either of the two manometers described.

Fig. 86.

Besides these, there is a metallic manometer, invented by M. BOURDON, and known as BOURDON'S Metallic Manometer. It is not so reliable as those described.

III.—APPLICATION TO PUMPS AND OTHER MACHINES.

The Air-pump.

125. An AIR-PUMP is a machine for exhausting the air from a closed space. The air-pump was invented by OTTO VON GUÉRICKE, in 1650.

A perspective view of one of the most common forms of the air-pump is given in Fig. 87. The details of its construction will be best studied from Figs. 88 and 89; the former represents a longi-

Illustrate. How is this manometer used? (**125.**) What is an Air-pump? When invented, and by whom?

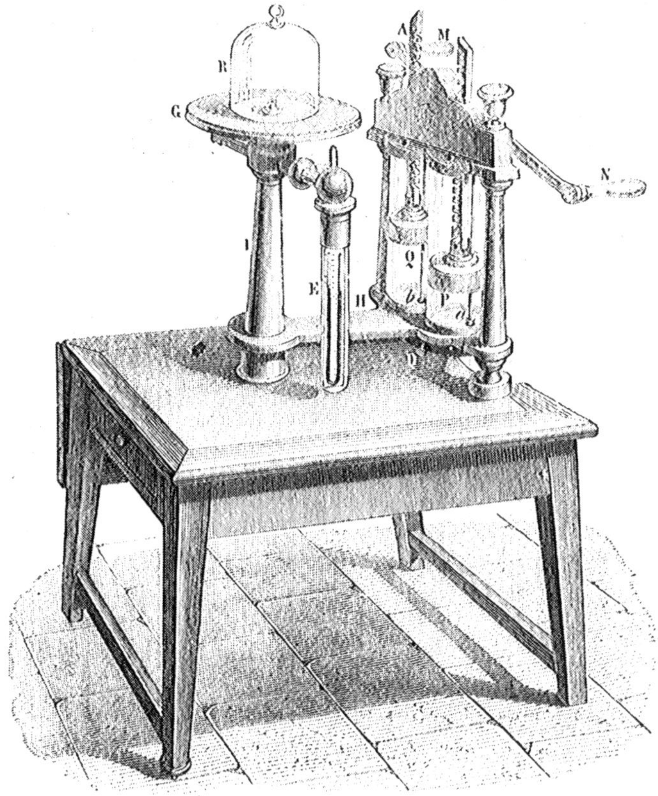

Fig. 87.

tudinal, and the latter a transverse section. In all of the figures, the same letters indicate corresponding parts.

The air-pump consists of two glass cylinders, called *barrels*, in which are pistons, *P* and *Q*, made of leather, thoroughly soaked in oil. The pistons are attached to rods, and are elevated and depressed by a lever, *NM*, Fig. 89, which imparts an oscillating motion to a pinion, *K*. The teeth of this pinion engage with corresponding ones

Give a complete description of the air-pump. Barrels. Pistons Rods.

on the inner sides of the piston rods, *A* and *B*. The machine is so arranged that one rod ascends whilst the other descends. The cylinders rest upon and are firmly attached to a platform, *H*, Fig. 88. On the same platform, *H*, is a column, *I*, which supports a plate, *G*. Resting upon the plate *G*, is a bell glass, *R*, called a *receiver*. The receiver communicates with both cylinders by a pipe, shown in Fig. 88.

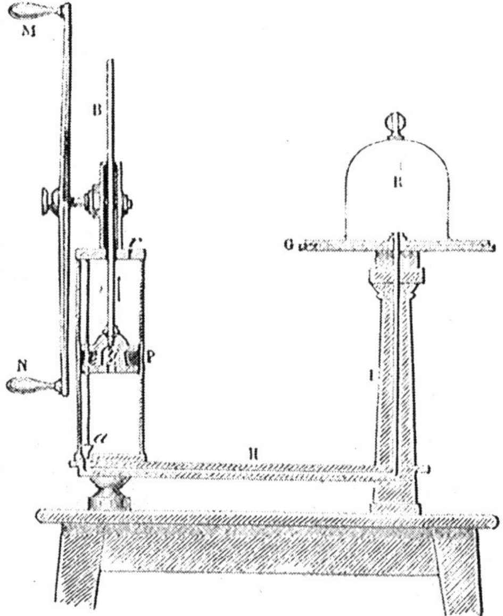

Fig. 88.

This pipe branches near the cylinders, one branch leading to each cylinder, as shown in Fig. 89. The pipe communicates with the cylinders by openings, which may be closed by conical valves, *a* and *b*. The valves *a* and *b* are attached to rods which pass through the pistons, and fitted to slide with gentle friction as the pistons move up and down. In the pistons are valves, *s* and *t*, which are gently

Receiver. Pipe. Valves. Valve rods.

pressed by spiral springs so as to permit the condensed air to escape and then to close the orifices in the valves. All of the valves, *a*, *b*, *s*, and *t*, open upwards.

In explaining the action of the air-pump, it will be sufficient to consider a single barrel, as shown in Fig. 88. The piston, *P*, being at the bottom of the barrel, the valves *a* and *t* are closed. If the piston be raised, the valve *a* is opened, whilst the valve *t* is kept closed by the spiral spring and the pressure of the atmosphere. The valve *a* is soon arrested by its rod coming in contact with the top of the barrel, and it then remains open during the ascent of *P*. The air in the barrel above the piston is driven out at the opening, *r*, and that in the receiver and pipe expands so as to fill the receiver, pipe, and barrel. If the piston, *P*, be depressed, it at once closes the valve *a*, and compresses the air in the barrel till its elastic force becomes great enough to force open the valve *t*, when it escapes into the atmosphere.

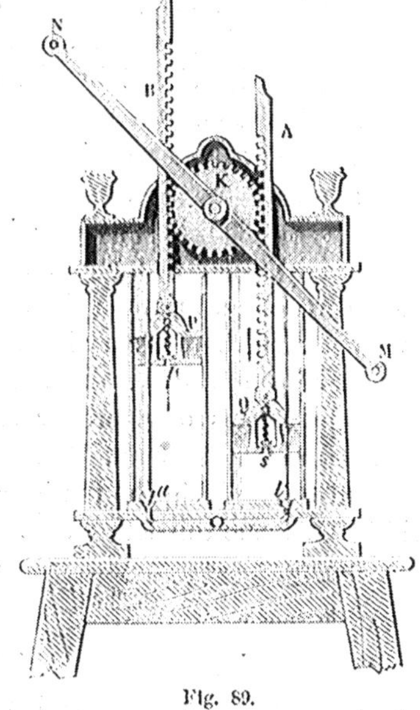

Fig. 89.

By this double stroke of the piston, *P*, a portion of the air is exhausted from the receiver, and if a second double stroke be made, a portion of what remains may in like manner be exhausted, and so on until nearly a perfect vacuum is formed in the receiver,

Describe the action of the air-pump in detail.

R, or in any other closed vessel attached to the pipe of the machine.

What has been said of one barrel, is equally true of the other; in fact, the instrument, as figured, is a double pump.

Measure of the Rarefaction produced.

126. In order to measure the degree of rarefaction produced, a glass cylinder, *E*, Fig. 87, is connected with the pipe by means of an opening through the column *L*. In this cylinder, is a glass tube bent into the form of the letter U, one branch being closed at the top, and the other open. The tube has its closed branch filled with mercury, and is called a *siphon gauge.*

The mercury, under ordinary circumstances, is kept in the closed branch by the atmospheric pressure, but as the air becomes rarefied in the receiver, the tension of the air becomes less and less, and finally the mercury falls in the closed branch and rises in the open one. The difference of level between the mercury in the two branches, is due to the tension of the rarefied air, and if this difference be determined by means of a proper scale attached to the gauge. the tension can be found. Thus, if the difference of level is reduced to one inch, the tension of the air in the receiver will be only one thirtieth part of the tension of the external atmosphere.

Experiments with the Air-pump.

127. We have already described several experiments requiring the employment of the air-pump, such as the shower of mercury, Fig. 1: the fall of bodies in a vacuum, Fig. 2; the bladder in a vacuum, Fig. 70; the bursting membrane, Fig. 72; and finally, the hemispheres of Magdebourg, Fig. 73.

(126.) How may the degree of rarefaction be measured? What is the siphon gauge? Explain its action and use.

The machine may be used to show that the air is necessary to the support of combustion and animal life. If a lighted taper be placed under the receiver, and the air exhausted, the light will grow dim, and finally will go out entirely. If an animal or bird be placed under the receiver, and the air exhausted, it will struggle and soon die. This experiment is shown in Fig. 90.

Fig. 90.

Animals and birds die as soon as they are placed in a vacuum; reptiles support life longer when deprived of air; as to certain insects, they live for many days under an exhausted receiver. They are enabled to live on the small supply of air which remains in the receiver, after as much of it as possible is extracted.

Preservation of Food in a Vacuum.

128. It has been discovered that articles of food which would soon perish if exposed to the air, may be preserved fresh for a long time if kept in a vacuum.

If fruits, vegetables, and the like, be placed in a bottle with water, and then heated gradually till ebullition takes place, all of the air will be driven out, being replaced by steam. If the bottle is corked and sealed in this condition, the fruit will remain fresh for years. On this principle, vast quantities of meat, fruit, vegetables, and the like, are prepared for naval and other purposes. Instead of bottles, tin canisters may be employed, which, after expelling the air, are hermetically sealed by soldering.

(**127.**) *How is it shown that air is necessary to combustion and animal life? What animals support life longest in a vacuum?* (**128.**) *How are articles of food preserved in vacuo? What applications are made of this principle?*

The Condenser.

129. A Cóndenser is a machine for condensing air, by forcing large quantities into a small space.

Such a machine is represented in Fig. 91. It is similar to the air-pump in its general construction, but differs in some of its details. The receiver is of very thick glass, and is

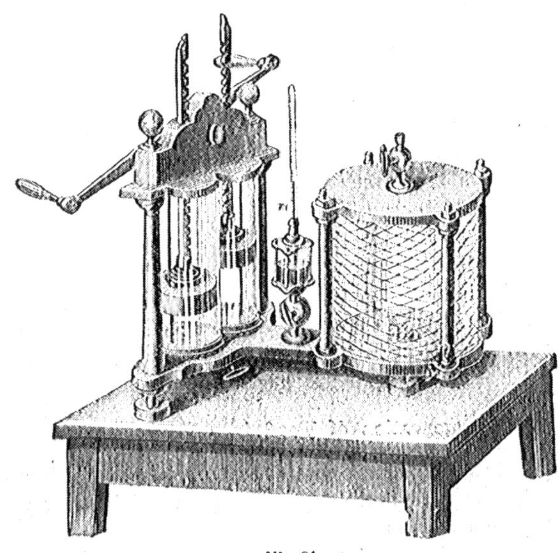

Fig. 91.

confined upon the plate by a second plate at the top, which is connected with the bottom plate by four brass rods with suitable screws and nuts. To prevent danger in case of rupture, the glass receiver is surrounded by a netting of strong wire. The four valves open in a direction contrary to that of the valves in the air-pump, so that air is forced

(**129**) What is a Condenser? Difference between it and the air-pump? How is the receiver guarded?

into the receiver at every double stroke, instead of being exhausted, as in that instrument. Finally, a closed manometer, *m*, is employed to indicate the tension of the compressed air. The machine is worked in the same way as the air-pump.

A taper burns more freely in compressed air than in the air under the ordinary pressure. Animals placed in compressed air do not experience any extraordinary inconvenience. In many submarine operations, it becomes necessary for men to work in an atmosphere of compressed air, and it has been found that no other inconvenience is felt under a pressure of three atmospheres, than a painful sense of compression in the ears. This feeling takes place only at the beginning and end of the operations, disappearing when an equilibrium is established between the tension of the air in the internal ear and that without.

Artificial Fountains.

130. An Artificial Fountain, is a machine by means of which water is forced upward in the form of a jet by the tension of compressed air. The most interesting instrument of this class, is that known as Hero's fountain, so named from its inventor, Hero, of Alexandria, born 120 B. C.

Hero's Fountain.

131. An ornamental form of Hero's Fountain is shown in Fig. 92. It consists of two globes of glass, connected by two metallic tubes. The upper globe is surmounted by a brass basin, connected with the globes by tubes, as shown in the figure.

To use the instrument, the tube which forms the jet is withdrawn, and through the opening thus made, the upper globe is nearly filled with water, the lower one containing air only. The jet tube is then replaced, and some water is poured into the basin.

How is the degree of condensation measured? *What effect has condensed air on combustion? On animal life? On divers?* (**130.**) What is an Artificial Fountain? (**131.**) Describe Hero's Fountain. How is it prepared for use?

Fig. 92.

The water in the basin, acting by its weight, flows into the lower globe, through the tube shown on the left of the figure, as indicated by the arrow head. This flow of water into the lower globe forces out a part of the air in it, which, ascending by the tube shown on the right of the figure, accumulates in the upper globe. The pressure of the air in the upper globe, acting upon the water in that part of

Explain its action.

the instrument, forces a part of it up through the *jet tube*, giving rise to a jet of water, which may be made to play for several hours without re-filling the instrument.

Intermittent Fountain.

132. An INTERMITTENT FOUNTAIN is one in which the flow is intermittent, that is, in which the flow takes place at regular intervals. Such fountains exist in nature. Fig. 93 represents an artificial fountain of this character.

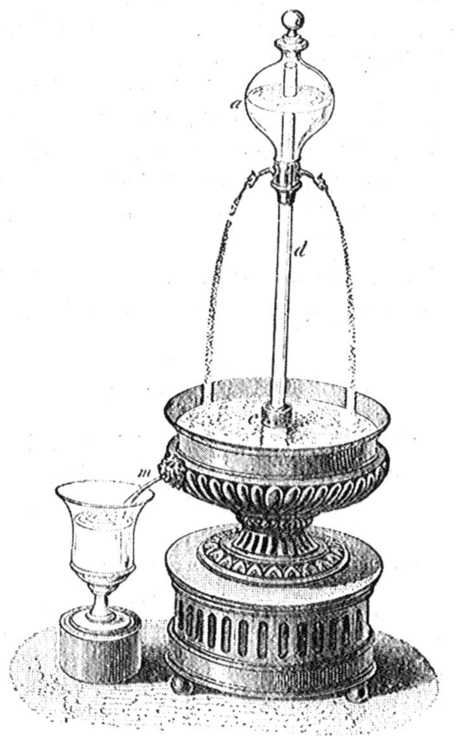

Fig. 93.

(**132.**) What is an Intermittent Fountain?

It consists of a glass globe, *a*, closed above by a glass stopper, and having two small tubes below, through which water can flow without interruption. The globe *a*, is supported by a hollow glass stem, *d*, which, rising from a metallic basin, enters the globe and reaches nearly to the top of it. Around the bottom of the tube *d*, are small holes, *c*, through which air can enter it, and thus reach the upper part of the globe *a*. A small spout, *m*, serves to draw off the water from the basin.

To use the instrument, the globe *a*, and the metallic basin, are nearly filled with water. So long as the holes, *c*, are covered with water, no flow will take place from the globe *a*, but as soon as the basin is emptied by the spout *m*, so as to expose the holes, *c*, the air enters the tube *d*, and reaching the globe *a*, the flow from the two tubes commences. The flow will continue until the holes, *c*, are again submerged, when it will cease, and so on as long as any water remains in the globe.

Of course the capacity of the two tubes, attached to the globe *a*, must be greater than that of the spout *m*.

The Atmospheric Inkstand.

133. An inkstand has been devised in accordance with the principles of atmospheric pressure, which, whilst preserving the ink from evaporation, is extremely simple in its construction.

The inkstand, partially filled with ink, is represented in Fig. 94. The body of the inkstand is air-tight. Near the bottom is a tube for supplying the ink as wanted, and also for filling the inkstand when necessary. The inkstand is filled by turning it until the tube is at the top, when the

Describe the artificial one shown in Fig. 93. Explain its action. (**133.**) Explain the construction and use of the Atmospheric Inkstand.

ink can be poured in through the tube. The pressure of the atmosphere prevents the ink from flowing out. When the ink has been used till its level falls below *o*, where the tube joins the main body of the inkstand, a bubble of air enters, and rising to the top, acts by its pressure to fill the tube again, and so on until the ink is exhausted.

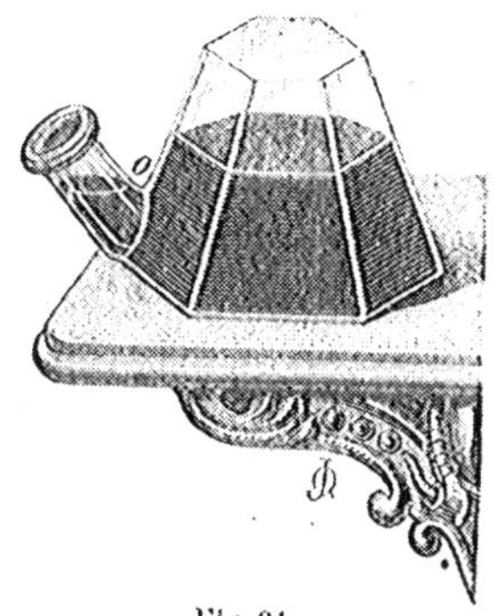

Fig. 94.

Water Pumps.

134. A WATER PUMP is a machine for raising water from a lower to a higher level, generally by the aid of atmospheric pressure. Three separate principles are employed in working pumps: the *sucking*, the *lifting*, and the *forcing* principles. Pumps are often named according as one or more of these principles are employed.

The Sucking and Lifting Pump.

135. A SUCKING AND LIFTING PUMP is represented in Fig. 95, in which a portion of the barrel is removed, to show more clearly the relative position of the parts.

It consists of a cylinder, usually of cast iron, called the *barrel* of the pump. The barrel communicates with a reservoir by a narrow pipe, called the *sucking pipe*, a part of which is shown in the figure. At the top of the sucking pipe is a valve opening upwards, called the *sleeping valve.* Within the barrel is a disk of metal or wood, packed with leather, called the *piston.* The piston is attached to a rod, *b*, called the *piston rod*, and is moved up and down through

(**134.**) What is a Water Pump? How many principles may be employed? What are they? How are pumps named? (**135.**) Describe the Sucking and Lifting Pump. Its barrel. Sucking pipe. Sleeping valve. Piston.

a certain space, called the *play* of the piston, by a lever, *B*, called the *pump-handle.* To cause the rod to work vertically, it is connected with the handle by a forked piece,

Fig. 95.

which is united to the piston rod by a hinge joint. This arrangement permits the rod, *b*, to glide up and down through a *guide*, as shown in the figure. Finally, the piston

Play of the piston. Piston rod. Guide.

itself is pierced in its centre, and carries a second valve, also opening upward, called the *piston valve*.

In explaining the action of this pump, we refer to Figs. 96, 97, and 98, which represent sections of the pump in different states of action. In all of the figures, a is the sleeping valve, c the piston valve, and B the sucking pipe.

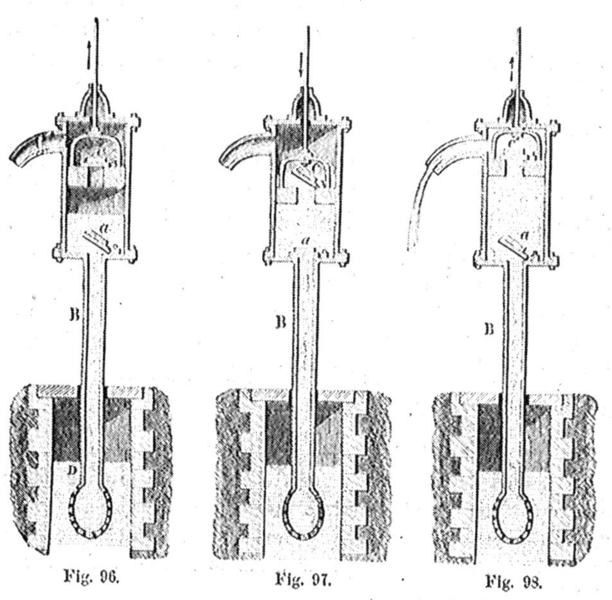

Fig. 96. Fig. 97. Fig. 98.

Suppose the piston to be at the lowest point of its play; there will then be an equilibrium between the pressure of the air within the pump and that without. When the piston is raised to the highest point of its play, the air beneath it is rarefied, and its tension diminished; the tension of the air in the sucking pipe then forces up the sleeping valve, and a portion of it escapes into the barrel. The tension of the air in the sucking pipe being less than that of

Piston valve. Explain the action of this pump.

the external atmosphere, a quantity of water rises in the pipe, to restore the equilibrium. The water continues to rise till its weight, increased by the tension of the air in the pump, is just equal to the tension of the external air. When the equilibrium is restored, the sleeping valve closes by its own weight.

Now, if the piston be depressed, the air in the barrel is condensed, forces open the piston valve, and a portion escapes into the external atmosphere. If the piston be raised again, an additional quantity of water will be forced into the pump, and after one or two strokes of the piston, it will begin to flow into the barrel, as shown in Fig. 96.

When the water rises above the lowest limit of the play of the piston, the latter in its descent will act to compress the water in the barrel. This pressure forces open the piston valve, and a portion of the water passes above the piston, as shown in Fig. 97. By continuing to elevate and depress the piston, the water will be raised higher and higher in the pump, till at length it will flow from the spout, as shown in Fig. 98.

As the water is raised in the pump by atmospheric pressure, it is necessary that the lowest limit of the play of the piston should not be more than 34 feet above the surface of the water in the reservoir, even at the level of the sea. To provide against barometric fluctuations and other contingencies, it is usual to make this distance considerably less than 34 feet.

The Forcing Pump.

136. In the Forcing Pump, the sucking pipe may be dispensed with, and the barrel plunged directly into the reservoir, as shown in Figs. 99 and 100, or a sucking pipe may be employed, as will be explained hereafter. We

---

What is the lowest limit of the play of the piston? (**136.**) What two forms may be given to the Forcing Pump?

shall first consider the case in which the sucking pipe is omitted.

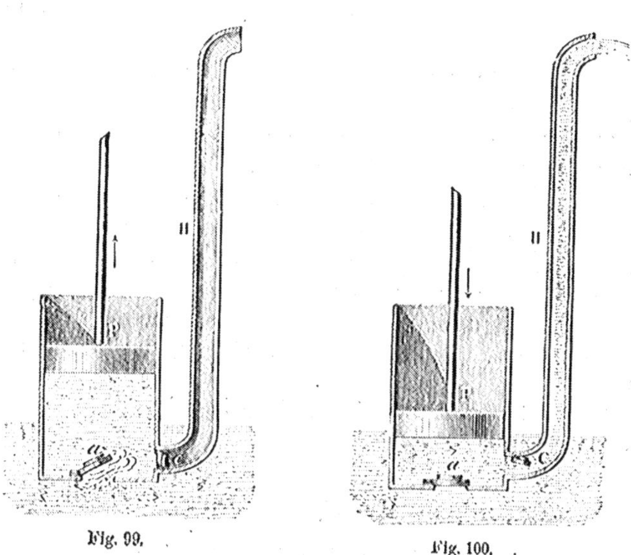

Fig. 99.

Fig. 100.

In this case the piston is solid, and a lateral pipe, *H*, called the *delivery pipe*, is introduced below the level of the lowest position of the piston. There are two valves, both fixed, the sleeping valve *a*, as in the sucking pump, and a valve *c*, opening into the delivery pipe.

When the piston is raised to its highest position, as shown in Fig. 99, the pressure of the atmosphere on the water in the reservoir forces open the sleeping valve, and the barrel is filled with water up to the bottom of the piston, when the sleeping valve closes by its own weight. On depressing the piston, the valve *c*, is forced open, and a portion of the water in the barrel is forced into the delivery pipe. When

Describe the piston. The delivery pipe. Explain the action of the forcing pump in detail.

the piston reaches its lowest position, the weight of the water in the delivery pipe closes the valve *c*, and prevents the water in the delivery pipe from returning into the barrel.

By continually raising and depressing the piston, additional quantities of water are forced into the delivery pipe, which finally escape from the spout at the top of the delivery pipe, as shown in Fig. 100.

To regulate the flow of the water through the delivery pipe, and to facilitate the working of the pump, an air-vessel is generally introduced, as will be explained in the next article. Sometimes the working is rendered uniform by combining two forcing pumps in such a manner, that the piston of the one ascends, whilst that of the other descends. This combination is also explained in the next article.

The oil in a carcel-lamp is forced up into the wick by a double forcing pump, moved by clock-work.

The Fire Engine.

137. A Fire Engine is a double forcing pump, having its delivery pipe composed of leather or other flexible material. It is used, as its name implies, for extinguishing fires.

Fig. 101 shows a section of the essential parts of a fire engine. In this figure, *PQ* is the lever to which are attached the piston rods, that move the pistons *m* and *n*; *R* is an air-vessel with two valves, one admitting water from each barrel; *Z* is the entrance to the hose or delivery pipe; *M* and *N* are rods sustaining the framework of the machine.

The two barrels are plunged into a reservoir which is kept supplied with water. This water flows into a space

---

How is the flow regulated? How is the working rendered uniform? How is the oil raised in a carcel-lamp? (**137.**) What is a Fire Engine? Describe it in detail.

beneath the barrels through holes represented on the right and left of the figure, and from thence is forced into the air-vessel in a manner entirely similar to that explained in

Fig. 101.

the last article. When the water is forced into the air-vessel *R*, the air is at first compressed, after which it acts by its tension to force a continuous current through the hose.

The lever is provided with long handles at right angles to its length, so that it may be worked by several men acting together. Within a few years many improvements have been introduced into the fire engine, one of the most important being the application of steam as a motor.

How is it supplied with water? *How is it maneuvered?*

The Sucking and Forcing Pump.

138. This differs from the simple forcing pump described in Art. 136, in having a sucking pipe and an air-vessel. It consists of a barrel, *A*, a sucking pipe, *B*, a sleeping valve, *G*, and a solid piston, *C*, worked by a lever, *E*, and piston rod, *D*. A pipe leads from the bottom of the barrel, through a sleeping valve, *F*, into an air-vessel, *K*. The delivery pipe, *H*, enters the air-chamber at its top and extends nearly to the bottom.

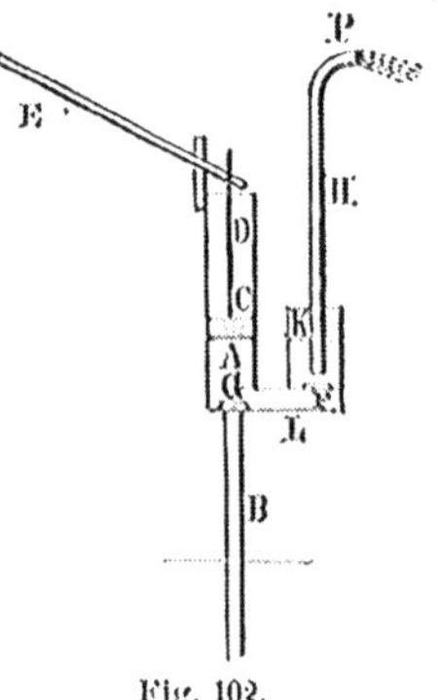

Fig. 102.

To explain the action of the pump, suppose it empty and the piston at its lowest position; when it is raised to its highest position, the air in the barrel is rarefied, the tension of the air in the sucking pipe forces open the valve, *G*, and a portion of it escapes into the barrel; the water is then forced up the sucking pipe by the tension of the external air acting on the surface of the water in the reservoir until an equilibrium is produced, when the valve, *G*, closes by its own weight. If the piston be again depressed to its lowest limit, the air in the barrel is condensed until its tension exceeds that of the external air, when it forces open the valve, *F*, and a portion escapes into the air-vessel. After a few double strokes of the piston the water rises through the valve, *G*, and the action becomes the same as in the pump described in Art. 136; with the exception of the air-vessel, which serves to keep up a continuous stream through the delivery pipe. The piston ought not to be more than 34 feet above the reservoir. The spout, *P*, may be at any height above *K*.

(**138.**) Describe the sucking and forcing pump and its mode of action.

The Siphon.

139. The SIPHON is a bent tube, by means of which a liquid may be transferred from one reservoir to another, over an intermediate elevation. The siphon may be used with advantage when it is required to draw off the upper portion of a liquid without disturbing the lower portion. This operation is called *decanting.*

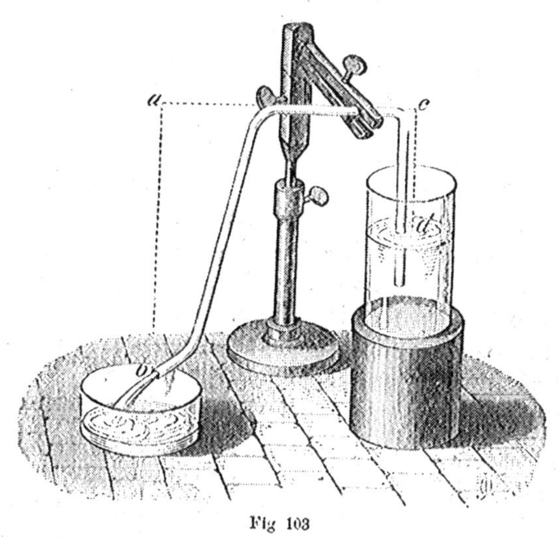

Fig 103

The siphon consists of two branches of unequal lengths, as shown in Fig. 103. The shorter one is plunged into the liquid to be decanted, and the flow takes place from the longer one.

To use the siphon, it must first be filled with the liquid. This operation may be effected by applying the mouth to the outer end of

(**139.**) What is a Siphon? When may it be used with advantage? What is decanting? Explain the operation. *How is the siphon prepared for use?*

the siphon, and exhausting the air by suction, or it may be inverted and filled by pouring in the liquid, and stopping both ends, after which it is again inverted, care being taken to open both ends at the same instant. Sometimes a sucking pipe is employed to exhaust the air and fill the siphon.

When the flow commences, it will continue until the liquid in the first reservoir falls below the level of the end of the siphon.

To understand the action of the siphon, we must consider the forces called into play. The water is urged from *d* towards *b*, by the pressure of the atmosphere on the fluid in the reservoir, together with the weight of the water in the outer branch of the siphon; that is, by the weight of a column of water whose height is *ab*. This motion is retarded by the pressure of the atmosphere at *b*, together with the weight of the fluid in the inner branch; that is, by the weight of a column whose height is *cd*. The difference of these forces is the weight of a column of the liquid whose height is the excess of *ab* over *cd*, and it is by the action of this force that the flow is kept up. The greater this difference the more rapid will be the flow, and the less this difference the slower the liquid will escape. When this difference becomes zero, the flow ceases altogether.

The siphon is used for conveying water over hills, but for this purpose the highest point of the tube should not be more than thirty feet above the level of the water in the reservoir, this being about the height at which the atmospheric pressure will sustain a column of water.

If a siphon be mounted on a piece of cork, so as to sink as the level of the fluid falls, the flow will be constant. Such a siphon is called a *siphon of constant flow*.

How long will the flow continue? Explain the principle and action of the siphon in detail. *How high can water be raised by a siphon? Describe a siphon of constant flow.*

IV.—APPLICATION TO BALLOONING.

Buoyant Effort of the Atmosphere.

140. It has been shown that a body plunged into a liquid is buoyed up by a force equal to the weight of the displaced liquid. That a similar effect is produced upon a body in the atmosphere, may be shown by means of an instrument called a *baroscope*, which is represented in Fig. 104.

The BAROSCOPE consists of a beam like that of a balance, from one extremity of which is suspended a hollow sphere of copper, and from the other extremity a solid sphere of lead. These are made to balance each other in the atmosphere.

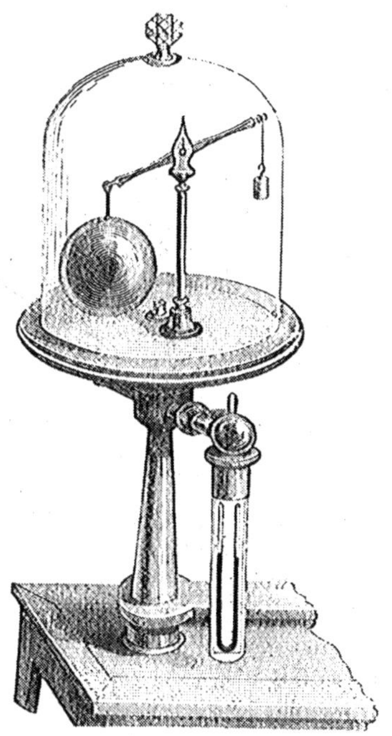

Fig. 104.

If the instrument be placed under the receiver of an air-pump and the air exhausted, the copper sphere will descend. This shows that in the air it was buoyed up by a force greater than that exerted upon the leaden sphere. If, now, the leaden sphere be increased by a weight equal to that of a volume of air equal to the bulk of the copper

(**140.**) What instrument is used to show the buoyant effort of the air? Describe the Baroscope. Explain its use.

sphere diminished by that of the leaden sphere, it will be found, after the air is exhausted, that the balance is in equilibrium. This shows that the buoyant effort is equal to the weight of air displaced. Hence we have the following principle, entirely analogous to the principle of ARCHIMEDES:

When a body is plunged into a gas, it is buoyed up by a force equal to the weight of the displaced gas.

If the buoyant effort is greater than the weight of the body, the latter will rise; if it is less, the body will fall; if the two are equal, the body will float in the atmosphere without either rising or falling.

Smoke, for example, rises, because it is lighter than the air which it displaces. It continues to rise until it reaches a stratum of air where its weight is just equal to that of the displaced air, when it will come to rest and remain suspended. A soap-bubble filled with warm air floats for a considerable time in the atmosphere, being nearly of the same weight as the displaced air.

The Balloon.

141. A BALLOON is a spherical envelope filled with some gas lighter than the air.

Balloons are of very different sizes, and are filled with gases of very different specific gravities, and consequently capable of raising very different weights in ascending to the upper regions of the atmosphere.

The first balloon was constructed by STEPHEN and JOSEPH MONTGOLFIER, two brothers, in 1783. It was made of linen, lined with paper. It was about forty feet in diameter, and weighed 560 lbs. It was filled with heated air and smoke, furnished by burning wet straw, paper, and the like, under the balloon, the lower part of which was left open to receive it. The balloon rose to a height of more than a mile, but it soon became cooled in the upper regions of the air and fell to the earth.

Give the law of buoyancy. *When will a body rise in the atmosphere? When fall? When remain neutral? Examples.* (**141.**) What is a Balloon? *Give an account of the early history of ballooning.*

In the following August, two brothers, named ROBERT, constructed a balloon of silk saturated with india-rubber, and filled it with hydrogen gas. The ascensional power of this balloon was very great, and being set loose in Paris, it rose with great rapidity, and at the end of four minutes had reached a height of nearly a thousand yards, when it was lost sight of by entering a cloud. It descended fifteen miles from Paris, to the astonishment of the people who saw it.

Manner of filling a Balloon and making an ascent.

142. Balloons may be filled either with hydrogen or with illuminating gas, which is a compound of carbon and hydrogen. On account of the readiness with which the latter gas can be obtained, together with its cheapness, it is generally employed. The envelope is made of silk, rendered air-tight by some kind of varnish, and is strengthened by a network of cords. This network also serves to sustain a wicker basket, or car, in which the aeronaut is seated.

Fig. 105 represents the method of filling a balloon, and preparing it for an ascension. Two masts are erected at a suitable distance from each other, at the tops of which are pulleys. A rope passing through a loop at the top of the balloon, also passes over the pulleys, and serves to raise the balloon during the process of filling.

When the process of filling commences, the balloon is raised till it is three or four feet above the ground, when the gas is introduced by means of a pipe or hose which connects with a gasometer. As the balloon fills with gas it is held down by ropes, and when completely filled, the opening is closed, and the car attached. Care should be taken not to fill the balloon completely, as the gas expands in rising, and unless an allowance is made for this increase of volume, the balloon might be ruptured.

To regulate the ascensional power, the car is ballasted by sand, contained in small bags. Everything being ready, the ropes are detached, and the balloon ascends with greater or less velocity, according to the ascensional force, that is, the excess of the buoyant effort over the weight of the entire balloon and its cargo.

When the aeronaut finds that he does not ascend fast enough, he increases the ascensional force by emptying one or more of the sand

(**142.**) *With what are balloons filled? Explain the method of filling a balloon. How is the ascensional power regulated?*

bags. In like manner, in descending, if the velocity is too great, or if the balloon tends to fall in a dangerous place, the weight of the balloon is diminished by emptying some of the sand bags.

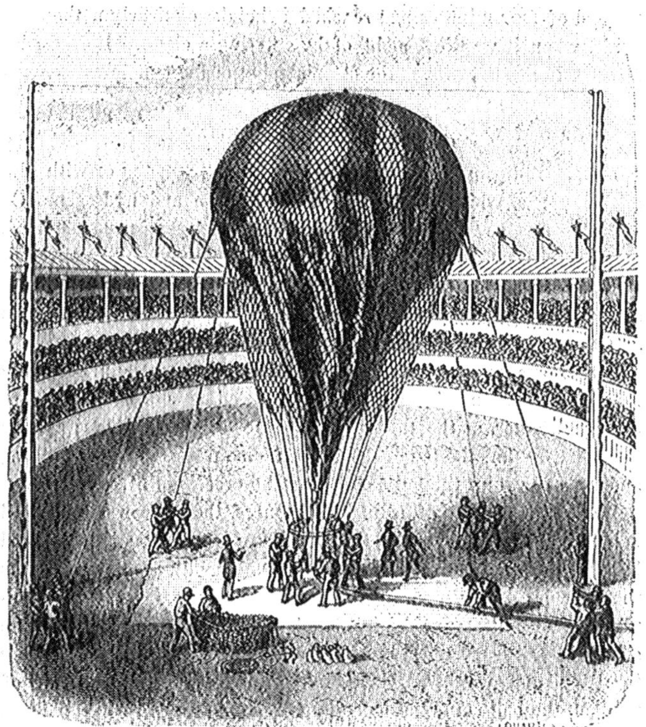

Fig. 105

To render the descent less difficult, the aeronaut is provided with an anchor or grapple, suspended from a cord, by means of which he can seize upon some terrestrial object when he comes near the earth. When the anchor is made fast, the aeronaut draws down the balloon by pulling upon the cord. The anchor, the sand bags, and the wicker car, are represented on the ground in Fig. 105.

How does the aeronaut make fast to the earth in descending?

At the top of the balloon is a valve kept closed by a spring; it can be opened by means of a string descending through the balloon to the car of the aeronaut. When he wishes to descend, he opens the valve, and allows a portion of the gas to escape. To ascertain whether he is ascending or descending, the aeronaut is provided with a barometer; when ascending, the barometric column falls, and when descending, it rises. By means of the barometer the height at any time may be determined.

The Parachute.

143. A PARACHUTE is an apparatus by means of which an aeronaut may abandon his balloon, and descend slowly to the earth.

The form and construction of a parachute is shown in Fig. 106. It consists of circular piece of cloth, 15 or 16 feet in diameter, presenting, when spread, the form of a huge umbrella. The ribs are made of cords, which, being continued, are attached to a wicker car, as shown in the figure.

When the aeronaut wishes to descend in the parachute, he enters the car and detaches the parachute from the balloon. At first he descends with immense rapidity, but the air soon spreads the cloth, and then acting by its resistance, the velocity is diminished, and the aeronaut reaches the earth without injury. A hole is made at the centre of the parachute, which, by allowing a part of the compressed air to escape, directs the descent and prevents violent oscillations that might prove dangerous.

The parachute was first tried by BLANCHARD, who placed a dog in the car, and detached it from the balloon. A whirlwind arrested its descent and carried it up above the clouds, where BLANCHARD soon after fell in with it, to the great joy of the poor animal. A current again separated the two voyageurs, but both reached the earth in safety, the dog being the last to descend.

J. GARNERIN was the first man who ventured to descend in a parachute, which he did by detaching himself from a balloon at the

---

What is the use of the valve at the top? What is the use of the barometer? (143.) What is a Parachute? Describe it. *Explain its use and action.*

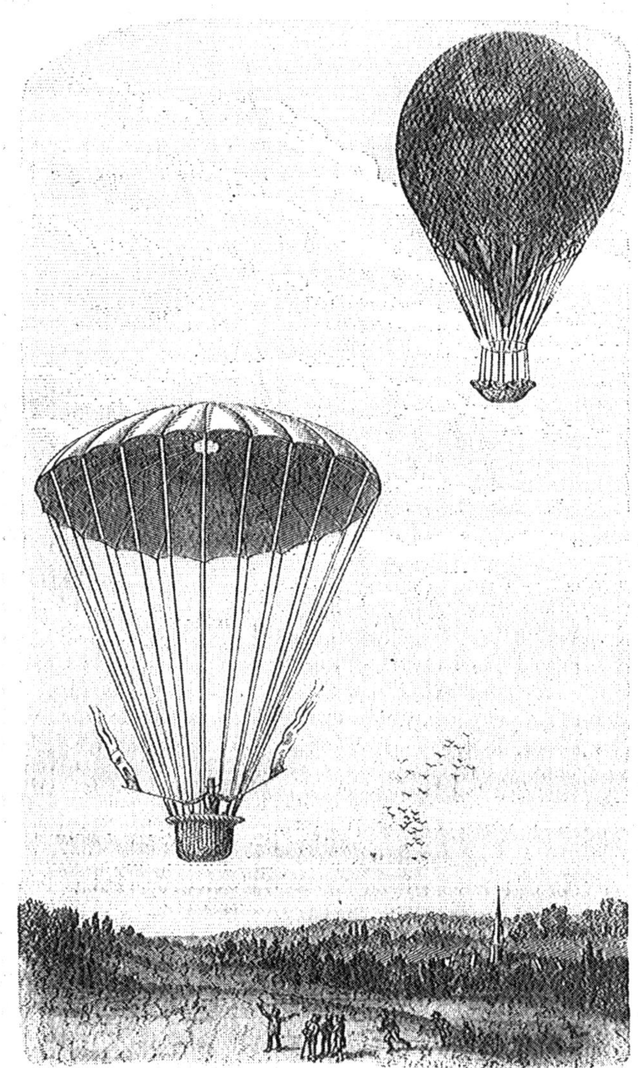

Fig. 106.

height of a thousand yards above the surface of the earth. He descended in safety.

Remarkable Balloon Ascensions.

144. The first ascension was made in October, 1783, by DE ROZIER. His balloon was filled with heated air, and was confined by a rope, so that he only rose to a height of about a hundred feet. In the following year DE ROZIER and D'ARLANDES ascended in a fire balloon from the Bois de Boulogne, and after a voyage of twenty-five minutes they descended on the other side of Paris. In a subsequent ascent DE ROZIER lost his life in consequence of his balloon taking fire. In 1785, BLANCHARD and JEFFRIES crossed the English Channel from Dover to Calais. During the voyage they had to throw overboard all of their ballast, then their instruments, and finally their clothing, to lighten the balloon. In 1804, GAY LUSSAC ascended to the height of 23,000 feet above the level of the sea. At this height the barometric column fell to 12.6 inches, and the thermometer, which at the surface of the earth was 31°, fell to 9½° below 0.

At such heights, substances which absorb moisture, like paper and parchment, become dry and crisp as if heated in an oven, respiration becomes difficult, and the circulation is quickened on account of the rarefaction of the air. GAY LUSSAC relates, that his pulse rose from 66 to 120. The sky becomes almost black, and the silence that prevails is frightful. After a voyage of six hours, GAY LUSSAC descended, having travelled about ninety miles.

On the 1st of July, 1859, Messrs. WISE, LA MOUNTAIN, GAGER, and HYDE, ascended from St. Louis, Mo., and descended at Henderson, Jefferson Co., N. Y., having travelled 1150 miles in a little less than twenty hours, or about fifty-seven miles per hour. This is the most celebrated voyage on record.

During the recent siege of Paris balloons were successfully employed as a means of communication between the forces within the city and those without the lines of the enemy. Balloons have also been used for making observations in the higher regions of the atmosphere.

(**144.**) *Describe some of the most remarkable Balloon Ascensions. That of* ROZIER. *Of* BLANCHARD *and* JEFFRIES. *Of* GAY LUSSAC. *What effect has the atmosphere at great elevations? Describe the great American voyage. Uses.*

CHAPTER IV.

ACOUSTICS.

I.—PRODUCTION AND PROPAGATION OF SOUND.

Definition of Acoustics.

145. ACOUSTICS is that branch of Physics which treats of the laws of generation and propagation of sound.

Definition of Sound.

146. SOUND is a motion of matter capable of affecting the ear with a sensation peculiar to that organ.

Sound is caused by the vibration of some body, and is transmitted by successive vibrations to the ear. The original vibrating body is said to be *sonorous*. A body which transmits sound is called a *medium*. The principal medium of sound is the atmosphere; wood, the metals, water, &c., are also media.

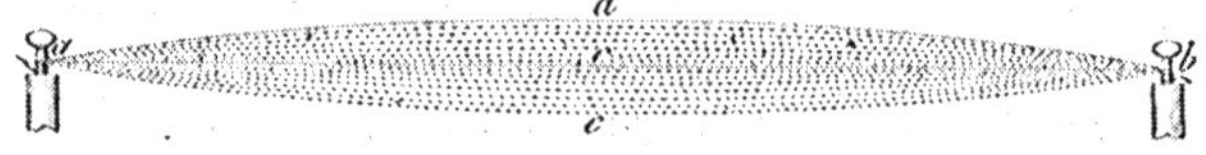

Fig. 107.

Let us take, for illustration, a stretched cord which is made to vibrate by a bow, as in a violin, for example. When the cord is drawn from its position of rest, *acb*, Fig. 107, to the position *adb*, every point of the cord is drawn from its position of equilibrium;

(**145.**) What is Acoustics? (**146.**) What is Sound? What is its cause? How is it transmitted? What is a sonorous body? A medium? Examples. *Explain the vibrating cord.*

when it is abandoned, it tends, by virtue of its elasticity, to return to its primitive state. In returning to this position, it does so with a velocity that carries it past *acb* to *aeb*, from which it returns again nearly to *adb*, and so on vibrating backward and forward, until, after a great number of oscillations, it at length comes to rest.

Sound-waves in Air.—Mode of Propagation.

147. Sound-waves are produced in the air by the vibration of some sonorous body. When the body moves forward it strikes the air in front of it and condenses a stratum whose thickness depends on the rapidity of vibration; the particles of this stratum impart the condensation to those

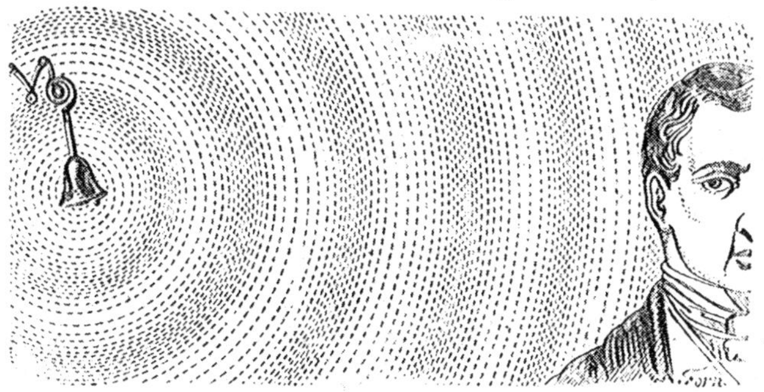

Fig. 108.

of the next, and these in turn to those of the next, and so on; the condensation thus transmitted outward is called the *condensed pulse*. When the body moves backward, the air in front of it follows and produces rarefaction in a stratum whose thickness depends on the rapidity of vibration; this causes a backward movement and consequent rarefaction in the next stratum, which is transmitted to the next, and so on; the rarefaction thus propagated outward is called the *rarefied pulse*. Each complete vibration of the sonorous body generates a condensed and a rarefied pulse, and these taken together constitute a *sound-wave*.

(147.) Describe the mode of sound propagation in the air.

If the vibrations are continuous, a series of sound-waves are generated travelling outward in the form of spherical shells, as shown in Figure 108.

The rate at which the sound-wave travels is the *velocity of sound;* the distance through which it travels in the time of one vibration of the sonorous body is the *wave length;* hence the wave length is always equal to the velocity of sound divided by the number of vibrations in one second. The form of the sound-wave is transmitted through the air, but the individual particles of air simply oscillate to and fro in the direction of wave propagation, moving forward on the passage of the condensed and backward on the passage of the rarefied pulse; the distance through which each particle oscillates is called the *amplitude of vibration* of the particle.

Any two particles situated on a line in the direction of propagation, and a distance from each other equal to a wave length, are always moving in the same direction and with equal velocities; such particles are said to be in the same *phase.* All the particles of any wave that are in the same phase are on the surface of a sphere, which is called a *wave front.*

Superposition of Sound-waves.

148. It is to be remarked that many sounds may be transmitted through the air simultaneously. This shows that the sound-waves cross each other without modification. In listening to a concert of instruments, a practiced ear can detect the particular sound of each instrument.

Sometimes an intense sound covers up or drowns a more feeble one; thus, the sound of a drum might drown that of the human voice. Sometimes feeble sounds, which are too faint to be heard separately, by their union produce a sort of murmur. Such is the cause of the murmur of waves, the rustling sound of a breeze playing through the leaves of a forest, and the indistinct hum of a distant city.

It has been shown that two sound-waves may, under certain circumstances, neutralize each other, producing silence.

What is the velocity of sound? The wave length? The amplitude of vibration of a particle? Its direction? A wave front? (**148.**) Do sound-waves interfere with each other's progress? How shown? *Explain the murmur of leaves. Waves.*

Sound is not propagated in a Vacuum.

149. That some medium is necessary for the transmission of sound, may be shown by the following experiment.

In a glass globe with a stop-cock, is suspended a bell, as shown in Fig. 109. When the globe is shaken, the sound of the bell is distinctly heard. If the air be exhausted from the globe, no sound is heard when the globe is shaken.

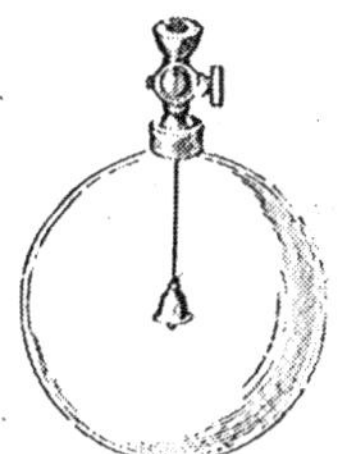
Fig. 109.

This experiment may be performed otherwise as follows:

A bell is placed under the receiver of an air-pump, provided with a striking apparatus set in motion by clock-work. Before the air is exhausted, the strokes of the hammer on the bell are distinctly heard, but as the air is exhausted the sound becomes fainter and fainter, till at last it ceases to be heard.

For the complete success of this experiment, the bell and clock-work should be placed upon a cushion, of some substance which does not readily transmit sound.

In ascending high mountains, the air becomes rarefied, and a corresponding diminution in the intensity of sounds is observed. SAUSSURE, on firing a pistol on the summit of Mt. Blanc, reports that it produced only a feeble sound, like that heard on breaking a stick.

Propagation of Sound in Liquids and Solids.

150. Sound is transmitted, not only by gases, but also by liquids and solids. Divers hear sounds from the shore when

(**149.**) How is it shown that sound is not transmitted in a vacuum? Another method of showing the same thing. *Effect of elevation on sound.* (**150.**) How is it shown that liquids and solids transmit sounds?

under water, and sounds made under water are heard on shore. A slight sound made at one end of a long stick of timber is distinctly heard by an ear at the other end, even when it might be inaudible at an equal distance through the air.

The earth transmits sounds, and by placing the ear in contact with it, sounds may be distinguished at a great distance. This method of hearing approaching footsteps of men or animals, is well understood by hunters. In the construction of subterranean galleries for mining purposes, the miner is often guided, as to the direction he should take, by sounds transmitted through large masses of earth and rock.

Velocity of Sound in the Air.

151. That sound occupies an appreciable time in passing from point to point may be shown by many familiar examples. If we notice a man cutting wood at a distance, we perceive that his axe falls some time before the sound of the blow reaches the ear. If a gun is discharged, we see the flash before we hear the report. In like manner the flash of lightning is seen before we hear the thunder.

In 1822, a number of scientific men undertook a series of very nice experiments to determine the velocity of sound. They placed a cannon on the hill of Montlery, near Paris, and another on a plain near Ville-Juif, the distance between them being 61,047 feet. At each station twelve discharges were made at intervals of ten minutes; the discharges alternating between the stations at intervals of five minutes. Observers placed at each station observed the intervals of time that elapsed between seeing the flash and hearing the report of the cannon at the other station. The average

How is it shown that the earth transmits sound? Illustrate. (**151.**) How is it shown that sound requires an appreciable time to pass from place to place? Illustrate. Explain the experiments made near Paris.

interval was 54.6 seconds, and the temperature was 61° F.; the actual velocity was found to be 1118 feet per second, which, after correcting for temperature, gave 1090 feet per second for the temperature 32° F. The velocity increases about 1 foot per second for each degree of Fahrenheit.

Velocity of Sound in Liquids and Solids.

152. Liquids and solids transmit sound more rapidly than air. Experiments made by transmitting sound across the Lake of Geneva, in Switzerland, show that the velocity of sound in water is about 4700 feet per second, which is more than four times its velocity in air.

That sound travels faster in iron than in air, may be shown by placing the ear at one extremity of a long iron bar or tube, whilst it is struck on the other end with a hammer. Two sounds will be heard, the first transmitted through the iron, and the second through the air.

Reflection and Refraction of Sound.

153. Sound-waves in air emanating from a point move in concentric spheres; if they meet an obstacle, they are turned back, forming a new set of waves whose common centre is as far behind the obstacle as the generating point is in front of it. This phenomenon is called *reflection*, and the two sets of waves are called *incident* and *reflected* waves. A ray of sound is a line along which sound acts; it is normal, or perpendicular to the wave fronts, and consequently its direction passes through their centre. A line drawn to any point of the reflecting surface from the centre of incident waves, is an *incident ray;* a line drawn through the same point from the centre of reflected waves, is a *reflected ray;* the point itself is the *point of incidence.*

What is the velocity of sound at 32° F.? At what rate does it increase with the temperature? (**152.**) How was it shown that sound travels faster in water than in air? In iron than air? (**153.**) What is reflection? Incident and reflected waves? Incident and reflected rays?

Let *I*, *I*, be incident waves whose centre is *C*; *W*, *W*, reflected waves whose centre is *C'*; and *PN*, a normal to the reflecting surface, *AB*, at the point *P*. Then is *CP* an incident ray; *PR* a reflected ray; and *P* is the point of incidence. The angle, *CPN*, is called *the angle of incidence*, and *NPR* is called *the angle of reflection*. The incident and reflected rays lie in a plane normal to the reflecting surface at the point of incidence, and the angles of incidence and reflection are equal.

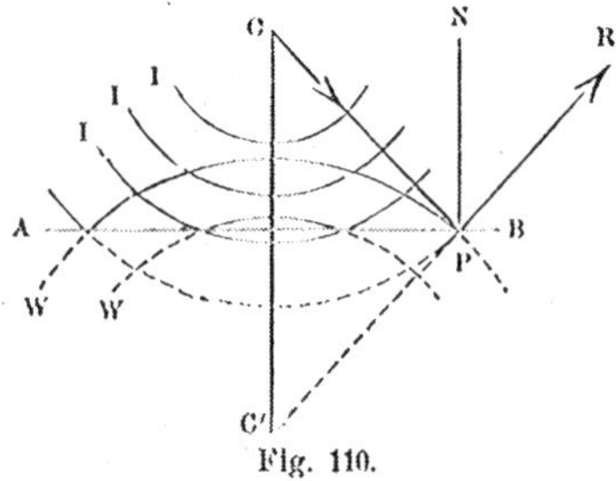

Fig. 110.

Refraction is the change of direction experienced by a ray when it passes obliquely from one medium to another.

Let *I*, *I*, be waves whose centre is *C*, incident on the surface *AB*, which separates two media, and let *W*, *W*, be modified waves in the second medium. If sound travel at different rates in the two media, the modified waves in the second medium will have their centre at some point, *C'*, on the perpendicular to *AB* through *C*. If the velocity in the second medium be less than in the first, *C'* will be further from *AB* than *C*, otherwise the point *C'* will be between *C* and *AB*. Suppose the former case, and let *NN'* be a normal to the deviating surface *AB* at *P*; then is *CP* an *incident ray*; *PR* a refracted ray; *CPN* the *angle of incidence*; *N'PR* the *angle of refraction*; and *CPC'* is the *amount of refraction*. The angles of incidence and refraction are in a plane normal to the deviating surface at the point of incidence, and the sine of the angle of incidence is equal to the sine of the angle of refraction multiplied by a constant quantity, called the *refractive index*.

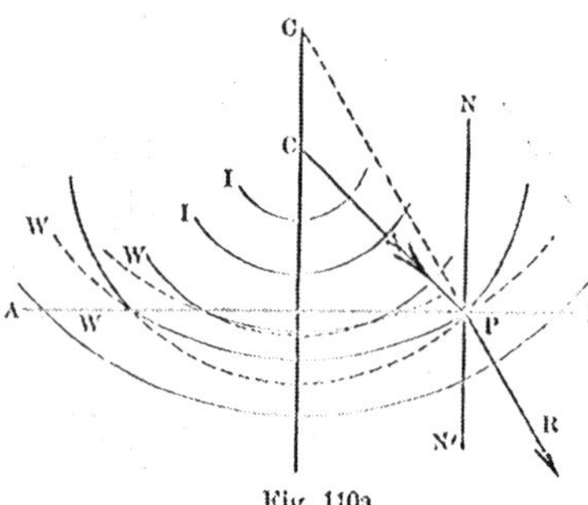

Fig. 110a.

What relation exists between the incident and reflected rays? Define refraction. What is the incident ray? The refracted ray? The relation between them?

Echoes.

154. An echo is a sound repeated by reflection. In order that the echo of any sound may be clearly heard, the reflecting surface ought to be so far distant from the listener as to require at least the fifth of a second for sound to travel to it and return.

It is not possible to pronounce or to hear distinctly more than five syllables in a second. The velocity of sound being 1090 feet per second, it follows that sound travels 218 feet in one fifth of a second. If, then, an obstacle be placed at the distance of 109 feet, sound will go to it and return in one fifth of a second. At the distance of 109 feet, the last syllable only of the echo will reach the ear after the sentence is pronounced. Such an echo is called *monosyllabic*. If the echo takes place from an obstacle at a distance of 218 feet, we hear two syllables; that is, the echo is *dissyllabic*. At distances of 327 feet, the echo is *trisyllabic*, and so on.

Sound may be reflected from several objects situated in different directions and at different distances. Such echoes are called *multiple echoes*. It is said that at a place three leagues from Verdun, a multiple echo formed by parallel walls fifty or sixty yards apart, repeats a sound twelve times. At the chateau of Simonnetta, in Italy, there is an echo which repeats the report of a pistol from forty to fifty times.

Echoes modify the tones of sound. Some repeat sounds with a roughened, others with a softened tone; some with a sneering, others with a plaintive accent.

Resonance.

155. When sounds are reflected from obstacles at a less distance than 109 feet, the reflected sound is superposed

---

What is an echo? Its cause? *Explain the monosyllabic, dissyllabic, and trisyllabic echoes.* What are multiple echoes? Examples. What effect have echoes on the tone of a sound? (**155.**) What is a Resonance?

upon the direct one, giving rise to a strengthened sound, which is called a *Resonance.*

It is the resonance from the walls of a room that makes it easier to speak in a closed apartment than in the open air. The resonance is more clearly perceived when the walls are elastic. In rooms where there are carpets, curtains, stuffed furniture, and the like, the sound-waves are broken up, and the resonance is diminished; but in houses where there is no furniture, the resonance is strengthened. Hence it is, that the sound of voices, footsteps, and the like, is so strongly marked in deserted and unfurnished buildings.

Intensity of Sound.

156. The intensity of sound depends on the force with which it strikes the ear. It varies very nearly as the square of the amplitude of vibration of the aerial particles. Some of the causes that modify the intensity of sound are noticed in the following article.

Causes that modify the Character of Sound.

157. The following are some of the causes that modify the intensity and rate of propagation of sound:

1. It is shown by theory and confirmed by experiment, that the intensity of sound diminishes as the square of the distance from the sonorous body increases.

This is expressed by saying that, *the intensity of sound varies inversely as the square of the distance from the sonorous body.*

2. The intensity of sound diminishes with the *amplitude* of the vibration of the aerial particles.

When a cord vibrates, the sound is observed to diminish as the vibrations become smaller, and when the vibrations cease, the sound

Illustrate by examples. (**156.**) What is Intensity? On what depend? (**157.**) What are the laws of Intensity? 1. Effect of distance? 2. Amplitude of vibration?

is no longer heard. The amplitude of vibration of the sonorous body determines *the length*, or *amplitude* of the vibrations of the ærial particles.

3. The density of the air modifies sound. When the air is rarefied, the intensity is diminished. This fact has been shown by the experiment of a bell in an exhausted receiver.

The presence of watery vapor in the air also modifies sound, that substance being a good conductor of sound. When the air is cooled, it becomes more dense, hence, sounds are louder in cold than in warm weather.

4. The wind modifies sound. The velocity of sound is increased or diminished by the velocity of the wind, according as the direction of the wind conspires with or opposes the propagation.

The effect of the wind is to move the whole mass of air, carrying along the sound-waves unaltered.

5. Sound is increased in intensity when the sonorous body is in contact with, or even in the neighborhood of another body capable of vibrating in unison with it.

Hence, the sound of a vibrating cord is reinforced or strengthened by stretching it over a thin box filled with air, as in the violin. In this case the air in the body of the violin vibrates in unison with the cord. The ancients placed in their theatres vessels of brass, to reinforce and strengthen the voices of the actors.

Intensity of Sounds in Tubes.

158. When a sound is transmitted through a tube, the sound-waves can not diverge laterally, and consequently the

---

Illustrate. 3. Density of the air? *Illustrate.* 4. How does wind modify sound? 5. Effect of a neighboring sonorous body? *Illustrate.* (158.) What effect has a tube on sound?

sound is transmitted to a great distance without much loss of intensity.

M. Biot was able to carry on a conversation in a low tone through a tube a thousand feet in length. He says that the sound was transmitted so well, that there was but one way to avoid being heard, and that was not to speak at all.

Fig. 111.

This property of tubes is utilized in hotels and dwelling-houses, for transmitting messages from one story to another. The tubes employed for this purpose are called speaking tubes. The method of employing the speaking tube, is illustrated in Fig. 111.

The Speaking Trumpet.

159. The Speaking Trumpet, as its name implies, is a conical tube employed to transmit the voice to a great

Biot's *experiment. Practical applications.* (**159**.) What is a Speaking Trumpet?

distance. It is used by firemen and by mariners, as shown in Fig. 112.

Fig. 112.

By means of the speaking trumpet, the voice of the captain can be heard above the noise of the winds and waves in a tempest. According to Father KIRCHER, ALEXANDER THE GREAT employed a speaking trumpet in commanding his armies.

The effect of the speaking trumpet has been explained by successive reflections of sound-waves from the sonorous material of which the instrument is composed, by virtue of which the voice is transmitted only in the direction of the tube.

But the fact is, that sound is transmitted in all directions, which would indicate that its effect should be attributed to a reinforcement of the voice by the vibration of the column of air contained in the trumpet, according to the principle that sound is reinforced by an auxiliary vibrating body.

How is the effect of the speaking trumpet explained?

The Ear Trumpet.

160. The Ear Trumpet is a trumpet employed by persons whose hearing is defective, as shown in Fig. 113.

Fig. 113.

It is simply the speaking trumpet reversed. It serves to collect and concentrate the sound-waves, which are thus enabled to produce a more powerful impression on the drum of the ear. The shape of the ear in man and in animals is such as to perform the function of the trumpet.

II.—MUSICAL SOUNDS.

Difference between a Musical Sound and a Noise.

161. A Musical Sound results from a succession of vibrations of equal duration. Such vibrations are called *isochronal.*

(**160.**) What is an Ear Trumpet? How does it differ from the speaking trumpet? What is its use? (**161.**) What is a Musical Sound?

NOISE results from a single impulse, or from a succession of vibrations of unequal duration. Thus, the crack of a whip, the discharge of a pistol, the rattling of thunder, or the roar of the waves of the ocean, are destitute of musical value, and are simply noises.

Pitch of Sounds.—Music.

162. The PITCH of a musical sound depends upon the frequency of the vibrations. Those sounds which result from very rapid vibrations, are called *acute*, whilst those which arise from very slow vibrations, are called *grave*.

The terms acute and grave are relative; thus, a given sound may be acute with respect to a second, whilst it is grave with respect to a third; thus, a sound which corresponds to 160 vibrations, is acute with respect to one corresponding to 80 vibrations, and grave with respect to one corresponding to 320 vibrations per second. A well arranged and happy combination of grave and acute sounds according to the principles of harmony, constitutes *music*.

Limits of perceptible Sounds.

163. M. SAVART investigated the subject of sound with respect to the number of vibrations, corresponding to the most grave and acute sounds perceptible by the human ear, by means of an apparatus devised for that purpose.

As the result of his investigations, he concluded that the gravest perceptible sound was produced by 16 vibrations per second, and the most acute by 48,000 vibrations per second. Allowing 1090 feet as the velocity of sound, we find for the length of the waves, corresponding to the gravest sounds, 68 feet, and for the length corresponding to the most acute sounds, a little more than a quarter of an inch.

What is a Noise? **162.** What does Pitch depend upon? What is an acute sound? A grave one? *Illustrate by examples. What is music?* (**163.**) Who investigated the limits of audible sounds? Give the results of his investigation.

The limits of sounds employed in music are much narrower, especially in singing. SAVART gives for the gravest sounds of the male voice, 190 vibrations per second, and for the female voice, 572. For the most acute sounds of the male voice he gives 678 vibrations per second, and for the female voice, 1606.

Two sounds, corresponding to the same number of vibrations pe second, are in *unison*.

Musical Scale.—Gamut.

164. The ear not only distinguishes between given sounds—which is most grave, and which is most acute—but it also appreciates the relations between the number of vibrations corresponding to each. We can not recognize whether for one sound the number of vibrations is precisely two, three, or four times as great as for another, but when the number of vibrations corresponding to two successive or simultaneous sounds have to each other a simple ratio, these sounds excite an agreeable impression, which varies with the relation between the two sounds.

From this principle there results a series of sounds characterized by relations which have their origin in the nature of our mental organization, and which constitute what is called a *Musical Scale.*

In this scale sounds recur in the same order in groups of seven. Each group constitutes what is called a *gamut* of seven notes. The notes are named, *do, re, mi, fa, sol, la, si,* but they are usually denoted by the letters, C, D, E, F, G, A, B. The relation between these notes is given in the table below, n denoting the number of vibrations corresponding to the note C:

C,	D,	E,	F,	G,	A,	B,	C;
n,	$\frac{9}{8}n$,	$\frac{5}{4}n$,	$\frac{4}{3}n$,	$\frac{3}{2}n$,	$\frac{5}{3}n$,	$\frac{15}{8}n$,	$2n$,

What are the limits in singing? When are sounds in unison? (164.) What is a Musical Scale? *What is a gamut?* How are the notes named?

The number of vibrations corresponding to any note may be measured by an instrument called a siren.

Intervals.—Accords.

165. An interval is the ratio of the number of vibrations corresponding to any note to the number corresponding to some higher note.

The intervals between consecutive notes, called seconds, is given in the following table:

C to D,	D to E,	E to F,	F to G,	G to A,	A to B,	B to C;
$\frac{9}{8}$,	$\frac{10}{9}$,	$\frac{16}{15}$,	$\frac{9}{8}$,	$\frac{10}{9}$,	$\frac{9}{8}$,	$\frac{16}{15}$;

If the interval comprise *two*, *three*, *four*, &c., *seven* notes, it is called, *a third*, *a fourth*, *a fifth*, &c., *an eighth*, or an *octave;* thus, the interval between C and E is a *third*, and is equal to $\frac{5}{4}$; the interval from C to F is a *fourth*, and is equal to $\frac{4}{3}$; the interval from any note to the next note of the same name is an octave, and is always equal to 2.

The coexistence of several sounds is called an *accord.* When the ear can distinguish, without fatigue, the relation between two sounds, which is the case when this relation is simple, the coexistence of these sounds is called a *consonance;* when the ear is painfully affected by the coexistence, it is called a *dissonance.*

The most simple accord is *the unison*, in which the number of vibrations are equal; then comes *the octave*, in which the number of vibrations, corresponding to one sound, is double that corresponding to the other; then *the fifth*, in which the numbers are as 3 to 2; then *the fourth*, in which the numbers are as 4 to 3; and finally *the third*, in which the ratio is that of 5 to 4.

When the numbers of vibrations corresponding to three simultaneous sounds, are as 4, 5, and 6, the combination is called a *perfect accord.* For example, the notes C, E, G, form a perfect accord, as

---

(165.) What is an Interval? *What is a third, fourth, fifth, sixth, seventh, octave?* What is an Accord? A Consonance? A Dissonance? *What is the simplest accord? The next simplest? Next three in order? What is a perfect accord? Example.*

do the notes G, B, D. These accords produce upon the ear the most agreeable sensation.

The Tuning Fork.

166. A TUNING FORK is an instrument used in tuning musical instruments of fixed sounds, like the piano.

It consists of a plate of steel, bent into the shape of the letter U, mounted upon a wooden box, as shown in Fig. 114. The wooden box is open at one extremity, and serves to

Fig. 114.

reinforce the sound, which would otherwise be feeble. The fork is made to sound by drawing across one of its branches a violin bow, or by straining the branches apart by a wedge of wood or metal, and then suddenly withdrawing it, or finally, by striking one of the branches with a solid body. The tuning fork is usually constructed so as to sound the A, which corresponds to 856 vibrations per second.

(**166.**) What is a Tuning Fork? Describe it. How is it made to sound?

Transverse Vibrations of Cords.

167. We have already seen (Art. 146), that when a stretched cord is drawn from its position of equilibrium and abandoned, it returns to its position of rest by a succession of continually decreasing vibrations.

Cords used in musical instruments are generally made of catgut, or of twisted wires. They are made to vibrate by drawing a bow across them, as in the violin; by drawing them aside, as in the harp; or by percussion with little hammers, as in the piano. In all of these cases, the vibrations are *transversal*, that is, the movements take place perpendicularly to the direction of the cord.

Laws of Transversal Vibrations of Cords.

168. The number of vibrations of a stretched cord in any given time, as in one second, for example, depends upon its length, its thickness, its tension, and its density. The following are the laws that govern the number of vibrations in a fixed time:

1. *The tension being constant the number of vibrations varies inversely as its length.*

If a given cord makes 18 vibrations per second, it will make 36 if its length be reduced to one half, 54 if its length be reduced to one third, and so on. This property is utilized in the violin. By applying the finger, we virtually reduce the length of the vibrating portion at pleasure.

2. *The tension and length being the same, the number of vibrations varies inversely as its diameter.*

Small cords vibrate more rapidly than large ones, and consequently render more acute sounds. A cord of any given size

(**167.**) Of what are musical cords made? How set in vibration in different instruments? (**168.**) Upon what does the number of vibrations of a cord depend? What is the first law? *Illustrate.* The second law? *Illustrate.*

makes twice as many vibrations as one of double the size. Other things being equal, the notes rendered differ by an octave.

3. *The length and size being the same, the number of vibrations varies as the square root of the tension.*

If a cord renders a given note, it will, if its tension be quadrupled, render a note an octave higher, and so on. This property is utilized in stringed instruments by means of an apparatus for increasing or diminishing the tension at pleasure.

4. *Other things being equal, the number of vibrations varies inversely as the square root of the density.*

Dense cords render graver notes than those of less density. Small, light, and short cords, strongly stretched, yield acute notes. Large, dense, and long cords, not strongly stretched, yield grave notes.

Verification of the Laws of Vibration.

169. The laws enunciated in the preceding article may be verified by means of an instrument called a *Sonometer*, shown in Fig. 115.

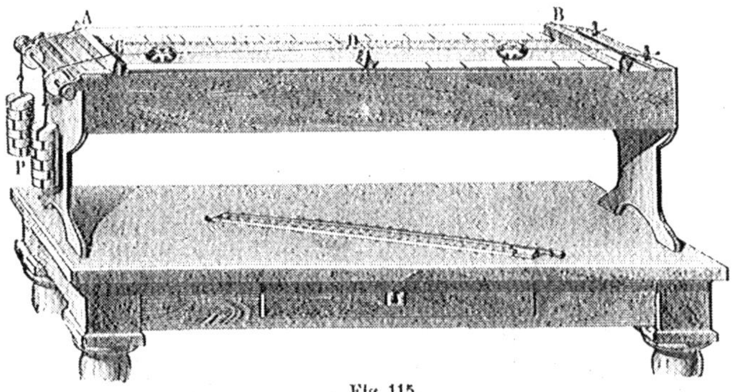

Fig. 115.

The sonometer is said to have been invented by PYTHAGORAS, about 600 years before our era. In its present form,

The third law? *Illustrate.* The fourth law? *Illustrate.* (**169.**) How may the preceding laws be verified? What is a Sonometer?

it consists of a wooden box about four feet in length, upon which are mounted two fixed bridges, *A* and *B*, and a movable one, *D*. On these bridges, two cords, *CD* and *AB*, fastened firmly at one end and passing over pulleys at the other end, are stretched by means of weights, *P*.

Let the cords be exactly alike and stretched by equal weights. If the bridge *D*, be moved so as to render *CD* equal to one half of *AB*, the notes of the two cords will differ by an octave; that is, *CD* will vibrate twice as fast as *AB*. If *CD* be made equal to one third of *AB*, by moving the bridge *D*, the former will vibrate three times as fast as the latter, and so on. This verifies the *first* law.

To verify the *second* law, we remove the bridge *D*, and use two cords, one of which is twice as large as the other. It will be found that the notes yielded will differ by an octave. If one cord be taken three times as large as the other, the latter will be found to vibrate three times as fast as the former.

To verify the *third* law, let the two cords be alike, and stretch one by a weight four times as great as that employed to stretch the other. The notes will differ by an octave. If the stretching force in one, is nine times that in the other case, the former will vibrate three times as fast as the latter, and so on.

To verify the *fourth* law, we make use of cords equal in length, size, and equally stretched, but of different densities. It will be found that the law is verified in each case

Stringed Instruments.

170. All stringed instruments of music are constructed in accordance with the preceding laws. They are divided into instruments with *fixed sounds*, and instruments with *variable sounds*.

Describe it. How is the first law verified? The second? The third? The fourth? (**170.**) How are stringed instruments classed?

To the former class belong the piano, the harp, &c. They have a cord for each note, or else an arrangement is made so that by placing the finger at certain points, as in the guitar, the same cord may be made to render several notes in succession.

To the latter class belong the violin, the violoncello, &c. They are provided with cords of catgut, or sometimes of metal, put in vibration by a bow. Various arrangements are made for regulating the notes, such as increasing the tension, placing the finger upon the cords, and the like. These instruments are difficult to play upon, and require great nicety of ear, but in the hands of skillful players they possess great power. They are the soul of the orchestra, and it is for them that the finest pieces of music have been composed.

Sound from Pipes.

171. When the air in a *pipe*, or hollow tube, is put into vibration, it yields a sound. In this case, it is the air which is the sonorous body, the nature of the sound depending upon the form of the pipe and the manner in which the vibrations of its contained air are produced.

To produce a sound from a pipe, the contained air must be thrown into a succession of rapid condensations and rarefactions, which is effected by introducing a current of air through a suitable *mouth-piece.* Two principal forms are given to the mouth-piece, in one of which the parts remain fixed, and in the other there is a movable tongue, called a *reed.*

Pipes with fixed Mouth-pieces.

172. Pipes with fixed mouth-pieces are of wood or metal, rectangular or cylindrical, and always of considerable

Examples of each class. Which are most difficult to play upon? (**171.**) What is the sonorous body in the case of a pipe? How thrown into vibration? What is a mouth-piece? How many forms? (**172.**) What are the characteristics of pipes with fixed mouth-pieces?

length compared with their cross section. To this class belong the flute, the organ pipe, and the like. Some of the forms given to pipes of this class are shown in Figs. 116, 117, 118, 119, and 120.

Fig. 116 represents a rectangular pipe of wood, and Fig. 117 shows the form of its longitudinal section. *P* represents the tube

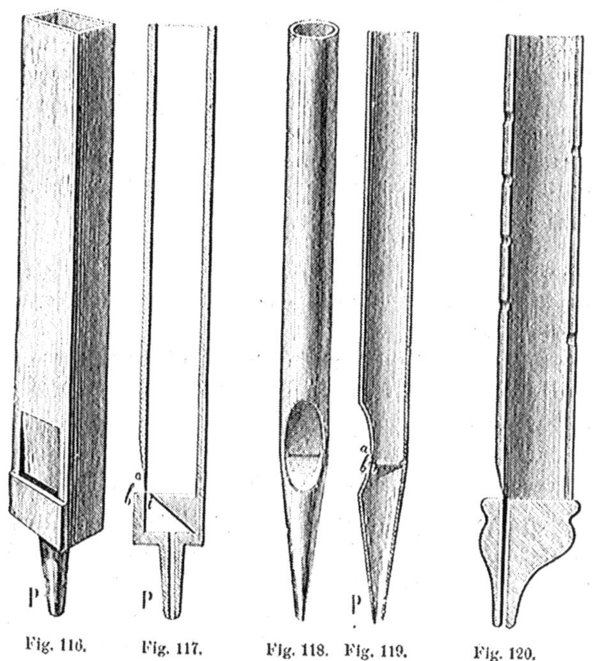

Fig. 116. Fig. 117. Fig. 118. Fig. 119. Fig. 120.

through which air is forced into it. The air passes through a narrow opening, *i*, called the *vent*. Opposite the vent is an opening in the side of the pipe, called the *mouth*. The upper border, *a*, of the mouth, is bevelled, and is called the *upper lip*, the lower border is not bevelled, and is called the *lower lip*.

Describe the mouth-piece. The vent. The mouth. The lips.

The current of air forced through the vent strikes against the upper lip, is compressed, and by its elasticity, reacts upon the entering current, and for an instant arrests it. This stoppage is only for an instant, for the compressed air finds an outlet through the mouth, again permitting the flow. No sooner has the flow commenced than it is a second time arrested as before, again to be resumed, and so on.

This continued arrest and release of the current gives rise to a succession of vibrations, which are propagated through the tube, causing alternate and rapid condensations and rarefactions, which result in a continuous sound. The vibrations are the more rapid as the current introduced is stronger, and as the upper lip approaches nearer the vent

Fig. 118 represents a second form of organ pipe, which is shown in section in Fig. 119. This is but a modification of the pipe already explained. The letters indicate the same parts as in the preceding figures.

Fig. 120 represents the form of the mouth-piece of the flageolet, and it will be seen that it bears a close resemblance to the pipes already explained.

In the flute, an opening is made in the side of the pipe, and the arrest and flow of the current are effected by the arrangement of the lips of the player.

Reed Pipes.

173. In Reed Pipes the mouth-piece is provided with a vibrating tongue, called a *Reed*, by means of which the air is put in vibration. To this class belong the clarionet, the hautboy, and the like. The reed may be so arranged as to beat against the sides of the opening, or it may play freely through the opening in the tube.

Figs. 121 and 122 show the arrangement of a reed of the first kind. A piece of metal, *a*, shaped like a spoon, is fitted with an elastic tongue, *l*, which can completely close the opening. A piece of

---

Explain the action in detail. How is the mouth-piece in the flute? (**173.**) What is a reed? What are some of the reed instruments? *Explain the arrangement of a reed of the first kind.*

metal, r, which may be elevated or depressed by a a rod, b, serves to lengthen or shorten the vibrating part of the reed. This arrangement enables us to diminish or increase the rapidity of vibration at pleasure.

The mouth-piece, as described, connects with the tube T, and is set in a rectangular box, KN, which is in communication with a bellows, from which the wind is supplied. For the purpose of class demonstration, the upper part of the tube KN, has glass windows on three sides to show the motion of the reed.

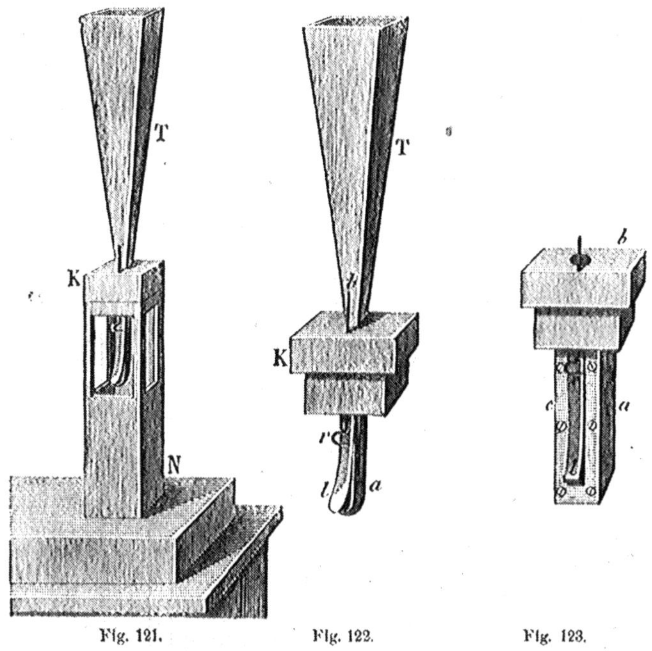

Fig. 121. Fig. 122. Fig. 123.

When a current of air is forced into the tube KN, the reed is set in rapid vibration, causing a succession of rarefactions and condensations in the air of the pipe T, and causing it to emit a sound. The air entering the tube KN, first closes the opening by pressing the

Explain its action.

reed against it; the reed then recoils by virtue of its elasticity, permitting a portion of condensed air to enter the pipe, when the reed is again pressed against the opening, and so on as long as the current of air is kept up. It is evident, that the rapidity of vibration will be increased by increasing the tension of the air from the bellows, and also by shortening the vibrating part of the reed.

Fig. 123 shows the arrangement of the free reed. The vibrating plate, *l*, is placed so as to pass backwards through an opening in the side of the tube *ca*, alternately closing and opening a communication between the tube and the air from the bellows. The regulator, *r*, is entirely similar to that shown in Figs. 121 and 122, as are the remaining parts of the arrangement. The explanation of the action of this species of reed is entirely similar to that already described.

Fig. 124.

Explain the arrangement of the Free Reed. What is its mode of action?

The Bellows.

174. Fig. 124 represents one form of the *Bellows*, used in causing pipes to sound. It is worked by a lever. The air enters a valve, *S*, through which it passes to a leathern reservoir, *R*. The top of the reservoir is weighted so as to force the air into a box, from which it is admitted to the pipes by means of valves, which are opened and shut at the will of the player.

Wind Instruments.

175. WIND INSTRUMENTS of music consist of pipes, either straight or curved, which are made to sound by a current of air properly directed.

In some, the current of air is directed by the mouth upon an opening made in the side, as in the flute. In others, the current of air is made to enter through a mouth-piece, as in the flageolet. In others, a reed is used, as in the clarionet. In the organ, there is a collection of tubes, similar to those shown in Figs. 116 and 118. In some instruments, as the trumpet and the horn, a conical mouth-piece is used, of the form shown in Fig. 125, within which the lips of the musician vibrate in place of the reed. The rapidity of vibration can be regulated at will.

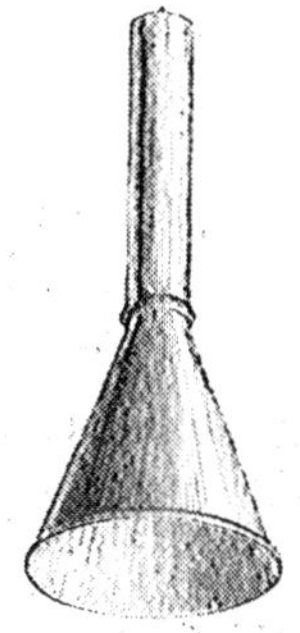
Fig. 125.

(174.) Describe the Bellows used with wind instruments. (175) What are Wind Instruments? Explain their different varieties.

CHAPTER V.

HEAT.

I.—GENERAL PROPERTIES OF HEAT

Definition of Heat.

176. HEAT is the physical agent that produces the sensation we call warmth; the term *heat* is also applied to the sensation itself. *Cold* is a negative term used to express the absence of heat.

Theories of Heat.

177. Two theories have been advanced to explain the phenomena of heat: the *emission* theory and the *undulatory* theory.

According to the emission theory, heat is a fluid, destitute of weight and capable of passing from one body to another with great velocity. Its particles repel each other, but are attracted by the particles of all other bodies. A body becomes heated by receiving more of this fluid than it gives out; it becomes cooled by giving out more than it receives.

According to the undulatory theory, the heat of a body is caused by a rapid vibration of its molecules; this motion may be transmitted from one body to another through an elastic medium called *ether*, in the same way that sound is transmitted through the air. According to this view heat is a mode of motion, and those bodies are hottest whose molecules vibrate with greatest velocity, and through the greatest amplitudes.

(**176.**) What is heat? Cold? (**177.**) What two theories of heat have been advanced? Explain the emission theory. The undulatory theory.

The undulatory theory is the one generally adopted by physicists; it affords a better explanation of the phenomena and at the same time serves to show the intimate relation between heat and light. In what follows the phenomena will be explained, as far as possible, independently of both theories.

General Effects of Heat.

178. Heat, accumulating in bodies, penetrates into their substance, and acting upon their ultimate molecules, gives rise to repellent forces which counteract those of cohesion. Hence, the most noticeable phenomenon of heat is, that it causes bodies *to expand.* If applied in sufficient quantity, the particles of solids are so far repelled, as to move freely amongst each other, becoming *liquid;* or if still greater quantities of heat are applied, the body passes into a *state of vapor.* When heat is abstracted from a vapor, it returns to a liquid state, and if still more heat be abstracted, it becomes solid, and if the process be continued, the solid goes on contracting under the influence of the molecular forces.

Hence we say, that heat *dilates* bodies, and cold contracts them. Heat also converts solids into liquids, liquids into vapors, and acting upon gases and vapors, causes them to expand.

Expansion of Bodies by Heat.

179. All bodies are expanded by heat, but in very different degrees. The most dilatable bodies are gases, then vapors, then liquids, and finally solids. In fluids we regard only increase of volume, but in solids we distinguish two kinds of expansion, *linear expansion*, that is, expansion in length, and *expansion of volume.*

(**178.**) Describe the general effects of heat on solids. On liquids. What effect has cold on vapors? On liquids? (**179.**) What bodies are most dilatable? The least dilatable? What is linear expansion? Expansion of volume?

Fig. 126 represents the method of showing and measuring the linear expansion of the metals. A rod of metal, *A*, passes through two metallic supports, being made fast at one extremity by a clamp-screw, *B*, and being free to expand at the other extremity. The free end abuts against the short end, *C*, of a lever, the long end, *D*, of which plays in front of a graduated arc.

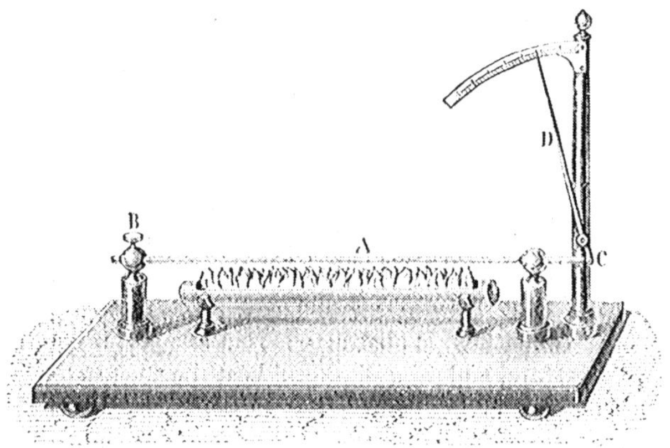

Fig. 126.

When the rod is heated, by placing fire beneath it, as shown in the figure, the rod *A* expands, and the expansion is shown by the

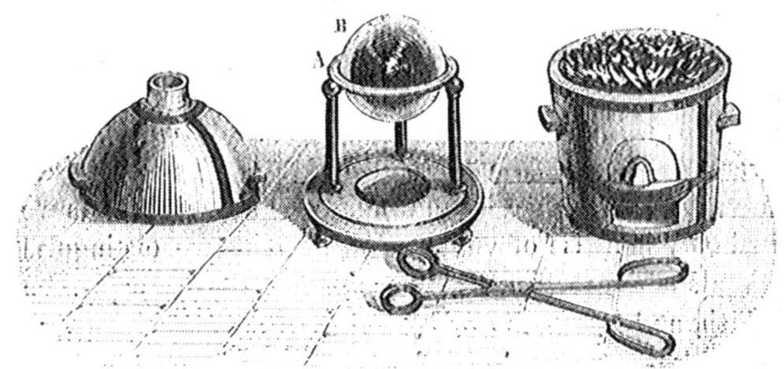

Fig. 127

How is the linear expansion of metals shown?

motion of the index, *D*. When the rod, *A*, is of steel, copper, silver, &c., the amount of expansion varies, as is shown by the different amounts of displacement of the index. Brass, for example, expands more, for the same amount of heat, than iron or steel.

Fig. 127 shows the method of demonstrating that bodies undergo an expansion in volume when heated. A ring, *A*, is constructed so that a ball, *B*, passes freely through it when cold. If the ball be heated in a furnace, it will no longer pass through the ring, but if allowed to cool, it again falls through the ring. The method of making the experiment is fully shown in the figure.

Liquids and gases being more expansible than solids, their expansion is more easily shown by experiment. For liquids, we take a hollow glass sphere, terminating in a narrow tube, open at the top, and fill the globe and a portion of the stem with some fluid, like mercury, as shown in Fig. 128. If heat be applied to the globe, the liquid will rise in the stem from *a* towards *b*, indicating an increase of volume; and if sufficient heat be applied, the liquid will fill the stem, and will ultimately be converted into vapor. If the liquid is allowed to cool, it again returns to its original volume.

An analogous experiment shows the expansion of gases and vapors. A bulb of glass is provided with a long and fine tube of the same material, which is bent twice upon itself, as shown in Fig. 129. An index of mercury is introduced into the stem in the following manner. The bulb is heated, and a portion of the air which it contains is driven

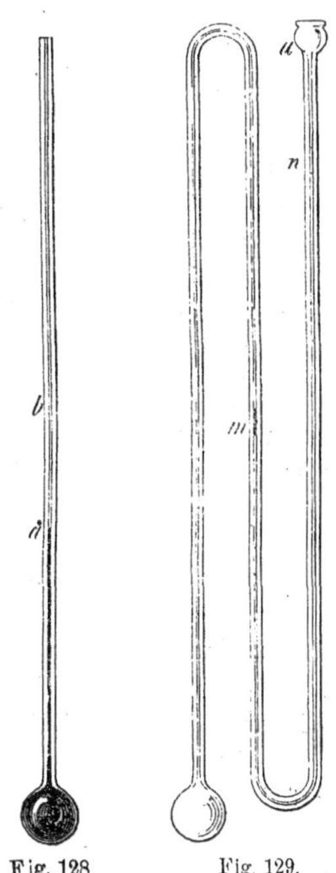

Fig. 128 Fig. 129.

How is expansion in volume shown? How is the expansion of liquids shown? Of gases?

out, when a drop of mercury is poured into the funnel, *a*. If the instrument is allowed to cool, the air in the bulb contracts, and the pressure of the atmosphere drives the drop of mercury along the tube to some position, *m*.

The instrument having been prepared in this manner, if the bulb is held in the hand for a few minutes, the air becomes heated and expands, the expansion being indicated by the index moving to some new position, as *n*. If allowed to cool, the index returns to *m*.

From what precedes, we infer that heat expands all bodies, and that cold contracts them. There are apparent exceptions to this law, but they are only apparent. Thus, bodies capable of absorbing water, like paper, wood, clay, and the like, contract on being heated. This contraction is only apparent; it arises from the water which they contain being vaporized and driven off, which produces an apparent diminution of volume; after they are thoroughly dried, they follow the general law.

The property just explained is used for bending absorbent bodies. To effect this they are heated on one side only, which drives out the water from that side, and causes them to bend in that direction. It is this principle that causes wooden articles to warp, and therefore demands that articles of furniture, and wooden parts of buildings, be coated with oils, paints, or varnishes, to prevent the absorption of water.

The principle of expansion and contraction is often utilized in the arts.

A familiar example, is the process of setting the tire of a wagon-wheel. The tire is made a little smaller than the outer periphery of the wooden part of the wheel. It is then heated, and placed around the wheel; on cooling, it contracts powerfully, and draws the felloes firmly together. The same principle has been applied

---

What is the general conclusion with respect to the action of heat and cold? Explain the apparent exceptions to the law. *Explain the process of warping Are the principles of contraction and expansion utilized? Explain the operation of setting a tire. Of drawing walls together.*

in bringing the walls of a building back to their original position after they had begun to separate from each other.

Sensible and Latent Heat. — Temperature.

180. Heat may act on a body in two ways. *First,* it may act to increase the warmth of the body; in this case it is said to be *sensible.* *Secondly,* it may be absorbed and act solely to produce a change of state of the body, without becoming manifest to the senses; in this case it is said to be *latent.* Thus, when ice melts, it absorbs an immense amount of heat without appearing to become any warmer; this heat acts to change the body from a solid to a liquid state.

The *temperature* of a body is the amount of its sensible heat.

II. — THERMOMETERS.

The Thermometer.

181. A THERMOMETER is an instrument for measuring temperatures.

The thermometer depends upon the principle that bodies expand when heated, and contract when cooled. Thermometers have been constructed of a great variety of materials. For common purposes, the mercurial thermometer is preferred, on account of the uniformity with which both mercury and glass expand when heated.

The mercurial thermometer consists of a bulb of glass, at the upper extremity of which is a narrow tube of uniform bore, hermetically sealed at its upper end. The bulb and a part of the tube are filled with mercury, and the whole is attached to a frame on which is a scale for measuring the rise and fall of the mercury in the tube.

(**180.**) What is sensible heat? Latent heat? Temperature? (**181.**) What is a Thermometer? On what principle does it depend? What is the best thermometer for common use? Describe a mercurial thermometer.

Method of making a Thermometer.

182. A capillary tube of glass is provided, of uniform bore, upon one end of which a bulb is blown, and upon the other a funnel, as shown in Fig. 130.

The funnel is nearly filled with mercury, which is at first prevented from penetrating into the bulb by the resistance of the air and the smallness of the tube. The bulb is therefore heated, when the air within expands, and a portion escapes in bubbles through the mercury. On cooling, the pressure of the external atmosphere forces a quantity of mercury through the tube into the bulb. By repeating this operation a few times, the bulb and a portion of the tube are filled with mercury.

The whole is then heated till the mercury boils, thus filling the tube, when the funnel is melted off and the tube hermetically sealed by means of a jet of flame urged by a blow-pipe. On cooling, the mercury descends to some point of the tube, as shown in Fig. 131, leaving a vacuum at the upper end. It only remains to graduate it, and attach a suitable scale.

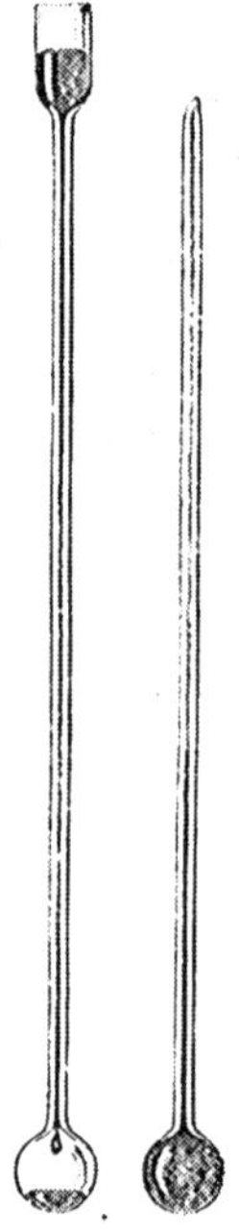

Figs 130. 131.

Method of Graduation.

183. Two points of the stem are first determined, the *freezing* and the *boiling point*. These are determined on the principle that the temperatures at which distilled water freezes and boils, are always the same, that is, when these changes of state take place under equal atmospheric pressures.

The instrument is first plunged into a bath of melting ice, as shown in Fig. 132, and is allowed to remain until it takes the

(182.) *Describe the process of filling a thermometer with mercury. How is the tube sealed?* (183.) *On what principle are the freezing and boiling points determined?*

temperature of the mixture, say twenty or thirty minutes. A slight scratch is then made on the stem at the upper surface of the mercury, and this constitutes the *freezing point.*

The instrument is next plunged into a bath of distilled water, in a state of ebullition, care being taken to surround it with steam by means of an apparatus like that shown in Fig. 133. After the mercury ceases to rise in the tube, which will be in a few minutes, the level of its upper surface is marked on the stem, by a scratch, as before, and this constitutes the *boiling point.*

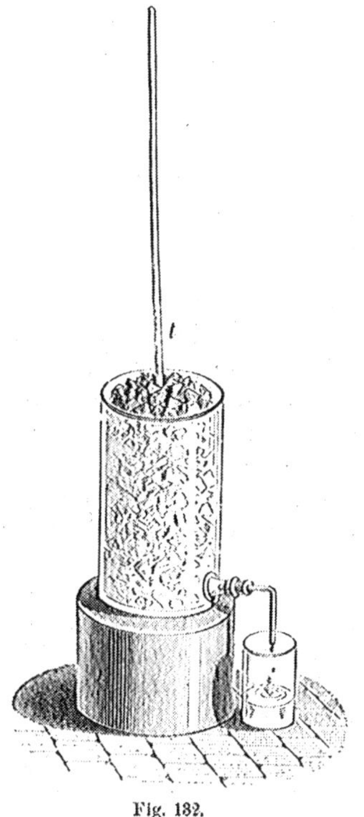

Fig. 132.

The space between the boiling and freezing points is then divided into a certain number of equal parts, and the graduation is continued above and below as far as may be desired. These divisions may be scratched upon the glass with a diamond, or, as is usually done, they may be made on a strip of metal, which is attached to the frame. The divisions are numbered according to the kind of scale adopted.

Thermometer Scales.

184. Three principal scales are used: the *Centigrade scale*, in which the space between the freezing and boiling

How is the freezing point determined? The boiling point? How is the intermediate space divided? (**184.**) What are the three principal scales used?

points is divided into 100 equal parts, called *degrees;* *Reaumur's scale*, in which the same space is divided into 80 equal parts, called *degrees;* and *Fahrenheit's scale*, in which this space is divided into 180 equal parts, also called *degrees.*

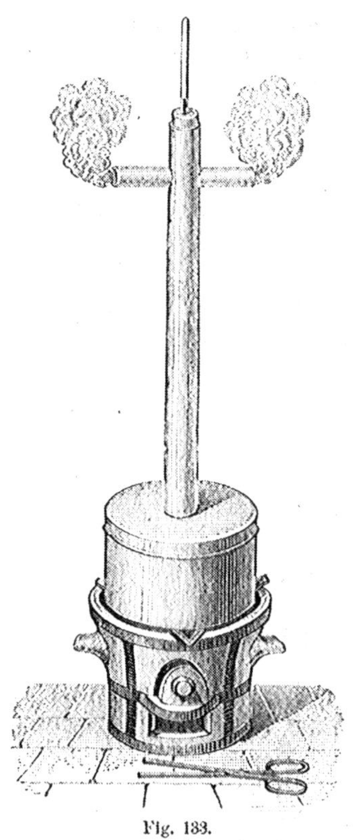
Fig. 133.

In the centigrade scale, the freezing point is marked 0, and the degrees are numbered both up and down, the former numbers being considered positive, and designated by the sign +, whilst the latter are considered negative, and designated by the sign —. Of course the boiling point is marked 100°.

Fig. 134 represents a thermometer mounted and graduated according to the centigrade scale. In it the mercury indicates 30° C.

In Reaumur's scale, the freezing point is marked 0, and the boiling point 80°. The degrees below freezing are marked as in the centigrade scale.

In Fahrenheit's scale, which is the one most used in the United States, the 0 point is taken 32° below the freezing

Where is the 0 point of the centigrade scale? Explain the signs + and —. What is the boiling point marked? Where is the 0 of the Reaumur scale? The boiling point? Where is the 0 of Fahrenheit's scale?

point, and the divisions are numbered from this point both up and down. The boiling point of distilled water is, 212° F.

Conversion of Centigrade and Reaumur's Degrees into Fahrenheit's.

185. A degree on the centigrade scale is equal to one and eight tenths of a degree on the Fahrenheit scale, and one on Reaumur's scale is equal to two and a quarter on Fahrenheit's. Hence, to convert the reading on a centigrade to an equivalent one on Fahrenheit's scale, multiply it by 1.8 and add to the result 32°. Thus, a reading of 25° centigrade, is equivalent to 25° × 1.8 + 32°, or 77° F. To convert a reading on Reaumur's scale to an equivalent one on Fahrenheit's, multiply by 2¼, and to the result add 32°. Thus, a reading of 24° Reaumur is equivalent to 24° × 2¼ + 32°, or 86° F.

By reversing the above processes, readings on Fahrenheit's scale may be converted into equivalent ones on the centigrade or Reaumur's scale.

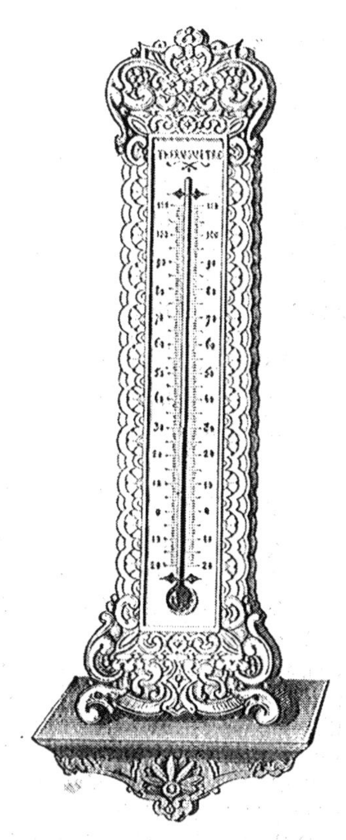

Fig. 134.

The boiling point? (**185.**) *Explain the method of converting readings from one scale to another.*

Alcohol Thermometers.

186. An ALCOHOL THERMOMETER is similar to a mercurial one in all respects, except that alcohol, tinged red, is used in place of the mercury.

Because alcohol does not expand regularly with a regular increase of temperature, the alcohol thermometer has to be graduated by experiment, comparing it degree by degree with a standard mercurial thermometer.

An alcohol thermometer is more easily filled than a mercurial one, no funnel being required. The bulb is heated until a portion of the contained air is driven off, and then the open end of the tube is plunged into a vessel of alcohol. As the air in the bulb cools, the pressure of the external atmosphere forces a portion of alcohol up into the bulb. If this be boiled, the vapor of alcohol will expel the remainder of the air, and by dipping the open end of the tube into the alcohol once more, the bulb will be completely filled, when it again becomes cool. The instrument is then treated like the mercurial thermometer.

Relative advantages of Mercurial and Alcohol Thermometers.

187. For ordinary purposes, the mercurial thermometer is to be preferred, on account of the uniformity with which the mercury expands with a uniform increase of temperature. But mercury congeals at 39° below 0 of the Fahrenheit scale, and where a lower temperature than this is to be observed, it becomes absolutely necessary to employ the spirit thermometer. In the severe cold of the polar regions, mercury often congeals, but no degree of cold has yet been obtained that will congeal absolute alcohol.

For high temperatures, mercury only is capable of being used; this liquid does not boil till raised to 662° F., whilst alcohol boils at 174° F. The latter liquid can not, therefore, be used to observe tem-

(**186**) How does the alcohol differ from the mercurial thermometer? *How is the alcohol thermometer graduated? Why? How is it filled?* (**187**.) *When is the alcohol thermometer preferable to the mercurial one? When must the latter be used?*

peratures higher than 174° F., nor can it be relied upon even for temperatures considerably lower than this.

It is to be observed, that mercury can not be relied upon for temperatures lower than 32° below 0, on account of irregularities in its rate of contraction below that limit.

Rules for using a Thermometer.

188. Before noting the height of the mercurial column, the instrument should be allowed to acquire the temperature of the medium in which it is placed. This, in general, will require some minutes.

In determining the temperature of a room, the thermometer should not be hung against the walls, but should be freely suspended, so as to take the temperature of the atmosphere. When hung against a wall, especially an outer wall, an error of several degrees may result. In like manner, if hung against a wall containing a flue, or adjoining another room of different temperature, a similar error of several degrees might result.

To determine the temperature of the atmosphere, the thermometer should be freely suspended in the air at some distance from any building or tree. It should be sheltered from the direct action of the sun's rays, as well as from the influence of reflecting substances. Furthermore, it should be protected from winds and currents of air.

The Differential Thermometers.

189. A Differential Thermometer is a thermometer contrived to show the difference of temperature between two places near each other. The two principal forms of the differential thermometer are Rumford's and Leslie's.

When can the former only be used? (**188.**) What precautions are to be taken in noting the temperature of a room? Why? In noting the temperature of the atmosphere? (**189.**) What is a Differential Thermometer? What are its two forms?

Rumford's Differential Thermometer.

190. RUMFORD'S DIFFERENTIAL THERMOMETER is represented in Fig. 135.

It consists of two bulbs of thin glass, *A* and *B*, connected by a fine tube bent twice at right angles, as shown in the

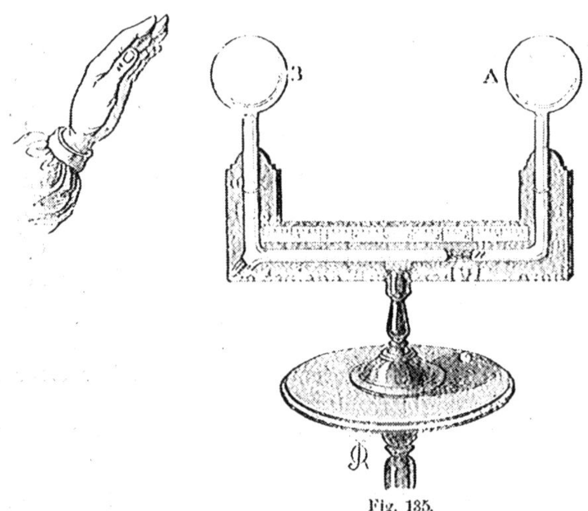

Fig. 135.

figure. The whole apparatus is attached to a suitable frame, which supports a scale parallel to the horizontal branch of the connecting tube. The 0 of the scale is at its middle point, and the graduation is continued from it in both directions. The bulbs and a large part of the connecting tube are filled with air; there is, however, in the tube a small drop of fluid which separates the air in the two extremities.

The instrument is so constructed that the index, *n*, is at the 0 of the scale when the temperature of the two bulbs is

(**190**) Describe RUMFORD'S form. Explain the scale. Explain its action. How is the scale graduated?

the same. When one of the bulbs is heated more than the other, the air in it expands and drives the index towards the other, until the tensions of the air in the two bulbs exactly balance each other.

The scale is divided by experiment by the aid of a standard mercurial thermometer.

Leslie's Differential Thermometer.

191. LESLIE'S DIFFERENTIAL THERMOMETER is shown in Fig. 136. It differs from RUMFORD'S, in having the bulbs smaller, and in containing a longer column of liquid in the tube. The scales are placed by the sides of the vertical portions of the tube, having their 0 points at the middle. There is, then, a double scale. The method of graduating and using this thermometer is the same as that described in the last article.

Fig. 136.

Pyrometer.

192. A PYROMETER is an instrument for measuring higher temperatures than can be observed by means of the mercurial thermometer.

The most important pyrometers are those of WEDGEWOOD and BROGNIART. The former is founded on the diminution of the volume of clay at high temperatures, and the latter on the principle of the expansion of metals. The indications of these instruments are very unreliable, and it yet remains

(**191.**) Describe LESLIE'S Differential Thermometer. (**192.**) What is a Pyrometer? What are the most important ones? What is the principle of each? Are they reliable?

to discover some accurate method of measuring temperatures higher than 600° F.

III.—RADIATION OF HEAT.

Propagation of Heat.

193. The ethereal medium that transmits heat extends through space, and is almost perfectly elastic. It penetrates all bodies and occupies the intervals between their molecules. The heat vibrations of bodies are thus imparted to the surrounding ether, and by it are propagated outward in spherical waves similar to sound-waves in air. Heat propagated in this way is called *radiant heat*. A line perpendicular to a wave front is called a *ray of heat*.

A ray of heat indicates a direction in which heat is propagated and along which it produces its effect. In a homogeneous medium heat-rays are straight lines radiating in every direction from a heated body. Rays of heat, like rays of sound, may be refracted and reflected. Radiant heat does not impart warmth to the medium that transmits it, but when intercepted by a body the motion of the particles of ether is imparted to the molecules of the body, and the phenomena of heat are developed.

Laws of Radiant Heat.

194. The radiation of heat takes place according to the following laws:

1. *Heat is radiated equally in all directions.*

This law may be verified by placing thermometers at equal distances and in different directions from a heated body.

2. *Rays of heat are straight lines.*

This law may be verified by interposing a screen anywhere in a right line joining the heated body and the thermometer, when the thermometer will cease to rise.

(193.) How is heat transmitted through space? What is radiant heat? What are rays of heat? **(194.)** What is the first law of radiant heat? *How verified?* What is the second law? *How verified?*

If a ray pass from one medium to another, it is bent from its course; this bending is called *refraction.*

The laws of refraction for heat are the same as for sound.

1°. *The plane of the incident and refracted rays is normal to the deviating surface at the point of incidence;* and

2°. *The sine of the angle of incidence bears a constant ratio to the sine of the angle of refraction.*

3. *The intensity of radiant heat varies directly as the temperature of the radiating body, and inversely as the square of the distance to which it is transmitted.*

The first part of this law is verified by exposing one of the bulbs of a differential thermometer to a blackened cubical box, filled with hot water, the other bulb being protected by a screen. If the water is in the first instance of a given temperature, and then falls to a half, or a third of that temperature, the differential thermometer will manifest a half, or a third of its original indication, and so on for any temperature.

The second part of the law may also be verified by means of the differential thermometer. In this case the heated body is kept always at the same temperature, and one bulb of the differential thermometer is placed at different distances from it. It will be found that at a double distance the indication is only a fourth of the original indication, at a triple distance only a ninth, and so on.

Exchange of Heat between bodies.

195. The process of radiation of heat between bodies is *mutual* and *continuous.* According to the laws given in the preceding article, those bodies which are most heated give off most heat; hence, the hottest bodies of a group give off more heat than they receive, and the coldest ones receive more than they give off. The consequence is that there is a continual tendency towards equalization of tem-

What is the third law? *How is the first part of the law verified? The second part?* Explain the laws of refraction of heat rays. **(195.)** Explain the action of radiation to produce uniformity of temperature.

perature. If all the bodies are of the same temperature, each will give off as much as it receives, and no further change of temperature can occur. The process of radiation, however, goes on as before.

All the bodies in a room, for example, tend to come to a uniform temperature. We say, tend to come to a uniform temperature, because this condition is never fully realized. Bodies nearest the walls are continually exchanging heat with the walls, and as these are in communication, either with the outer air, or with other rooms, their temperature will be influenced thereby, and will in turn exert an influence upon the remaining bodies in the room.

V.—REFLECTION, ABSORPTION, EMISSION, AND CONDUCTIBILITY.

Reflection of Heat.

196. When a ray of heat falls upon the surface of a body, it is divided into two parts, one of which enters the body and is absorbed, whilst the other is deflected or bent from its course. This bending is called *reflection*.

The point at which the bending takes place, is called *the point of incidence*. The ray before incidence is called *the incident ray*; after incidence it is called *the reflected ray*. If a perpendicular be drawn to the surface at the point of incidence, it is said to be *normal* to the surface at that point. The angle between the incident ray and the normal is *the angle of incidence*; the angle between the normal and the reflected ray is *the angle of reflection*. The plane of the incident ray and the normal is the plane of incidence; the plane of the reflected ray and the normal is the plane of reflection. These planes coincide.

Illustrate by the example of articles in a room. (**196.**) What is reflection of heat? What is the point of incidence? The incident ray? The reflected ray? The plane of incidence? The plane of reflection? The angles of incidence and reflection?

Laws which govern the Reflection of Heat.

197. The following laws, indicated by theory, have been confirmed by experience:

1. *The plane of the incident and reflected rays is normal to the reflecting surface at the point of incident.*

2. *The angles of incidence and reflection are equal.*

The apparatus, employed in establishing these laws, is shown in Fig. 137. *A* is a tin box with its faces blackened, in which hot water is placed. *B* is a reflecting surface, and *D* is a differential thermometer. *BC* is a normal to the reflecting surface.

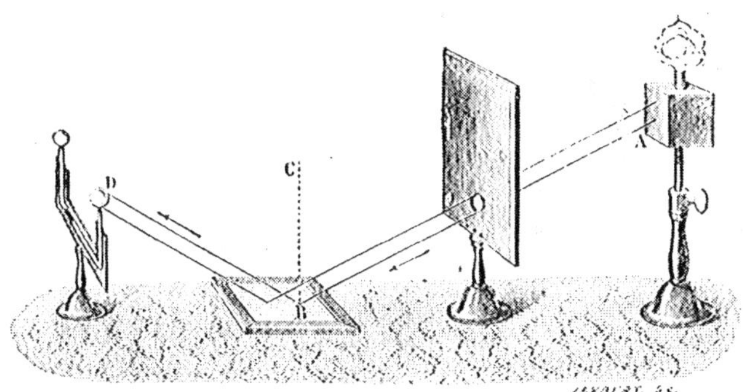

Fig 137.

The surface, *A*, radiates heat in all directions, but only a single ray is permitted to fall upon the reflector, *B*, the remainder being intercepted by a screen, having a small hole in it. By suitably arranging the thermometer, and other parts of the apparatus, it may be shown that the plane *ABD* is normal to the reflecting surface at *B*, and that the angles, *ABC* and *CBD*, are equal to each other.

(**197.**) What is the first law of reflection? The second law? *Explain the apparatus for verifying these laws. Explain the mode of verification.*

Reflection of Heat from Concave Mirrors.

198. A CONCAVE MIRROR is a polished spherical or parabolic surface, usually of metal, employed to concentrate rays of heat at a single point. For experimental purposes the parabolic mirror is generally used.

It is a property of such mirrors that all rays which before incidence are parallel to the axis, are after reflection converged to a single point, which point is the *focus* of the mirror. Conversely, if the rays proceed from the focus they will be reflected in lines parallel to the axis.

A and *B*, Fig. 138, represent two parabolic reflectors, having their axes coincident, and their surface turned to each other. In the focus, *n*, of the mirror, *A*, is placed a ball

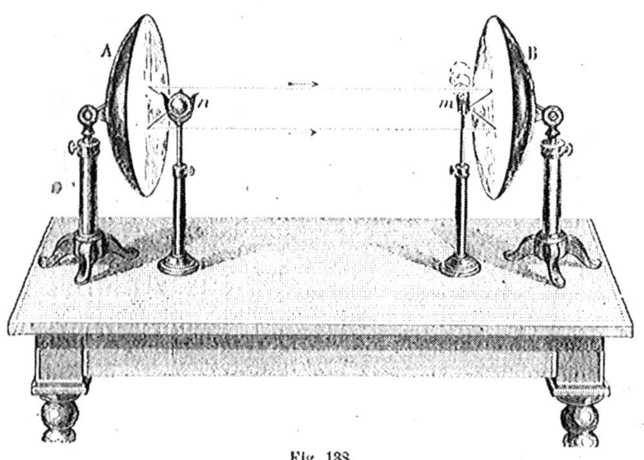

Fig. 138.

of hot iron, and in the focus, *m*, of the mirror, *B*, is placed an inflammable substance, as a piece of phosphorus. The

(**198.**) What is a Concave Mirror? What form is used for experiment? How are rays parallel to the axis reflected? What is the focus?

heat radiating from the ball, is reflected from A, parallel to the common axis of the mirror, and falling upon B, is again reflected to the focus m; the heat, concentrated at m, is sufficient to inflame the phosphorus, even when the mirrors are several yards distant from each other. If the mirror, A, alone is used, the phosphorus is not inflamed.

Fig. 139.

The property of parabolic mirrors, above explained, enables us to concentrate the heat of the sun's rays. In this case the reflector is called a *burning mirror*. Fig. 139 shows the manner of using a burning mirror. It is placed so that its axis is parallel to the rays of the sun, which, on falling upon it, are reflected to the focus, where they produce heat enough to set inflammable substances on fire.

It is said that ARCHIMEDES was enabled by means of mirrors to

How are rays from the focus reflected? Explain the experiment. *What is a burning mirror? Explain its use.*

set fire to the Roman ships in the harbor of the City of Syracuse. BUFFON showed the possibility of such an operation, by setting fire to a tarred plank, by means of burning mirrors, at a distance of more than 220 feet.

Reflecting Power of different substances.

199. It has been stated that a ray of heat which falls upon a body is divided into two parts, one being *absorbed* and the other *reflected*. The relative proportions between these two parts varies with the nature of the substance and the character of the reflecting surface.

Those bodies which reflect a large portion of the incident heat, are called *good reflectors;* those which reflect but little of the incident heat, are called *bad reflectors*. Good reflectors are *bad absorbers;* and bad reflectors are *good absorbers*.

Fig. 140 shows the method of determining the relative

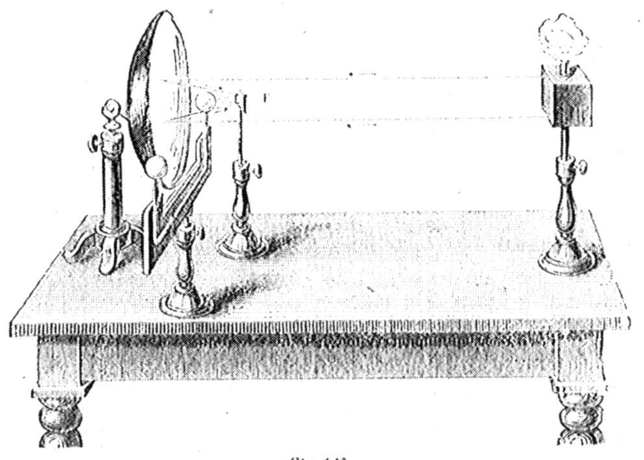

Fig. 140.

(199.) Into how many parts is an incident ray divided? What is a good reflector? A bad reflector? A good absorber? A bad absorber?

reflecting powers of different bodies, adopted by LESLIE. He placed a cubical tin box filled with water at the boiling point, in front of a parabolic reflector. The rays of heat, falling upon the reflector, are reflected and tend to come to a focus at *F*, but by interposing a square plate of some substance between the mirror and its focus, the rays are again reflected, and come to a focus as far in front of the plate, as *F* is behind it. The heat thus reflected is received upon one bulb of a differential thermometer, by means of which it is measured. By interposing plates of different substances in succession, their relative reflecting powers are determined.

In this way LESLIE showed, that polished brass possessed the highest reflecting power; silver reflects only nine tenths, tin only eight tenths, and glass only one tenth as much as brass. Plates blackened by smoke do not reflect heat at all.

Absorbing Power.

200. In order to determine the relative powers of absorption, LESLIE employed the apparatus shown in Fig. 141.

The source of heat and the reflector remaining as before, he placed the bulb of the differential thermometer in the focus of the reflector, covering it successively with layers of the substance to be experimented upon. In this way he showed, that those substances which reflect most heat absorb least, and the reverse.

When the bulb was blackened by smoke, the thermometer indicated the greatest change of temperature, and when covered with leaves of brass, it indicated the least change.

Explain LESLIE's method of determining the reflecting power of different bodies. What did LESLIE find to be the best reflector? The next in order? What of blackened plates? (**200.**) Explain LESLIE's method of determining the absorbing power of bodies. What was the result of his experiments?

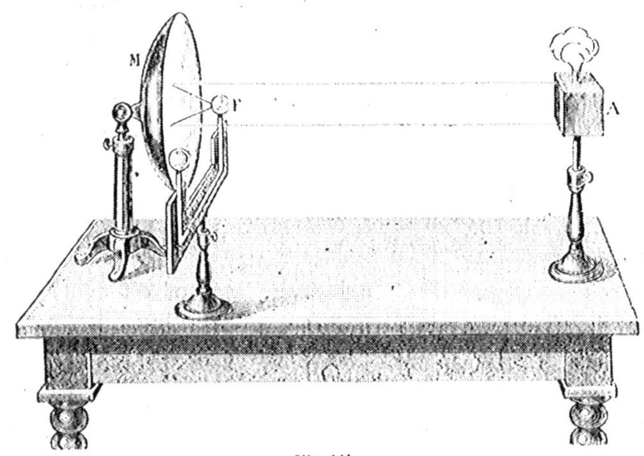

Fig. 141.

Radiating Power.

201. The RADIATING POWER of a body is its capacity to *emit*, or *radiate* the heat which it contains.

In determining the radiating power, LESLIE employed the apparatus shown in Fig. 141. In this case, instead of covering the bulb of the thermometer with layers of the substances to be experimented upon, he covered the different faces of the cubic box with layers of the different substances.

For example, let one face be made of tin, let a second be blackened by smoke or lamp-black, let a third be covered by a layer of paper, and a fourth by a plate of glass. On turning these different faces towards the reflector, the thermometer indicates different degrees of temperature. If the blackened face be turned towards the reflector, the thermometer rises, showing that this face is a good radiator; if the paper-covered face be next turned towards

(**201.**) What is the Radiating Power of a body? Explain LESLIE's method of determining it. Give an example of his process.

the reflector, the t'
poorer radiator tha
be turned towards t
lower, indicating that glas. poore
finally, if the tinned face is turned towards the refle
thermometer falls still lower, indicating the fact that tin is a poorer radiator than glass.

LESLIE found by this course of proceeding, that the radiating powers of bodies are the same as their absorbing powers; that is, a good radiator is also a good absorber, but a bad reflector, and the reverse.

Modifications of the Reflecting Powers of Bodies.

202. The principal causes that modify the reflecting and absorbing powers of bodies, are: *polish*, *density*, *direction of the incident rays*, *nature of the source of heat*, and *color*.

Other things being equal, *polished bodies are better reflectors and worse absorbers than unpolished ones.*

Other things being equal, *dense bodies are better reflectors and worse absorbers than rare ones.*

Other things being equal, *the nearer the incident ray approaches the normal, the less will be the portion reflected and the greater the portion absorbed.*

The nature of the source of heat sometimes modifies the reflecting and absorbing powers. Thus, if a body is painted with white lead, it absorbs more heat from a cubical box of boiling water, than though the same heat were emitted by a lamp. But if a body is painted with lamp-black, the amount absorbed is the same, whatever may be its source.

What relation did he find between the radiating and absorbing powers of bodies? (202.) What causes modify the reflecting and absorbing powers of bodies? Effect of polish? Of density? Of direction of rays? Of the source of heat?

' *bodies absorb less*
...red ones. White
...s the worst. White
...absorb... ...ad black ones the best.

Applications of the preceding principles.

203. Articles of clothing are intended to preserve uniformity of temperature in the human body by excluding the too violent heats of summer, and by preventing too rapid radiation of animal heat in winter.

Loose substances, like woollens and furs, are bad radiators, and therefore are suitable for winter clothing. Compact substances, like linens and cottons, are good reflectors, and therefore are suitable for summer clothing. As far as color is concerned, white is best adapted to both seasons, because white bodies are at once better reflectors and worse radiators, than those of dark colors.

The animals of the polar regions are generally of light colors, often becoming completely white in winter. This wise provision of Nature is calculated to adapt them to sustain more readily the severe cold of those inhospitable regions.

Oils and fats are good reflectors and bad radiators. Hence we find the Laplanders and Esquimaux rubbing their bodies with oils to prevent the too rapid radiation of animal heat, whilst the negroes of the tropical regions do the same thing to prevent the absorption of heat from without.

Snow is a good reflector but a bad absorber and radiator. Hence it is that a layer of snow in winter acts to protect the plants which it covers. Snow and ice, when exposed to the rays of the sun, melt but slowly, but if a branch of a tree or stone projects through the snow, it causes the latter to melt in its neighborhood, first by absorbing the heat of the sun, and then radiating it to the surrounding particles of ice or snow.

Of color? (**203**) *What is the object of clothing? Why are furs and woollens suitable to winter? Linens and cottons to summer? What color is best adapted to all seasons? Color of animals in Arctic regions? Effect of oils and fats on radiation and absorption? Examples. Effect of snow? Why do snow and ice melt slowly?*

If a stone is thrown upon a field of ice, it soon causes the ice around it to melt, forming a hole into which it sinks. A dark cloth spread upon snow acts in the same manner, and soon sinks under the influence of the sun's rays.

Water is soonest heated in a vessel whose surface is black and unpolished, because the vessel in this state is best adapted to absorb the heat which is applied to it, but on removing it from the fire, the water cools rapidly. To retain heat in liquids, they should be confined in dense and polished vessels, as these are poor radiators. Hence, for boiling and cooking, rough and black vessels should be employed, but to keep the articles warm, dense and polished vessels should be used. It is for this reason that a silver teapot is better than an earthen one. But as silver is a good conductor of heat, the handle should be *insulated* by interposing between it and the vessel some non-conducting substance, as ivory or bone.

Stoves, being intended to radiate heat, should be rough and black, but fire-places, being intended to reflect heat into the room, should be lined with white, dense, and polished substances, like glazed earthenware, or glazed fire-bricks.

Conductivity of Solid Bodies.

204. Conductivity is that property of bodies by virtue of which they transmit heat. Those bodies that transmit heat readily, are called *good conductors;* those that do not transmit it readily, are called *bad conductors.*

Ingenhousz showed that solid bodies possess different degrees of conductivity, by means of an apparatus shown in Fig. 142. It consists of an oblong vessel to contain water, from one side of which projects a system of short tubes for receiving rods of different kinds of solids, such as metals, marble, wood, glass, and the like.

Ingenhousz coated the different rods with a soft wax that

Explain the effect of a stone thrown upon ice. Of a dark cloth upon snow. Why is water soonest heated in black and unpolished vessels? In what vessels is it best kept hot? Of what material should stoves be constructed? Fire-places? Why? **(204.)** What is Conductivity? Good conductors? Bad conductors? Explain Ingenhousz' apparatus.

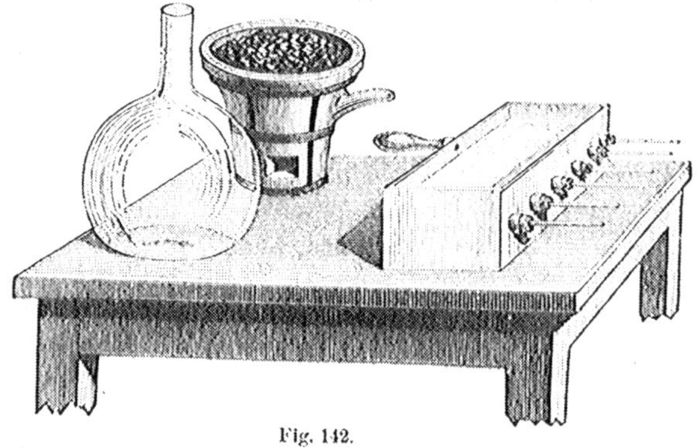

Fig. 142.

would melt at about 140° F., and then filled the vessel with boiling water. Upon some of the rods the wax melted rapidly, upon some more slowly, and upon others not at all. This showed that the rods varied in their conductivity.

It has been shown that metals are the best conductors, after which comes marble, then porcelain, bricks, wood, glass, resin, &c.

Conductivity of Liquids. — Convection.

205. Liquids are bad conductors of heat, except mercury, which is a metal. They are such bad conductors that RUMFORD asserted that water is not a conductor at all. More careful experiments have shown that all liquids are conductors, but all are extremely bad ones.

Liquids are heated by a process of circulation amongst their particles, called *convection*, the heat being applied from below, as shown in Fig. 143. When the particles at the bottom become heated, they expand, and as they are then lighter than the cooler particles above them, they rise to the

---

Explain his method of using it? *What are the best conductors? What bodies come next in order?* (**205.**) Are liquids good or bad conductors? How are liquids heated? Explain the illustration.

top of the vessel to give place to the heavier and cooler ones that supply their places. In this way a double current of particles is set up, as shown in the figure by the arrows, the hot ones rising and the cool ones descending. This process of circulation goes on till a uniform temperature is imparted to all of the liquid.

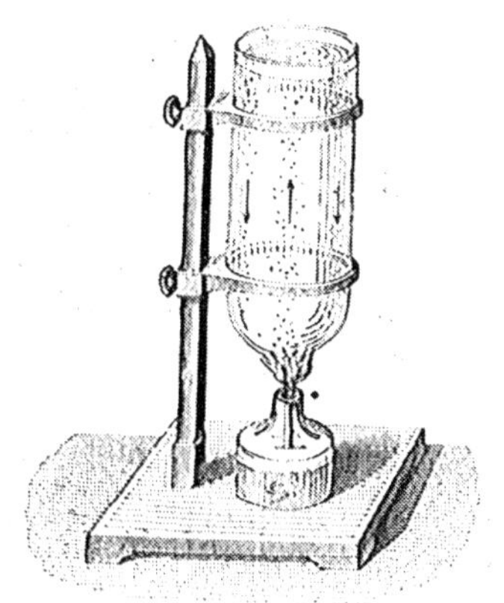
Fig. 143.

The circulation of particles may be shown by putting into the vessel particles of a substance of nearly the same density as the liquid; as, for example, oak sawdust. These particles will partake of the motion of the fluid, rising up in the centre, and descending along the walls of the vessel as shown in the figure

Conductivity of Gases.

206. Gases are bad conductors of heat, but on account of the extreme mobility of their particles, it is difficult to establish the fact by direct observation.

Gases are heated by convection, in the same manner as liquids.

Applications of the preceding principles.

207. If the hand be placed upon different articles in a cold room, they convey different sensations. Metals, stones, bricks, and the like, feel cold, whilst carpets, curtains, and the like, feel warm.

How may the circulation of particles be demonstrated? (**206.**) Are gases good or bad conductors? How are they heated? (**207.**) *Explain the different sensations experienced on touching bodies in a room.*

The reason of this is, that the former are good conductors, and readily abstract the animal heat from the hand, whilst the latter are bad conductors, and do not convey away the heat of the hand.

Wooden handles are sometimes fitted to metallic vessels which are to contain heated liquids. This is because wood is a bad conductor, and therefore does not convey the heat to the hand. For a similar reason, when we would handle any heated body, we often interpose a thick holder of woollen cloth, the latter being a bad conductor.

To preserve ice in summer, we surround it with some bad conductor, as straw, sawdust, or a layer of confined air. The same means are adopted to preserve plants from the action of frost. In this case, the non-conducting substance prevents the radiation of heat.

Cellars are protected from frost in winter by a double wall inclosing a layer of air, which is a non-conductor. It is the layer of confined air that renders double windows so efficient in excluding frost from our houses.

The feathers of birds and the fur of animals are not only in themselves bad conductors, but they inclose a greater or less quantity of air, which renders them eminently adapted to the exclusion of cold.

The bark of trees is a bad conductor, and so serves to protect them from the injurious effects of heat in summer, and cold in winter.

Our warmest articles of clothing are composed of non-conducting substances, inclosing a greater or less quantity of air. Such are furs, woollen cloths, and the like. It is not that these are warm of themselves, but they serve as non-conductors, preventing the escape of animal heat from our bodies.

V.—LAWS OF EXPANSION OF SOLIDS, LIQUIDS, AND GASES.

Laws of Expansion of Solids.

208. Numerous experiments have been made to determine the exact amount of expansion which bodies experience

Why are wooden handles attached to metallic vessels? How is ice preserved in summer? Why? How are plants protected? Why? How are cellars protected from frost? Why? Why are feathers adapted to exclude cold? Bark of trees? What substances form the warmest clothing? Why?

by the addition of a given amount of heat. As in a former article, it will be found convenient to consider first, *linear expansion*, and afterwards, *expansion in volume*.

1. *Linear expansion.* In order to compare the rate of linear expansion of different bodies, we take for a term of comparison, the expansion experienced by a unit of length of each body when heated from 32° F. to 33° F. This is called *the coefficient of linear expansion.*

The coefficients of linear expansion for a great number of bodies were determined in the latter part of the last century by LAVOISIER and LAPLACE. They reduced the substance to be experimented upon to the form of a rod or bar, then exposed it for a sufficient time to the temperature of melting ice, and measured its exact length. They next exposed the bar to a temperature of boiling water, and again measured its length. The increased length, divided by 180, gave the increase in length of the whole bar for 1° F. This result, divided by the length of the bar at 32° F., gave the linear expansion of a unit of length, and for an increase of temperature of 1° F., that is, *the coefficient of linear expansion.*

The following are some of the latest results:

SUBSTANCE.	COEFFICIENT.	SUBSTANCE.	COEFFICIENT.
Glass.......	0.00000474	Brass	0.00001044
Platinum...	0.00000483	Copper...	0.00000957
Steel.......	0.00000631	Silver	0.00001068
Iron	0.00000665	Lead.....	0.00001565
Gold	0.00000800	Zinc	0.00001653

From the above table, it is seen that the amount of expansion is always very small.

(208.) What is the coefficient of linear expansion of solids? How determined by LAVOISIER and LAPLACE? *Give some of the results.*

2. *Expansion in volume.* The *coefficient of expansion in volume* is the increment which a cubic unit of the substance experiences when its temperature is raised 1° F. This coefficient may be determined experimentally, or it may be found by multiplying the coefficient of linear expansion by 3.

Applications.

209. The principle of expansion explains many familiar phenomena, some of which are indicated below.

A cold tumbler is often broken when it is suddenly filled with hot water. The explanation is simple. Glass is a bad conductor of heat, hence the inside becomes heated by contact with the water more rapidly than the outside, and this inequality of heating produces an inequality of expansion that ruptures the glass. The thinner the glass, the less will be the inequality of expansion, and consequently the less will be the danger of rupture. In a metallic vessel such an accident is not to be apprehended, because metals are good conductors, and but little, if any, inequality of expansion can arise.

When a candle is held too near a pane of glass, the glass is often broken; the reason is the same as before.

Sometimes a glass vessel is broken by suddenly opening a door or window. This is due to a current of cold air which, falling upon the outer surface of the glass, causes an inequality of contraction that may produce rupture. All articles of glass should be guarded from sudden changes of temperature, would we avoid risk of breakage.

In the art of engineering, it is important to take into account the expansion and contraction of the metals. In laying the track of a railroad, for example, the rails should not be laid so as to touch each other, otherwise in warm weather the expansion, acting through a long line, might produce a force sufficient either to bend the rails or

---

What is the coefficient of expansion in volume? How determined? (209.) *Why does hot water break a cold tumbler? Which is more easily broken, a thin glass or a thick one? Why? Why is a pane of glass broken by the approach of a candle? Why may a glass vessel be broken by opening a door or window? Precautions? Explain the effect of expansion on a line of rails.*

to tear them from their fastenings. In employing iron ties in building, arrangements should be made by means of nuts and screws to tighten them in warm weather, and loosen them in cold weather, otherwise the forces of contraction and expansion would weaken and eventually destroy the building. Very serious accidents have occurred from omitting this precaution.

The principle of expansion and contraction of metals has been utilized in bringing the walls of a building together after they have commenced to separate. A system of iron ties is formed, passing through the opposite walls, on the outside of which they are secured by nuts. The alternate rods being heated, they expand, and the nuts are screwed up close to the walls. On cooling, the force of contraction brings the walls nearer together. The remaining rods are next heated, and the nuts screwed up. On cooling, a further contraction takes place, and so on until the walls are restored to their proper position. This method was successfully employed to restore the walls of a portion of the *Conservatoire des Arts et Metiers*, in Paris, which had begun to separate.

Compensating Pendulum.

210. The construction of the *Compensating Pendulum* depends upon the principle of contraction and expansion of metals. We have seen already that the time of oscillation of a pendulum depends upon its length, vibrating faster when shortened, and slower when lengthened. In consequence of variations of temperature, if a pendulum were suspended by a single metallic rod, its rate of vibration would be continually changing.

To obviate this defect and secure uniformity of rate, various devices have been employed, one of the most important of which is HARRISON'S Gridiron Pendulum, shown in Fig. 144. It consists of five parallel bars of metal, arranged as shown in the figure. The bars *a*, *b*, *c*, and *d*, are of steel, and when they expand, the effect is to lengthen the pendulum; the bar, *d*, passes freely through the cross piece, *or*, and is firmly attached to the piece, *mn*. The bars,

---

Precautions to be taken in building with iron? Explain the method of straightening walls. (**201.**) *What effect has heat upon a pendulum? How are its defects remedied? Explain the theory and construction of* HARRISON'S *Gridiron Pendulum.*

h and *k*, are of brass, firmly attached to both of the cross pieces, *mn* and *or*. When they expand, the effect is to raise the piece, *mn*, and thus to shorten the pendulum.

If the pieces are properly adjusted, the amount of shortening is exactly equal to he amount of lengthening before mentioned, and these two balancing each other, the length of the pendulum remains invariable. The adjustment requires that the lengths of the rods should be inversely as their coefficients of linear expansion.

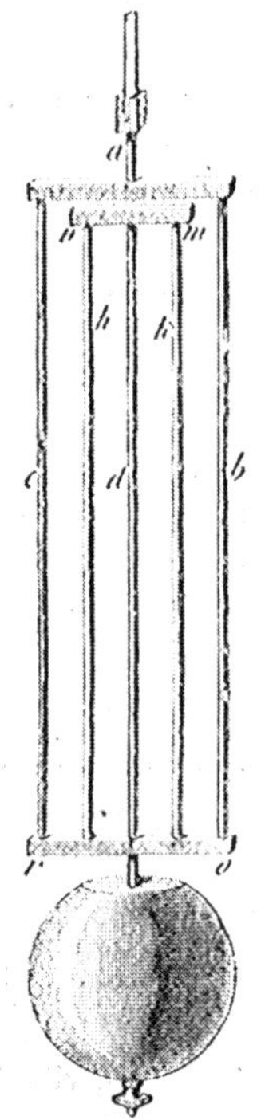

Fig. 144.

Laws of Expansion of Liquids.

211. Liquids are much more expansible than solids, on account of their feeble cohesion; their expansion is also much more irregular, especially when their temperature approaches the boiling point.

The expansion of a liquid may be *absolute* or *relative*. The absolute expansion of a liquid is its actual increase of volume; the relative expansion is its increase of volume with respect to the containing vessel. For example, in a thermometer the rise of the liquid in the stem is due to its relative expansion, with respect to that of the stem. Both expand, but the liquid more rapidly than the glass. The capacity of the bulb increases with an increase of heat, but the volume of its contained mercury increases more rapidly, and therefore rises in the stem. The absolute is usually greater than the relative expansion. It is the relative expansion that we generally observe.

(**211.**) Why are liquids more expansible than solids? What is absolute expansion? Relative expansion? Example. Which is generally observed?

The *coefficient of expansion* of a liquid is the expansion of a unit of volume, corresponding to an increase of temperature of one degree.

Taken with reference to glass, the coefficient of expansion for mercury is 0.000833; that of water is three times as great, and that of alcohol nearly eight times as great as that of mercury.

Maximum Density of Water.

212. If water is cooled down gradually, its volume continues to contract until it reaches the temperature of 39.°2 F., when it attains its maximum density. If it be still further cooled it begins to expand, and at 32° F. it becomes solid, or *freezes*.

This curious phenomenon may be shown by using a water thermometer in connection with a mercurial one. As the temperature is diminished, the liquids descend in the stems of both thermometers until the mercurial one shows 39.°2 F., after which, if the cooling process be continued, the mercury will continue to fall, whilst the water will begin to rise.

This apparent exception to the law of expansion and contraction is explained from the fact, that at the temperature of 39.°2 F., the particles begin to arrange themselves in a new order, preparatory to taking a crystalline form. Some other substances, such as melted iron, sulphur, bismuth, &c., exhibit a similar expansion of volume immediately previous to taking a solid crystalline form. It is this property of expanding at the time of crystallization, that renders iron so valuable a metal for casting. The expansion of the metal acts to fill the mould, thus giving sharpness and accuracy to the casting.

What is the coefficient of expansion of a liquid? What is its value for mercury with reference to glass? How do the coefficients of water and alcohol compare with mercury? (**212**) At what temperature has water the greatest density? When does it freeze? How may the phenomenon be shown? How explained? What other bodies exhibit similar phenomena? Why is iron so valuable for casting?

The fact that water has its greatest density at 39°.2 F., causes ice to form at the surface instead of at the bottom of rivers and lakes. Were it not that ice is lighter than water, it would sink to the bottom as fast as formed, or rather would form at the bottom, and in the colder regions of the globe would soon convert entire lakes into solid masses of ice. As ice and water are bad conductors of heat, the summer sun would not posesss the power to convert them again into water.

In Switzerland it is found by experiment that the temperature of the water at the bottom of deep and snow-fed lakes remains during the entire year at the uniform temperature of 39°.2 F., although the surface is frozen in winter, and in summer rises to 75° or 80° F.

It is because water has its maximum density at 39°.2 F., that it is taken at this temperature, as the standard of comparison for determining the specific gravity of bodies.

Law of expansion of Gases.

213. Gases are not only more expansible than solids and liquids, but they also expand more uniformly.

The *coefficient of expansion* of a gas, is the expansion which a unit of volume experiences when its temperature is increased one degree.

GAY LUSSAC supposed that all gases expand equally for equal increments of temperature; but more recent investigations show that the coefficients of expansion are slightly different for different gases. This difference is, however, so small, that for all practical purposes we may regard all gases as having the same coefficient. The value of the coefficient of expansion for gases is 0.00204, which is about eight times that of water.

Applications.

214. The law of expansion of gases when heated, has many important applications, some of which will be explained.

Explain the consequences of the expansion of water on freezing. Example of the lakes in Switzerland. Why is water taken at 39°.2 F. as a standard?

(213.) What bodies are most expansible? What is the coefficient of expansion? What was GAY LUSSAC's opinion? Was it strictly correct?

When the air of a room becomes warmed and vitiated by the presence of a number of persons, it expands and becomes lighter than the external air; hence it rises to the top of the room, and its place is supplied by fresh air from without, which enters through the cracks of the doors, or through apertures constructed for the purpose. Openings should be made at the upper part of the room to permit the foul air to escape. Such is the theory of *ventilation* of rooms.

In large buildings, like theatres, the spectators in the upper galleries often experience great inconvenience from the hot and corrupt air arising from below. To remedy this evil, large openings, called ventilators, should be constructed in the ceiling, and corresponding openings should be arranged near the bottom of the building, to supply a sufficient quantity of fresh air to keep up the circulation.

The principal of expansion gives a draft to our chimneys. The hot air ascends through the flue, and its place is supplied by a continued current of cold air from below, which keeps up the combustion in the fire-place or grate.

The same principle is applied in warming buildings by means of furnaces. Furnaces are placed in the lowest story of the building, and are provided with air chambers, which communicate with the external air by means of air-pipes. When the air becomes heated in the air chamber, it rises through pipes, or flues in the walls, to the upper stories of the building, and is admitted to or excluded from the different apartments by valves, called *registers*.

The principle of expansion of air explains many meteorological phenomena. When the air in any locality becomes heated by the rays of the sun, it rises and its place is supplied by colder air from the neighboring regions, thus producing the phenomena of winds. The circulation of the atmosphere in the form of winds, tends to equalize the temperature, and also, by transporting clouds and vapors, tends to equalize the distribution of water over the globe.

Winds also serve to remove the vitiated air of cities, replacing it by the pure air of the neighboring places, thus contributing to the preservation of life and health. Winds also act to propel vessels on

---

(**214.**) *How does the principle of expansion operate in ventilation? How are large buildings ventilated? What gives draft to chimneys? Explain the theory of heating by furnaces. How does the principle of expansion produce winds? Their effect on distribution of warmth and moisture?*

the ocean, thus contributing to the spread of commerce and civilization.

Without winds, our cities would become centres of infection, the clouds would remain motionless over the localities where they were formed, the greater portion of the earth would become arid and desert, without rivers or streams to water them, and the whole earth would s·on become uninhabitable.

Density of Gases.

215. The density of a gas depends upon the pressure to which it is subjected, and also upon its temperature.

It is for this reason that we select as a term of comparison the density at some particular pressure and temperature. The standard pressure is that of the atmosphere when the barometer stands at 30 inches, and the standard temperature is 32° F., or the freezing point of water. To determine the density at any other pressure, we apply MARIOTTE'S law; to determine it at any other temperature, we apply the coefficient of expansion, as explained in preceding articles.

Suppose it were required to determine the density of air when the barometer indicates 20 inches, and the thermometer 62° F., the density being equal to 1 at the standard temperature and pressure. The pressure being only two thirds the standard pressure, the air in the case considered would occupy once and a half its primitive volume, supposing the temperature to remain at 32° F. But the temperature being 62° F., or 30° above the standard, we multiply 1.5 by 30 times 0.00204 for the expansion. This product, added to 1.5, gives for a result, 1.5918. That is, a unit of volume at the standard pressure and temperature becomes 1.5918 units of volume at the given pressure and temperature. Because the density varies inversely as the volume, we shall have for the required density, $\frac{1}{1.5918}$, or 0.6282.

Other effects of winds? (**215.**) On what does the density of a gas depend? What do we take as a standard? How do we determine the density at any other pressure and temperature? *Example.*

The following table exhibits the density of some of the most important gases, air being taken as a standard:

TABLE.

GAS.	DENSITY.	GAS.	DENSITY.
Air.........	1.0000	Oxygen.....	1.1056
Hydrogen....	0.0692	Carbonic acid	1.5290
Nitrogen	0.9714		

Hydrogen is the lightest known body, its density being fourteen and a half times less than that of air.

VI.—CHANGE OF STATE OF BODIES BY THE ACTION OF HEAT.

Fusion.

216. It has been stated that heat not only causes bodies to expand, but that it may in certain circumstances cause them to change from the solid to the liquid state, or from the liquid to the gaseous state.

When a body passes from a solid to a liquid state, it is said to *melt*, or *fuse*, and the act of changing state in this case is called *fusion*.

If a melted body is suffered to cool, it becomes solid at the same temperature at which it melted. Hence the melting point is the same as the freezing point.

Fusion takes place when the force of cohesion, which holds the particles of a body together, is exactly balanced by the heat which tends to separate them. The temperature at which fusion takes place is different for different bodies. For some bodies it is very low, and for others very high, as is shown in the following

What is the lightest body? Give the densities of some other gases. (216.) *What is melting or fusion? When does fusion take place? Is the melting point the same for all solids?*

TABLE.

BODY.	TEMPERATURE OF FUSION.	BODY.	TEMPERATURE OF FUSION.
Mercury	− 39° F.	Bismuth.......	500° F.
Ice	32°	Lead	627°
Tallow	91°	Antimony	842°
White wax	149°	Zinc	932°
Sulphur........	232°	Silver.........	1832°
Tin............	455°	Gold..........	2282°

All bodies are not melted by the action of heat. Some are decomposed, such as paper, wood, bone, marble, &c. Simple bodies, that is, bodies which are composed of but one kind of matter, always melt, if sufficiently heated, with a single exception. Carbon has thus far resisted all attempts to fuse it.

Latent Heat of Fusion.

217. Bodies which can be melted always present the remarkable phenomenon, that when they are heated to the temperature of fusion, they can not be heated any higher until the fusion is complete. For example, if ice be exposed to heat, it begins to melt at 32° F., and if more heat be applied, the melting is accelerated, but the temperature of the mixture of ice and water remains at 32° until all the ice is melted.

The heat that is applied during the process of fusion, enters into the body without raising its temperature, and is said to become *latent*. When the body returns to its solid state, all the latent heat is again given out, and once more becomes *sensible*.

The phenomenon of latent heat may be illustrated by the following experiment. If a pound of pulverized ice, at 32° F., be mixed

Examples. Are all bodies melted by the action of heat? Examples. (**217.**) What is latent heat? Sensible heat? *How may the phenomenon of latent heat be illustrated.*

with a pound of water at 174° F., the heat of the water will be just sufficient to melt the ice, and there will result two pounds of water at the temperature of 32° F. During the process of melting, 142° of heat have been absorbed and become latent; hence, we say that the heat required to melt ice at 32° F. is 142°, or, in other words, the latent heat of water at 32° is 142°.

The enormous amount of heat which becomes latent when ice melts, explains why it is that large masses of ice remain unmelted for a considerable time after the temperature of the air is raised above 32° F. Conversely, the immense quantity of heat evolved when water passes to the state of ice, explains why it is that ice forms so slowly in extremely cold weather. The absorption of heat in melting, and production of heat in freezing, tend to equalize the temperature of climates in the neighborhood of large masses of water, like lakes and rivers.

Congelation.—Solidification.—Regelation.

218. Any body that can be melted by the application of heat, can be brought back to a solid state by the abstraction of heat. This passage from a liquid to a solid state is called *congelation*, or *solidification*.

In every body, the temperature at which congelation commences, is the same as that at which fusion begins. Thus, if water be cooled, it will begin to congeal at 32° F., and conversely, if ice be heated, it will begin to melt at 32° F. Furthermore, the amount of heat given out, or rendered sensible in congealing, is exactly equal to that absorbed, or rendered latent in melting.

Some liquids can not be congealed by the greatest cold to which we can subject them; such are alcohol and ether. Pure water congeals at 32°; the salt water of the ocean congeals at 27°; olive oil at 21°; linseed and nut oils at 17°.

Explain the action of latent heat on melting masses of ice. Also on freezing masses of water. (**218.**) What is congelation? How does the point of congelation compare with that of fusion? Illustrate. How does the heat given out in solidifying compare with that taken up in melting? *What liquids have never been frozen?*

Water reaches its maximum density at 38°.75, and as its temperature is diminished from this limit, its volume continues to increase until congelation is completed.

If two smooth pieces of melting ice be pressed against each other they are soon frozen together. This phenomenon is called *regelation*.

Regelation is explained by supposing the interior of the ice colder than the outer layer just passing into the state of water. When the pieces are pressed together the layer of water at 32° F. has a colder body on each side. The latent heat of fusion of this layer is soon absorbed and conducted away, and the water is converted into ice. The formation of a snow-ball depends on regelation. Below a temperature of 32° F. the particles of snow are dry and regelation cannot take place. Hence a coherent snow-ball can only be made of melting snow.

Crystallization.

219. When bodies pass slowly from the liquid to the solid states, their particles, instead of arranging themselves in a confused manner, tend to group themselves into regular forms. These forms are called *crystals*, and the process of forming them is called *crystallization*.

Flakes of snow, sugar candy, alum, common salt, and the like, offer examples of crystallized bodies. The forms of the crystals are best seen under a magnifying glass.

Bodies may be crystallized in two different ways. In the first case, we melt them, and then allow them to cool slowly. If a vessel of sulphur be melted and allowed to cool slowly, it will commence crystallizing about the surface, and if we break the crust thus formed, and pour out the interior liquid sulphur, we may obtain beautiful crystals of sulphur.

In the second case, we dissolve the body to be crystallized and then allow the solution to evaporate slowly. The dissolved body is then deposited at the bottom and on the

---

Explain the phenomenon of regelation. Illustrate. **(219.)** What are crystals? What is crystallization? Examples. How many methods of crystallization? Explain the first method. The second method.

sides of the vessel in the form of crystals. The slower the process, the finer will be the crystals. It is in this manner that we crystallize candy and various salts.

Freezing Mixtures.

220. The absorption of heat which takes place when a body passes from a solid to a liquid state, is often utilized in the production of intense cold. This result is best obtained by mixing certain substances, and these mixtures are then called *freezing mixtures.*

A mixture of one part of common salt and two parts of pounded ice forms a mixture that is used for freezing cream. The salt and ice have an affinity for each other, but they can not unite until they pass to the liquid state; in order to pass to this state they absorb a great quantity of heat from the neighboring bodies, and this causes the latter to freeze. By means of a mixture of salt and snow, the thermometer may be reduced to 0.

VII.—VAPORIZATION.—ELASTIC FORCE OF VAPORS.

Vaporization.—Volatile and Fixed Liquids.

221. When sufficient heat is applied to a liquid, it is converted into a gaseous form and is called a *vapor.* The change of state from a liquid to a gaseous state is called *vaporization.*

Conversely, if heat be abstracted from a vapor, it will return to a liquid form. The change of state from a vaporous to a liquid form is called *condensation.*

Vapors are generally colorless, and are endowed with an *expansive force,* or *tension,* which, when heated, may become very great.

(**220.**) What is a freezing mixture? Example. Explain its action? (**221.**) What is vaporization? Condensation? General properties of vapors?

The number of vapors that exist at ordinary temperatures is very small. Of these, watery vapor is the most familiar, as well as the most important, on account of the part which it plays in many natural phenomena.

Liquids are divided into two classes, with respect to the readiness with which they pass from the liquid to the vaporous state, viz.: *volatile liquids* and *fixed liquids.*

VOLATILE LIQUIDS are those which have a natural tendency to pass into a state of vapor even at ordinary temperatures, such as ether, alcohol, and the like. If a vessel of water, alcohol, ether, or chloroform be left exposed to the air, the liquid is slowly converted into vapor, and disappears; in other words, it evaporates. To the class of volatile liquids belong essences, essential oils, volatile oils, amongst which may be mentioned spirits of turpentine, oil of lavender, attar of roses, oil of orange, and the like.

FIXED LIQUIDS are those which do not pass into vapor at any temperature, as, for example, fish oils, olive oils, and the like. At high temperatures they are decomposed, giving rise to various kinds of gases, but to no true vapors that can be condensed into the original form of the liquid. Some oils, like linseed oil, harden on exposure to the air, but it is not by evaporation, but by absorbing oxygen from the air, and thus passing to a solid state. Some solids are capable of passing directly to a state of vapor without first becoming liquid. To this class belong camphor, musk, and odorous bodies generally. Snow and ice may, under certain circumstances, evaporate without melting.

Evaporation under pressure.

222. The influence of evaporation by pressure may be illustrated

The most important vapor? What two classes of liquids have we? What are volatile liquids? Examples. Illustrate. What are fixed liquids? Examples. Effect of high temperatures upon them? Give examples of solids that vaporize?

by means of an apparatus shown in Fig. 145. It consists of a curved tube, the short branch of which is closed and filled with mercury; the mercury also fills a portion of the long branch. A small quantity of ether is introduced into the short branch, when it at once rises to the top, *B*, of this branch. At ordinary temperatures, the pressure of the external atmosphere exerted through the mercury, is sufficient to prevent the ether from forming vapor

If, however, the tube is plunged into a vessel of water heated to 112°, the ether will be converted into vapor and will occupy a certain portion, *AB*, of the tube, holding in equilibrium the pressure of the atmosphere, together with the weight of the mercurial column whose height is *AC*.

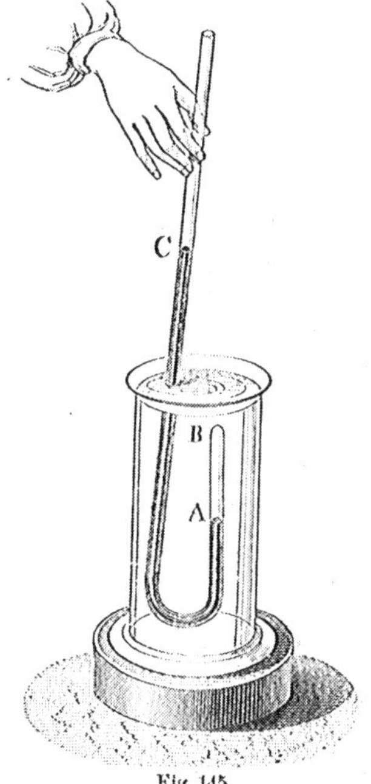

Fig. 145.

If the tube be withdrawn and allowed to cool, the vapor of ether will be condensed, and will appear as a liquid at *B*. If more heat be applied, it will again be converted into vapor, and the mercury will rise in the branch, *C*, as long as any ether remains to be evaporated. This shows that the tension of the vapor augments with the temperature. This principle holds true for all kinds of vapor.

The tension acquired by the vapor of water, or *steam*, often becomes so great by being heated as to burst the strongest vessels, and thus is the cause of frightful accidents. The cause of wood snapping when burned in a fire-place, is the expansion of the water in the pores, giving rise at last to an explosion. When a chestnut is roasted in the ashes, the moisture within the shell expands into

(222.) *Explain the experiment showing the influence of pressure on vaporization. Why does wood snap when burned? Why does a chestnut snap when roasted?*

steam, and explodes with sufficient force to throw the nut from the fire. Hence it is that a small puncture is usually made in the shell, which permits the escape of the steam and prevents explosion

Instantaneous Evaporation in a Vacuum.

223. Vapors formed upon the surface of a liquid escape by virtue of their tension. Under ordinary circumstances, the pressure of the air prevents a very rapid escape of vapor at ordinary temperatures, but when the atmospheric pressure is diminished in any way, evaporation takes place with great rapidity. If the pressure is entirely removed, the evaporation is instantaneous, like the flash of gunpowder, especially if the liquid is very volatile.

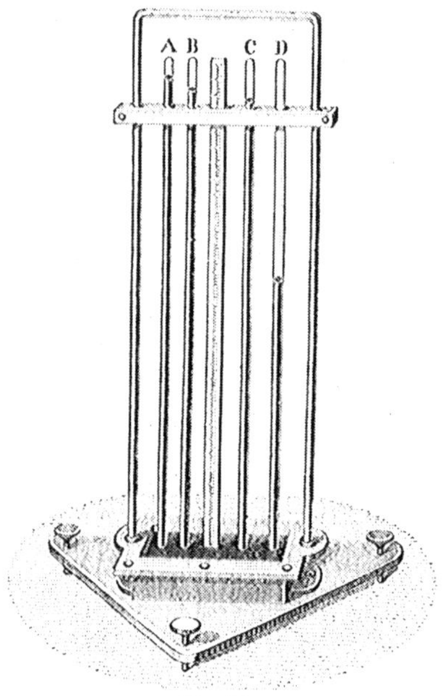

Fig 146.

This principle may be illustrated by means of the apparatus shown in Fig. 146. It consists of several barometer tubes, *A*, *B*, *C*, *D*, filled with mercury, and inverted in a common cistern of mercury, as shown in the figure. The whole apparatus is supported by a frame, to which is

How remedied? (**223.**) Why do vapors escape from the surfaces of liquids? When the pressure is removed, what happens? *How may the principle be illustrated? Explain the experiment in detail.*

attached a graduated scale. The mercury will stand at the same height in all of the tubes, at the height in A, for example.

If a few drops of water be introduced into the tube B, they will rise through the mercury in the tube, and on reaching the vacuum, will be instantly converted into vapor, as is shown by the depression that takes place in the column of mercury. If a little alcohol be introduced into the tube C, it will, in like manner, be converted into vapor, and will produce a still greater depression of the column. If a small quantity of ether be introduced into the tube D, a still greater depression of the mercury will be observed.

This experiment shows that the tension of the vapor of ether is greater than that of alcohol, and that of alcohol greater than that of water. By careful measurement, it is found that the tension of the vapor of ether is twenty-five times as great as that of water, and six times as great as that of alcohol.

Limit of the Tension of Vapors.

224. If a sufficient quantity of each of the liquids in the last experiment be introduced into the tubes, vapor will finally cease to form, and a portion will remain in the liquid state. In this case the tension of the vapor already formed is sufficient to balance the tendency of the liquid to pass into a state of vapor. In this state of affairs no more vapor can form without a change of temperature. This is the case supposed in the last article.

Saturation.

225. When a given space has taken all of the vapor that it can contain, it is said to be *saturated*. For example, if water be poured into a bottle filled with dry air, and the bottle be hermetically sealed, a slow evaporation will go on until the tension of the vapor given off is equal to the tendency of the remaining water to pass into vapor, when it

What does the experiment show? (**224.**) When does vapor cease to form? (**225.**) When is a space saturated with vapor? Example?

will cease. In this case, the space within the bottle is saturated.

It is a remarkable fact, established by numerous experiments, that for the same temperature, the quantity of watery vapor necessary to saturate a given space is always the same, whether that space is a vacuum, or whether it contain air or any other gas. The only point of difference in these cases is the rapidity with which the saturation takes place.

If the temperature varies, the amount of vapor required to saturate a given space will vary also. The higher the temperature, the greater will be the quantity of vapor required to saturate the given space, and the lower the temperature, the less the quantity required for saturation.

The quantity of watery vapor in the atmosphere is very variable, but notwithstanding the continued evaporation that is taking place from lakes, rivers, and oceans, the air in the lower regions of the atmosphere is never saturated. The reason is, that the vapor being less dense than the air at the surface, rises into the higher regions, where it is condensed by the greater cold existing there, and falls to the earth in the form of rain.

Causes that accelerate Evaporation.

226. The slow evaporation of water on the surface of our globe is accelerated by many causes, some of which are indicated below:

1. *Temperature.*—Increase of temperature also increases the tension of the vapor formed, and accelerates evaporation.

This property is utilized in the arts in the manufacture of extracts. The evaporation is carried on in chambers kept at temperatures of

What is the law of saturation at a given temperature? What effect has a change of temperature? Why is the amount of vapor in the atmosphere variable? (**226.**) What effect has increase of temperature on evaporation? *How is this property utilized?*

from 80° to 140° F., the air being continually renewed to carry off the vapor as fast as formed.

2. *Pressure.*—Diminution of pressure facilitates evaporation.

This principle has been utilized in the arts for the concentration of syrups. This application is illustrated by the method of concentrating syrups in sugar refining. The syrups are placed in large spherical boilers, from which the air is extracted by means of air-pumps worked by steam.

3. *Change of air.*—A continual change of the air in contact with the liquid facilitates evaporation, by carrying off the vapor which would otherwise saturate the layer in contact with the liquid, and effectually check the formation of additional vapor.

It is for this reason that the surface moisture of our fields and roads disappears more rapidly when there is a breeze than in calm weather. In the arts, the principle is applied by keeping a current of air playing across the surface of the liquid to be evaporated, by means of blowers, or otherwise.

4. *Extent of the liquid.*—A large surface is favorable to rapid evaporation, by affording a great number of points from which vapor may be formed.

This principle is utilized in the arts by employing shallow and broad evaporating pans. This application is illustrated by the process of making salt from sea-water. The water is spread out in large pans, which are very shallow, and then exposed to the influence of the sun's rays, when the water slowly evaporates, leaving the salt in the form of crystals.

Ebullition.

227. EBULLITION, OR BOILING, is a rapid evaporation, in which the vapor escapes in the form of bubbles. The

---

What effect has pressure? *How is this utilized?* What effect has change of air? *Application of this principle?* What effect has the extent of liquid? *How utilized in the arts? Example.* (227.) What is Ebullition?

bubbles are formed in the interior of the liquid, and rising to the surface, they collapse, permitting the vapor to pass into the air.

In heating water, the first bubbles are due to the small quantities of air contained in the liquid, which expand and rise to the surface. Afterwards, as the heat is kept up, particles of water are converted into vapor and rise through the liquid, becoming condensed by the colder layers of water above them. When all of the layers become suitably heated, the bubbles are no longer condensed, but rise to the surface, and escape with a commotion that we call boiling, as shown in Fig. 147.

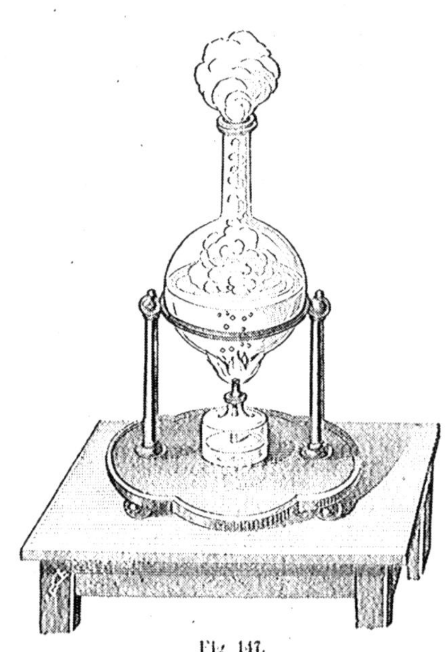

Fig 147.

The following are the laws that govern the phenomena of ebullition:

1. *Under the same pressure, each liquid enters into ebullition at a fixed temperature.*

The temperature at which a liquid boils is called its *boiling point.* When the barometer stands at 30 inches, the boiling point of pure water is 212° F.; the boiling point of ether is 108° F.; the boiling point of alcohol is 174° F., and the boiling point of mercury is 660° F.

Explain the phenomena of boiling. What is the first law of ebullition? Illustrate.

2. *The pressure remaining the same, a liquid can not be heated higher than the boiling point.*

For example, if water be heated to 212°, it will begin to boil, and no matter how much heat may be applied, it will continue to boil, but will never become hotter than 212°; all the applied heat passes into the vapor and becomes latent. It becomes latent, because it does not heat either the water or the steam above 212°. This will be explained hereafter.

Causes that modify the boiling point of Liquids.

228. The principal causes that influence the boiling point of liquids, are: *the presence of foreign bodies, variations of pressure*, and *the nature of the vessels* in which the boiling is effected.

1. *Presence of foreign bodies.*—Matter in solution generally raises the boiling point of a liquid. Thus, a solution of salt does not boil so readily as pure water. If, however, the body dissolved is more volatile than water, then the boiling point is lowered. Fatty matters combined with water, raise its boiling point. Hence it is, that boiling soup is hotter than boiling water.

2. *Variations of pressure.*—Increase of pressure raises, and diminution of pressure depresses, the boiling point. When the pressure is great, the vapor, in order to escape, must have a high tension, and this requires a high temperature. When the pressure is small, the reverse is the case.

This principle may be illustrated by the apparatus shown in Fig. 148. It consists of a bell-glass, connected with an air-pump. Beneath the glass is a vessel of water. If the air be exhausted from

What is the second law? Illustrate. (228.) What are the principal causes that modify the boiling point? What is the effect of impurities? Illustrate by examples. What is the effect of pressure? *Illustrate. Explain the experiment.*

the bell-glass, the water enters into ebullition, even at ordinary temperatures. This is because the pressure is diminished.

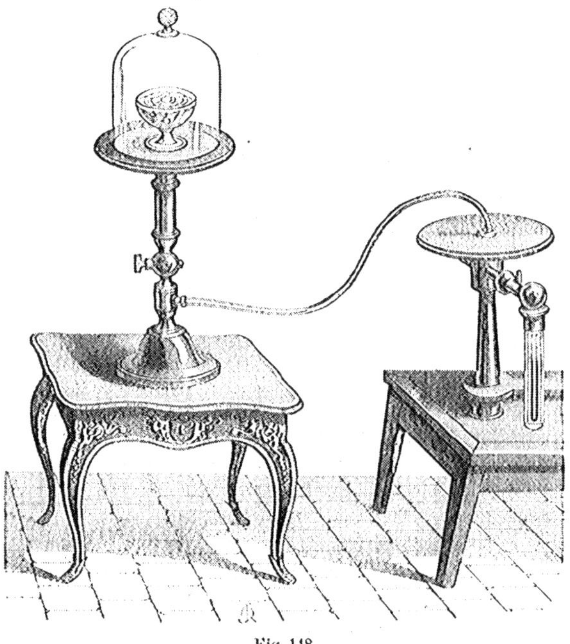

Fig. 148.

If it is desirable to continue the ebullition for some time, an arrangement must be made to remove the vapor as fast as formed. This can be effected by placing a dish of sulphuric acid under the bell-glass. The acid absorbs the vapor with great avidity. Furthermore, there is no increase of temperature in the water, but on the contrary the temperature continually falls, and the water may even be frozen.

The same principle may be further illustrated by a little instrument, shown in Fig. 149, called FRANKLIN's *Pulse Glass*. It consists

How may water be frozen by evaporation? Explain FRANKLIN's *Pulse Glass.*

of a glass tube, bent twice at right angles, and terminating at each extremity in a bulb, one of which is somewhat larger than the other. Before the larger bulb is sealed, a quantity of water is introduced,

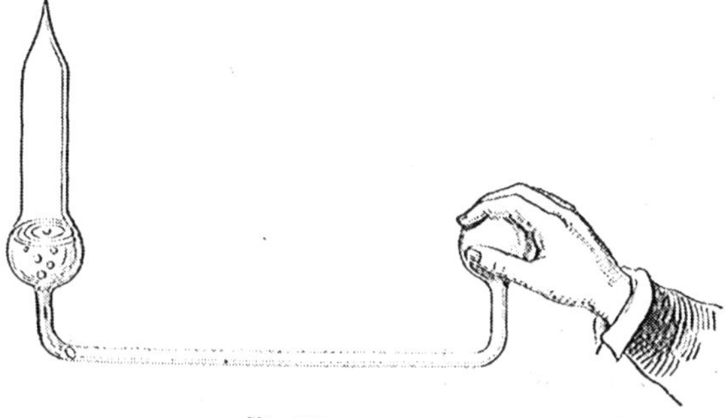

Fig. 149.

sufficient to fill the smaller one and a part of the larger one, and this is then made to boil over a spirit lamp until the air is driven out and the entire space is filled with steam. When this is effected, the large bulb is hermetically sealed by means of a jet of flame, directed across the open end of the tube. The space above the water is then filled with steam, which, as the instrument cools, is reduced to a low degree of tension. In this state of affairs the heat of the hand applied to the small bulb is sufficient to make the water boil, as indicated in the figure.

3. *Nature of the vessel.*—When the interior of the vessel is rough, the projecting points form centres for developing vapor, and the boiling point is lower than when the surface is smooth. Water boils at a lower temperature in an iron than in a glass vessel. In fixing the boiling point of thermometers, a metallic vessel should always be employed to boil the water in, on account of the fact just mentioned.

X What effect has the nature of the vessel on ebullition? Illustrate.

Papin's Digester.

229. When water is heated in open vessels, its temperature can not be raised beyond a certain limit, but in closed vessels both the water and its vapor may be raised to very high temperatures, so that the tension of the vapor may reach several atmospheres. The instrument employed to show this fact is called PAPIN'S *Digester*, so called because PAPIN invented it for extracting the nutriment from bones.

It is represented in Fig. 150, and consists of a thick bronze vessel, *M*, whose cover is held in place by a screw passing through a strong frame. To avoid danger of explosion, the instrument is provided with a safety-valve, similar to that used in steam-engine boilers. The safety-valve consists of a valve, *u*, fitting closely over an opening in the cover. This valve is held in place by a lever, *ab*, and a movable weight, *p*. One end of the lever is fastened at *a* by a hinge-joint. By moving the weight *p*, along the lever, we may vary the force with which the valve, *u*, is kept in place.

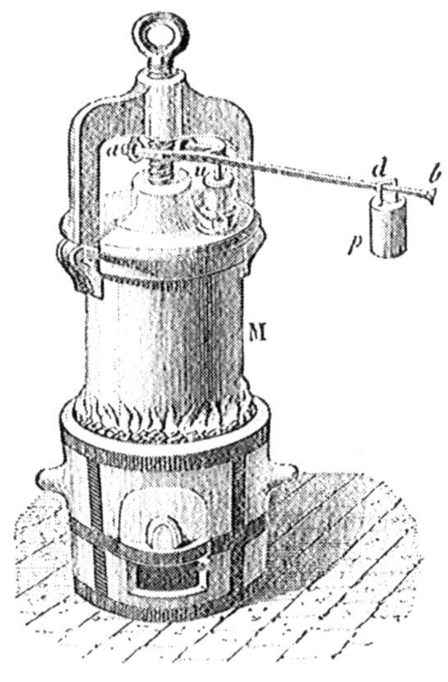

Fig. 150

Suppose the weight, *p*, to be 30 lbs., or 2 atmospheres, then if the distance *ad*, is made equal to four times the distance *a* , from the principle of the lever the pressure upon the valve will be that of the atmosphere increased by 120 lbs., that is, it will be equal to 7 atmospheres, and whenever the tension of the vapor within the

(**229.**) *What is* PAPIN'S *Digester? What principle does it illustrate? Explain its construction.*

digester exceeds this, the valve will be forced open, and a portion of the steam will escape with a whistling sound that indicates great compression.

If the valve be left open, the temperature can only be raised to 212°, and we have the phenomena of simple boiling. If water be heated in a well corked bottle, the tension of the vapor will finally cause the cork to spring from its place with a loud explosion.

It is the high tension of confined vapors that gives rise to the explosion of steam-boilers. Hence the necessity of constructing them of strong materials, and of providing them with proper safety-valves.

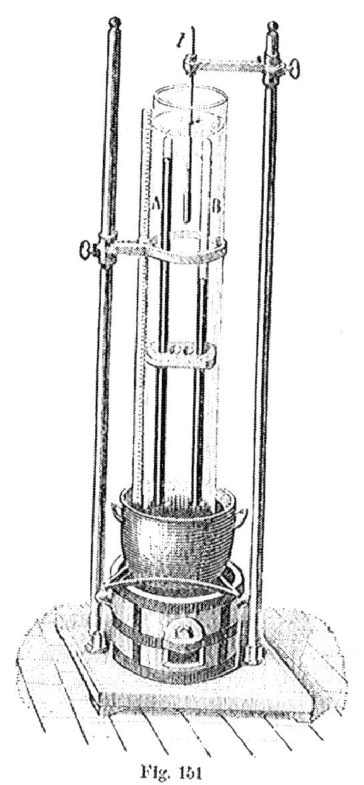

Fig. 151

Measure of the Elastic Force of Vapor.

230. DALTON measured the elastic force of watery vapor at every temperature, from 32° F., up to 212° F., by means of the apparatus shown in Fig. 151.

This apparatus consists of two barometer tubes, *A* and *B*, filled with mercury, and inverted in an iron boiler, also filled with the same liquid. The tube, *A*, contains mercury alone, whilst the tube, *B*, contains a small quantity of water

Illustrate its use by an example. What causes explosions of steam-boilers? Precautions to be taken. (**230.**) Explain DALTON's apparatus for measuring the tension of vapors, and the method of using it.

above the mercury. The tubes are kept in place by a wooden frame, placed in a long glass cylinder filled with water. A thermometer, *t*, is plunged into the water for the purpose of determining its temperature. When heat is applied to the boiler, the temperature of the whole apparatus is raised, and the water in the tube, *B*, is converted into vapor, whose tension is made known by the difference of level of the mercury in the tubes, *A* and *B*. This difference is measured by a scale attached to the cylinder.

For example, if, when the thermometer stands at 158° F., the difference of level in the tubes is 9 inches, we say that the tension of vapor at 158° is 9 inches of mercury, or 4.5 lbs., that is, it presses each square inch of surface, with which it is in contact, with a force of 4.5 lbs.

DALTON increased the temperature from 32° to 212°, noting at each degree the difference of level between the mercury in the tubes, and thus was enabled to form a table showing the elastic force of vapor at all temperatures within these limits.

DULONG and ARAGO have more recently extended DALTON'S table to temperatures above 212°. Their investigations show that the tension of watery vapor at 212° F. is 1 atmosphere; at 250° F. it is 2 atmospheres; at 273° F. it is 3 atmospheres; at 291° F. it is 4 atmospheres; at 306° F. it is 5 atmospheres.

From all of these results we infer that the tension increases very rapidly with the temperature.

Latent Heat of Vapors.

231. When a liquid begins to boil, all of the heat that is added enters into the vapor and becomes latent. The amount of heat that becomes latent, is different for different liquids. It is called the *latent heat of vaporization.*

It has been ascertained by experiment that the latent heat of watery vapor is about 990° F., that is, it takes 5½ times as much

---

Example. *Between what limits does* DALTON'S *table extend? What general inference may be drawn?* (**231.**) What is latent heat of vaporization? *What does it amount to for water?*

heat to convert any quantity of water into steam as is required to raise the same quantity of water from the freezing to the boiling point. This may be verified by mixing 1 lb. of steam at 212° with 5½ lbs. of water at 32°. The latent heat becomes sensible by the condensation of the vapor, and there results 6½ lbs. of water at 212°.

Examples of Cold produced by Heat becoming Latent.

232. If a few drops of ether be poured upon the hand and allowed to evaporate, a sensation of cold will be felt. The ether in evaporating extracts the heat from the hand, which becomes latent.

Damp linen feels cold when applied to the body, because the moisture in passing to a state of vapor extracts the animal heat, which entering the vapor, becomes latent.

The warm wind of summer is refreshing, because it causes a more rapid evaporation of the perspiration, which abstracts animal heat from the body to become latent in the vapor thus produced. The coolness that results from sprinkling the floor of an apartment in summer, arises from the passage of heat from a sensible to a latent state, in consequence of the evaporation of the water. For the like reason, a shower of rain is generally followed by a diminished temperature.

Water may be cooled by putting it in porous vessels. A small quantity escapes through the pores, and in evaporating abstracts a portion of heat from the remaining liquid, thus reducing its temperature. This is the process of cooling water employed in many tropical countries.

Congelation of Water and Mercury in a Vacuum.

233. When evaporation is rapidly increased, the absorption of heat is proportionally increased, and as it is taken from the surrounding objects, these are sometimes frozen. It has been stated that water may be frozen under

How is this shown? **(232.)** *Why does ether produce cold by evaporation? Why does damp linen feel cold? Why is warm wind refreshing in summer? Effect of sprinkling? Of a shower? How is water cooled in porous vessels?* **(233.)** Why does evaporation produce cold in surrounding objects?

the receiver of the air-pump by absorbing the vapor as rapidly as it is generated.

By operating with a liquid more volatile than water, a greater degree of cold is produced. By using sulphurous acid, which boils at 14° F., a sufficient degree of cold is produced to freeze mercury. This is effected by surrounding a thermometer bulb with cotton, saturated with sulphurous acid, and then placing it under a receiver and exhausting the air.

The rapid vaporization abstracts so much heat from the mercury that it freezes in a few minutes. If we break the bulb, the mercury is found in a solid mass like a leaden bullet. In this form mercury can be drawn out into sheets, or stamped like a coin, but it soon absorbs heat from neighboring bodies, and again passes to a liquid state

VIII.—CONDENSATION OF GASES AND VAPORS.—SPECIFIC HEAT.

Causes of Condensation.

234. The CONDENSATION of a vapor, is its change from a vaporous to a liquid state. This change of state may arise from *chemical action*, *pressure*, or *diminution of temperature.*

1. *Chemical action.*—The affinity of certain substances for the vapor of water is so strong that they absorb it from the air, even when the latter is not saturated; such, for example, are quick-lime, potash, sulphuric acid, and many others. When placed in a closed space, they in a short time abstract all of the moisture that is in it.

2. *Pressure.*—If a closed cylinder be filled with vapor, and

Explain the experiment with sulphurous acid. Can mercury be frozen? (**234.**) What is condensation of a vapor? Causes? Effect of chemical action? Examples. Effect of pressure?

this be compressed by a piston, as soon as the space occupied by the vapor is saturated, it will begin to condense, and if the pressure be continued, all the vapor will be reduced to the liquid state. Until the space becomes saturated, the pressure must be continually increased, on account of the augmented tension of the vapor, but after liquefaction begins, no further augmentation of tension takes place, and the pressure required to complete the liquefaction remains uniform.

3. *Diminution of temperature.*—When the temperature of any space is diminished, the amount of vapor required for saturation is diminished. After the point of saturation is reached, any further diminution of temperature causes a deposit of the vapor in a liquid form.

Steam is colorless, but when allowed to escape into the cold air, condensation takes place in the form of drops, which become visible. For the same reason, the moisture contained in the breath becomes visible in cold weather.

In winter the glass of our windows often becomes coated with drops like dew. This arises from the fact that the glass is colder than the air of the room, and thus acts continually to produce condensation of the vapor in the air. If the difference of temperature is sufficient, the particles of vapor are frozen as they are deposited, producing beautiful crystallizations. When the external air is warmer than that within, the deposit takes place on the outside of the glass. If a vessel of cold water be placed in a warm room, a deposition of moisture takes place on its exterior surface.

The nearer the air is to saturation, the more abundant is the deposit of dew. Hence, before a rain, the deposit is especially abundant. Stone walls, and the like, being cooler than the atmosphere, are often in summer covered with moisture, when they are said to sweat. The moisture in this case is condensed from the air,

Illustrate. How long must the pressure augment? Effect of diminution of temperature? *What is the color of steam? Why does it become visible? Explain the deposition of drops on glass. Explain frost-work crystals. Why is the deposition more abundant before rain? Deposition on stones and walls?*

and does not come from the stones. If the sweating of stones is indicative of rain, it is because the deposition is most abundant when the air is most nearly saturated.

Heat developed by Condensation.

235. When a liquid passes to a state of vapor, a great quantity of heat is absorbed from neighboring bodies, and becomes latent. When the vapor returns to a liquid state, an equal amount of heat is given out and becomes capable of affecting our senses; in other words, it becomes *sensible.*

Heating by Steam.

236. Buildings are heated by means of steam conveyed from a boiler in the lower story, through iron pipes in the walls. The steam, by its heat and by the heat given out on condensation, serves to warm the apartments through which it is made to pass. To this end, coils of pipes are placed in the rooms to be warmed.

Distillation.

237. DISTILLATION is the process of separating liquids from each other by means of heat.

The most volatile of the liquids is most easily evaporated, and its vapor is then condensed. The heat should be kept above the boiling point of the liquid that we wish to obtain, but below that which we wish to leave behind. The boiling point of alcohol being 174° F., and that of water 212°, if a mixture of alcohol and water be heated up to some temperature between these limits, the alcohol will all be vaporized, whilst most of the water will remain behind.

Why indicative of rain? (**235.**) Explain the development of heat by condensation? (**236.**) Explain the principle of heating buildings by steam? (**237.**) What is distillation? What degree of heat is required for distillation?

Distillation.

238. An ALEMBIC, or Still, is an apparatus for distillation.

The most usual form of an alembic is represented in Fig. 152. It is composed of a boiler, *A*, with a cover, *B*,

Fig. 152

called the *dome;* from the top of the dome a metallic tube, *C*, passes into a vessel, *S*, called the condenser, and is then bent into a spiral form. This tube is called the *worm*, and after passing through the condenser, *S*, it leads to a *receiver*, *D*. The condenser, *S*, is kept full of cold water by an arrangement shown in the figure.

The substance to be distilled is placed in *A*, and a suitable heat is then applied. The more volatile portion is converted

(238.) What is an Alembic? Describe the most usual form? How is distillation effected?

into vapor, rises into the dome, and passing through the worm, is condensed, and escapes in a liquid form into the receiver, *D*.

Wine is composed of water, alcohol, and a coloring matter. If this liquid be placed in the alembic and heated to any temperature between 174° and 212°, the alcohol is separated from the other ingredients. As a portion of water is evaporated, the alcohol thus obtained is not pure, and will require to be distilled again. At each distillation, the strength is increased, but no amount of distillation can render it absolutely pure.

By distillation, pure water may be obtained from the brine of the ocean, or from the impure water of our wells and springs.

Liquefaction of Gases.

239. Most of the gases have been liquefied, either by pressure alone, or by a combination of pressure with a diminution of temperature. An immense pressure may be had by utilizing the tension of the gases themselves, by generating large quantities in confined spaces.

One of the most interesting examples of the liquefaction of a gas is that of carbonic acid.

Carbonic acid is capable not only of liquefaction, but also of congelation. For this purpose, two immensely strong cylinders are fitted together, both being hermetically sealed, and communicating by a pipe. One of these cylinders is the *generator*, and the other the *receiver*. In the generator are placed the ingredients necessary to generate carbonic acid, usually carbonate of soda and sulphuric acid. After the opening is carefully closed, these materials are brought into contact, when an immense volume of carbonic acid is developed, and, being unable to expand, its tension becomes so great that a portion is condensed into a liquid form. The tension, at the temperature of 60° F., is equal to 50 atmospheres, or 750 lbs. on each square inch.

Explain the method of distilling alcohol? Water? (**239.**) How may gases be liquefied? *Example. Explain the apparatus for liquefying carbonic acid? The process of liquefaction?*

After liquefaction has ceased, if a stop-cock be turned so as to allow a part of the confined gas to escape, a portion of the liquid acid passes to a state of vapor with immense rapidity, and in doing so, absorbs so much heat from the remaining portion as to freeze it. The frozen acid is thrown out by the gaseous jet in flakes like snow. It is very white, and so cold as to freeze mercury instantly. It evaporates very slowly, and when tested with a spirit thermometer, its temperature is found to be 112° below the 0 of FAHRENHEIT'S thermometer. By using this solid with other substances for which it has an affinity, the greatest degree of artificial cold may be obtained.

Specific Heat of Solids and Liquids.

240. Experiment shows that different bodies require different amounts of heat to elevate their temperatures through the same number of degrees. The amount of heat required to heat any body a certain number of degrees, is called its *specific* heat.

If equal weights of water, iron, and mercury have the same amount of heat communicated to them, the mercury will be most heated, the iron next, and the water least of all. When heated to a certain temperature, water absorbs ten times as much heat as iron, and thirty-three times as much as mercury.

In order to compare bodies with respect to their specific heat, we take as a unit the amount of heat necessary to raise a given weight, say 1 lb., of water through 1° F. Two principal methods have been employed to ascertain the relative specific heat of bodies.

In the *first* method, the body to be experimented upon is brought to a standard temperature, say 212° F., and is then brought into contact with ice. The amount of ice melted makes known the quantity of heat given off by the body in

How may a portion be solidified? Describe the solid gas. How may intense cold be produced? What degree of Fahrenheit? (**240.**) What is specific heat? Illustrate. How do we compare bodies with respect to specific heat? Explain the first method of determining the specific heat of a body.

passing from 212° to 32°, from which the relative specific heat may be determined.

In the *second* method, the body to be experimented upon is heated to a certain temperature, and then plunged into water at a lower temperature. The two bodies interchange heat and come to a common temperature. Then, from a knowledge of the weights of the two bodies mixed, their original temperatures, and their common resulting temperature, their relative specific heats may be determined.

The following table shows the specific heat of a few of the most important substances:

TABLE.

SUBSTANCE.	SPECIFIC HEAT.	SUBSTANCE.	SPECIFIC HEAT.
Water	1.000	Copper	0.095
Glass	0.198	Silver	0.057
Iron	0.114	Mercury	0.033
Zinc	0.096	Platinum	0.032

Of all these bodies water has the greatest specific heat, and consequently it requires more heat to raise its temperature through any given number of degrees.

Water heats slowly, and mercury very rapidly. Of course mercury cools rapidly and water slowly.

The specific heats of gases have been determined with respect to air as a standard, but the results need not be given in this treatise.

The second method. What body has the greatest specific heat? What bodies heat fastest? Cool fastest? Examples.

IX.—HYGROMETRY.—RAIN.—DEW.—WINDS.

Hygrometry.

241. HYGROMETRY is the process of measuring the amount of moisture in the air with respect to the amount necessary to saturate it.

The object of hygrometry is not to determine the absolute amount of moisture in the atmosphere, but simply to find out its degree of saturation. The absolute amount of moisture remaining the same, the atmosphere might at one temperature be saturated, whilst at some other temperature it would be far from saturation.

In winter the air is generally damper than in summer, though in the latter season it generally contains a greater absolute amount of vapor than in the former. This is due to difference of temperature. For the same reason the air is damper at night than in the day time. A cold room is damper than a warm one for the same reason.

Moisture in the Air, and its Effects.

242. The quantity of moisture in the air varies with *the seasons*, with *the temperature*, with *the climate*, and with *different local causes.*

When the air is too dry, the exhalation by the pores of the skin, called insensible perspiration, is too abundant, the skin cracks, and exfoliates, and much suffering results. When the air is too moist, the insensible perspiration is retarded and often entirely stopped, resulting in many painful diseases.

Hence the importance, in a sanitary point of view, of regulating the amount of moisture in our dwellings so as to avoid both of these extremes. On this account it is that *evaporators* are attached to our

(**241.**) What is Hygrometry? *Illustrate.* Explain the difference between the hygrometrical state of the air in winter and summer. (**242.**) Under what circumstances does the quantity of moisture in the air vary? *Explain the effect of dryness and moisture on the system. Important sanitary precaution.*

furnaces, which, when properly regulated, keep up a suitable degree of moisture in the heated air, furnished to warm our apartments.

The Hygroscope.

243. A HYGROSCOPE is an instrument for showing the amount of moisture in the air.

Any hygrometric substance, that is, any substance capable of absorbing moisture, may be employed as a hygroscope. A great number of animal and vegetable substances, such as paper, parchment, hair, catgut, are elongated by absorbing moisture, and are shortened when dried, and are therefore adapted to the construction of a hygroscope. We shall explain the construction of a single instrument of this class in illustration of the principle employed in all.

It consists, as shown in Fig. 153, of a piece of wood cut out in the shape of a monk, having a cowl of pasteboard turning about an axis, *a*. The axis, *a*, passes through the neck of the figure, and connects with an apparatus shown in the section *A B*, on the left of the figure. The axis, *a*, is connected with a piece of twisted catgut kept tense by a spring. When the weather is dry, the catgut twists tighter, carrying with it the axis *a*, and the monk lays off his cowl, as shown in the figure. When the weather is damp, the catgut un-

Fig 153.

(243.) What is a Hygroscope? What substances may be used in the construction of a hygroscope? Examples. Explain the hygroscope shown in Fig. 153.

twists, and the monk puts on his cowl. In adjusting the instrument, care should be taken to have the cowl on the head when the catgut is damp.

Instruments of this kind are very uncertain in their action, and are therefore used as matters of curiosity rather than for any scientific value they may possess.

The Hair Hygrometer.

244. A HYGROMETER is an instrument for measuring the amount of moisture in the air. Several kinds have been invented, but the hair hygrometer is the most used.

This instrument is constructed from the principle that a hair elongates when moistened, and shortens when dried. The form usually given to it is shown in Fig. 154. A hair about eight inches in length is fastened at its upper end, and at its lower end it is wound around the axis of a small pulley, and then is made fast to it. A silk thread is wound around the pulley in an opposite direction, having a weight, *P*, attached to it to keep it tense. A needle attached to the pulley plays in front of a graduated arc, as the hair elongates and contracts.

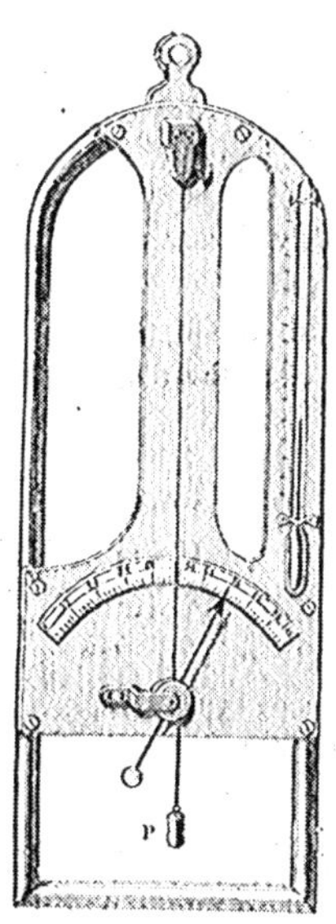

Fig. 154.

To graduate the instrument, it is placed under a bell-glass, and the air is thoroughly dried by some substance, such as quick-lime, which is capable of absorbing the moisture of the air. The point at which the needle then stands is marked 0. The air is then saturated

(244.) What is a Hygrometer? Explain the construction and use of the hair hygrometer. *How is it graduated?*

with moisture, and the point at which the needle stands is marked 100. The intervening space is divided into 100 equal parts, and these are numbered from 0 up to 100. The temperature is noted by a thermometer attached to the frame of the instrument.

To use the instrument, we note the reading of the needle and of the thermometer, and from these the exact amount of moisture in the air may be computed.

Hygrometric state of the Atmosphere.

245. By the HYGROMETRIC STATE of the atmosphere, we mean its relative degree of saturation. If we denote complete saturation by 1, and the air contain half the amount of vapor necessary to saturate it, its hygrometric state will be denoted by 0.5.

GAY LUSSAC has constructed a table, by means of which the hygrometric state of the air may be found when we know the reading of the hygrometer already described, together with that of the attached thermometer.

Formation of Fogs and Clouds.

246. FOGS and CLOUDS are masses of vapor condensed into drops, or vesicles, by coming in contact with colder strata of the atmosphere. The term fog, applies when these masses are in contact with the earth, and the term cloud, when they are suspended in the air.

The air at all times contains a greater or less quantity of invisible vapor, and if at any time the air becomes cooled below a certain limit, a portion is condensed and becomes visible; the result is either a fog or a cloud.

One of the most common causes of clouds is the cold generated by an ascending current of air. When the air becomes heated, it ex-

How is it used? (**245.**) What is meant by the hygrometric state of the atmosphere? (**246.**) What are Fogs? Clouds? How are fogs and clouds formed? *What is a common cause of a cloud?*

pands and ascends, and being continually subjected to a diminishing pressure, it expands rapidly, and a large amount of heat must become latent. This absorption of heat produces cold enough to condense the vapor into clouds. When a cloud floats into a warmer stratum of the atmesphere, it is often converted into invisible vapor and disappears. It is *dissolved*.

Mountains arrest the winds blowing from the plains, and force them to ascend their sloping sides. Coming in contact with the colder strata of the atmosphere, the moisture is converted into clouds and fogs. Hence we often see the mountain tops covered with fogs and clouds, when the other portions of the sky are clear. The condensation of water on the sides of mountains is the most fruitful source of our streams. When a cold wind meets with a warm and moist current of air, the cooling process is so great as to generate clouds.

Two theories have been advanced to explain the reason why clouds remain suspended in the air. According to the *first* theory, the particles of moisture are hollow spheres of water like soap-bubbles, filled with air less dense than that without. Consequently the little vesicles float in the air like so many minute balloons. According to the *second*, and favorite theory, the particles are extremely small, and float in the air in the same way that particles of dust and other small bodies are seen to be borne along by the atmosphere.

Fogs form over bodies of water and moist grounds, when the air above them is cooler than the water or earth.

Fogs are frequent along the course of rivers and upon inland lakes. The cause of the dense fogs that prevail in the neighborhood of Newfoundland, is the Gulf Stream. The water brought by the Gulf Stream is warmer than that of the surrounding ocean, and as the vapor rises from it, it is converted by the cold air from the neighboring regions into fog.

When does a cloud dissolve? Effect of mountains on clouds? Utility of mountain condensation? Explain the two theories of the formation of clouds. *Where are fogs most frequent? Why so many fogs on the banks of Newfoundland?*

Rain.

247. RAIN is a fall of drops of water from the atmosphere. When several particles of a cloud unite, the weight becomes too great to be supported by the air, and the drop thus formed falls to the ground.

When a cloud floats into a colder stratum of the atmosphere, it becomes more condensed, and we have a fall of rain. When it floats into a warmer stratum it dissolves. Hence we often see the clouds of the morning dissolve under the influence of the sun, which acts to heat the upper regions of the atmosphere.

The quantity of rain that falls in any country depends upon its neighborhood to the ocean or other bodies of water, upon the season; upon the temperature, and upon the prevailing direction of the winds. More rain falls near the coasts than in the interior; more rain falls in summer than in winter; more rain falls in tropical climates than in temperate and polar climates; and finally, more rain falls in those countries where the prevailing winds are from the ocean than where they are from the continents.

The following table indicates the number of inches of rain that fall during the year at the places named:

At	Copenhagen	18	inches.
"	Paris	22	"
"	Havana	90	"
"	Calcutta	81	"
"	Grenada	126	"

From this we see that the quantity of rain increases rapidly as we approach the equatorial regions.

(**247.**) What is Rain? *Explain the cause of rain.* Upon what does the amount of rain in any place depend? *Give examples of the amount of rain in different places. Inference.*

Dew and Frost.

248. DEW is a deposition of watery particles, that takes place upon the soil and plants during the calm nights of summer.

The true theory of dew was first established by WELLS. According to his theory, dew results from the earth and plants becoming cooled by radiation, thus producing a deposit of moisture from the neighboring strata of air. Good radiators are soonest covered with dew, whilst bad radiators have little or no dew formed upon them.

The state of the atmosphere influences the amount of dew. When the air is clear, the dew is abundant, when cloudy, little or no dew is formed. In this case the clouds radiate heat to the earth, and this prevents the latter from cooling so rapidly. A strong breeze prevents the formation of dew, by removing the strata of air next the earth before they have time to be cooled down to the point of saturation, or the *dew point*. A gentle breeze may facilitate the formation of dew, by replacing the layer of air from which the water has been deposited, by another which contains more moisture.

WHITE FROST is nothing more than frozen dew. It is often seen in autumn, and arises under the same circumstances as are favorable to the formation of dew. In order that frost may occur, the earth must be cooled below 32° F.

Snow and Hail.

249. SNOW is a collection of frozen particles of water, formed in the upper regions of the atmosphere, whence it falls to the ground in flakes.

(**248.**) What is Dew? What is WELLS' theory of dew? What bodies are soonest covered with dew? What ones have little dew upon them? What effect has the state of the atmosphere on dew? Why is there much dew on clear nights? Little on cloudy nights? What is the dew point? Effect of a gentle breeze? What is White Frost? (**249.**) What is Snow?

Snow flakes are made up of crystals, arranged in star-like forms with three or six branches, differently arranged, but always remarkable for their regularity and beauty. When snow falls, the temperature of the air is near 32° F. If the temperature is much lower, the snow is less abundant, because the amount of vapor in the air is less.

The quantity of snow that falls in any place is generally the greater as the place is nearer the pole, or as it is higher above the level of the ocean. At the poles, and on the summits of high mountains in all latitudes, snow remains through the entire year. As we approach the equator, the region of perpetual snow rises higher and higher above the level of the ocean. In the Andes, under the equator, the limit of perpetual snow is between 15,000 and 16,000 feet above the level of the ocean; in the Alps it is only 10,500 feet above the level of the ocean; towards the northern extremity of Norway it is but 3,000 feet above the ocean level.

HAIL is composed of layers of compact ice, arranged concentrically about nuclei of snow. Its formation is undoubtedly of electrical origin, and will be again treated of under the head of electricity.

Winds.

250. WINDS are currents of air, moving with greater or less rapidity. They are generally named from the quarter whence they blow; thus a wind that blows from the east is called an east wind, and so for other winds. Winds are sometimes named from some local peculiarity. Thus we have *trade winds*, *monsoons*, *siroccos*, and the like. The prevailing directions of the wind are different in different countries, for reasons that will be explained hereafter.

Causes of Winds.

251. Winds are caused by variations of temperature in the atmosphere; these variations produce expansions and

Describe a snow flake. *What law governs the fall of snow?* What is Hail? (**250.**) What are Winds? How named? (**251.**) What are the causes of winds?

contractions, thus disturbing the equilibrium of the atmosphere, causing currents. These currents are winds. For example, if the air is more heated over one country than over the neighboring countries, it dilates and rises, its place being supplied by the colder air which flows in from the surrounding regions. The surplus of air thus brought in flows over at the top of the ascending column. Hence there is a current near the earth in one direction, whilst at a higher elevation there is a current flowing in a contrary direction.

Regular, Periodic, and Variable Winds.

252. Winds are divided into three classes: Regular Winds, Periodic Winds, and Variable Winds.

1. *Regular winds.*—Regular winds are those which blow throughout the year in the same direction. They occur in the neighborhood of the equator, extending on each side about 30 degrees. From their advantage to commerce they are called *trade winds.* On the north side of the equator they blow from the north-east, on the south side they blow from the south-east.

The trade winds arise from currents of air flowing from the polar regions towards the equator; the velocity of the earth about its axis being greater as we approach the equator, these winds lag behind as it were, and become inclined to the westward, giving north-east winds on the north-side, and south-east ones on the south side of the equator.

2. *Periodic winds.*—Periodic winds are those which at regular intervals of time blow from opposite directions. Such are the *monsoons* that prevail in the Indian ocean,

(252.) How are winds divided? What are regular winds? Where do they occur? What are they called? What is their direction on the north side of the equator? On the south side? Explain the causes of the trade winds? What are periodic winds?

blowing one half of the year from north-east to south-west, and the other half in the opposite direction. When the sun is on the north of the equator, the southern portion of the Asiatic continent is warmer than the southern part of Africa, and the winds blow from south-west to north-east; when the sun is on the south side of the equator, the reverse is the case.

3. *Variable winds.*—Variable winds are those which blow sometimes in one direction and sometimes in another, without any apparent law of change. The further we recede from the equatorial regions, the more variable are the winds in their character.

The Simoon.—The Sirocco.

253. The SIMOON is a hot wind that blows from the deserts of Africa. It is felt in the northern and north-eastern parts of the African continent. During its prevalence the thermometer often rises to 120° F. In the desert this wind becomes suffocating from its heat and dryness. Travellers exposed to it cover their faces with thick cloths, and their camels turn their backs to escape its injurious effects.

The SIROCCO is a hot wind that sometimes is felt in Italy. When it blows, people remain in their houses, taking care to close every door and window. Some suppose this to be a continuation of the simoon from the African desert, others think that it has its origin in Sicily.

Velocity of Winds.

254. The velocity of winds is very variable. The velocity is measured by instruments called *anemometers.*

Explain the cause of the monsoons. What are variable winds? When are they most variable? (**253.**) What is the Simoon? Explain. What is the Sirocco? Explain. (**254.**) What is an anemometer?

These consist of a species of windmill attached to a train of wheel-work, by means of which the number of revolutions per minute can be registered. From the number of revolutions the velocity can be computed.

The velocity of the gentlest breeze, or zephyr, is not more than one mile per hour; a moderate wind travels at the rate of 4½ to 5 miles per hour, a brisk wind 20 miles per hour, a tempest 40 to 50 miles per hour, and a hurricane from 90 to 100 miles per hour.

X.—SOURCES OF HEAT AND COLD.

Sources of Heat.

255. The principal sources of heat, are: *the sun, electricity, chemical combination and combustion, pressure and percussion*, and *friction*.

1. *The sun.*—The sun is the most abundant source of heat. We are ignorant of the cause of heat in the sun's rays.

It has been computed that the heat received from the sun by the earth in a year is sufficient to melt a layer of ice extending over the entire globe, and 100 feet in thickness. Yet on account of the great distance of the earth from the sun, and its comparatively small size, it can receive only the minutest portion of the heat which the sun radiates in all directions.

2. *Electricity.*—The subject of heat due to electricity will be treated of under the head of Electricity.

3. *Chemical combination and combustion.*—Chemical combinations are generally accompanied by a disengagement of heat. When they take place slowly, the heat is inappre-

Describe it. *What are the velocities of some of the winds?* (**255.**) What are the principal sources of heat? What is the most abundant source? *What is the amount of heat received by the earth from the sun in a year?* Explain chemical combination as a source of heat.

ciable, but when they take place rapidly, there is often produced an intense heat, and sometimes a development of light.

Combustion is one form of chemical combination. The forms of combustion exhibited in our fire-places and our lamps, is a combination of the carbon and hydrogen of the wood and oil with the oxygen of the air. The products of such forms of combustion are watery vapor, carbonic acid, with gases and volatile products that appear under the form of smoke. Combustion is a decomposition of certain substances, accompanied by a composition of new products. In this change, no element is lost, simply a change of form takes place.

The flame produced in combustion, is a mixture of gaseous and volatile matters, heated red hot by the heat disengaged in the process of combustion.

The process of respiration is a species of slow combustion, in which the carbon and other matter of the blood unites with the oxygen of the air. This species of combustion gives rise to the heat of the body of men and animals. This heat is called *animal heat.*

Fermentation is a chemical process that gives rise to heat.

4. *Pressure and percussion.*—Whenever a body is compressed so as to reduce its volume, heat is developed. The greater the compression, the greater the amount of heat developed. If gas be suddenly and violently compressed, the heat generated is sufficient to set fire to inflammable bodies. This subject was referred to in the article on Compressibility, in which the instrument used for inflaming tinder is figured. (See Fig. 4.)

Percussion is a source of heat. If a body, like a piece of metal, for example, be hammered, it soon becomes hot. It is percussion that causes the heat when a flint is struck against a piece of steel. In this case there is a piece of the steel detached and rendered red hot by the collision.

Explain the phenomena of combustion. What is flame? What is respiration? What kind of heat comes from respiration? What is fermentation? Explain compression as a source of heat. Illustrate. Explain percussion as a cause of heat.

5. *Friction.*—Friction is the resistance which one body offers to another when they are rubbed together. This resistance is accompanied with a great development of heat. In many cases, the friction is so great that the rubbing bodies are set on fire. In this way many savage tribes procure fire. Pieces of ice when rubbed together, generate heat enough to melt them. In machinery, the friction on axles often sets them on fire, especially when lubrication has been neglected.

Sources of Cold.

256. The principal sources of cold are: *fusion*, *vaporization*, *expansion of gases*, and *radiation of heat.*

1. *Fusion.*—When a body melts, it absorbs heat from the surrounding bodies, which becomes latent in the melted body.

2. *Vaporization.*—When a liquid passes to a state of vapor, it absorbs heat, which becomes latent in the vapor. Both of these causes of cold have been considered already.

3. *Expansion of gases.*—When a gas is compressed, it gives out heat, and conversely, when it expands it absorbs heat. This heat, it is, that acts to keep the particles asunder, and the further apart the particles are kept, the greater the amount of heat required.

Heat is the repulsive force that keeps a body in a gaseous state at all, or even in a liquid state.

If air be compressed in a condenser and then allowed to escape into the atmosphere, a slight cloud will be formed; this is due to the cold generated by the expanding air, which condenses the vapor in the air. This experiment illustrates the manner in which clouds are formed in the upper regions of the atmosphere.

Explain friction as a source of heat. (**256.**) What are the principal sources of cold? Explain fusion as a source of cold? Vaporization. Expansion of gases. *Explain the formation of a cloud when compressed air expands.*

4. *Radiation.*—Radiation produces cold in the radiating body, because radiation is simply giving off heat.

The earth, and all bodies on its surface, are continually radiating heat. This is compensated during the day by the heat received from the sun; in fact, the amount received is greater than that given off. But at night the reverse holds true, and a greater amount is radiated than is received. This cooling of the earth's surface is, as has been stated, the cause of dew and frost.

It is often said that it freezes harder when the moon shines than when it is concealed by clouds. This is the case, but the moon has nothing to do with the freezing. The true explanation of the phenomenon is this: When the moon shines, it is generally cloudless, and the radiation goes on more rapidly, and of course a greater degree of cold is produced. On the contrary, when the moon is obscured, it is generally cloudy; now the clouds are good radiators of heat, and the heat that they send back to the earth is nearly or quite enough to compensate for that radiated from the earth; hence the process of freezing is either retarded or entirely prevented.

Plants are good radiators, hence they are more likely to be affected by frost than other objects. To protect them from frost, we cover them with mats, which prevent radiation, or rather radiate back the heat that the plants throw off.

Explain radiation as a cause of cold. *Illustrate. What effect has the moon on freezing? Why is it colder when the moon shines than when cloudy? Why are plants likely to be affected by frost? How are they protected?*

CHAPTER VI.

OPTICS.

I.—GENERAL PRINCIPLES.

Definition of Optics.

257. Optics is that branch of Physics which treats of the phenomena of light.

Definition of Light.

258. Light is that physical agent which, acting upon the eye, produces the sensation of sight.

Two Theories of Light.

259. Two theories have been advanced to account for the phenomena of light: *the Emission Theory*, and *the Undulatory, or Wave theory.*

According to the *emission theory*, light consists of infinitely small particles of matter, shot forth from luminous bodies with immense velocity, which, falling on the retina of the eye, produce the sensation of sight.

According to the *undulatory theory*, light, like heat, is caused by the vibrations of the molecules of bodies. It is transmitted by a highly elastic medium called *ether*.

(**257.**) What is Optics? (**258.**) What is Light? (**259.**) What two theories of light have been advanced? Explain the emission theory. Explain the wave theory.

This medium, which also transmits radiant heat, extends through space, penetrates all bodies, and exists in the intervals between their molecules. The molecular vibrations of a luminous body are imparted to the neighboring ether, and are propagated through it by a succession of spherical waves; these waves falling on the retina of the eye excite the sensation of sight.

Light and radiant heat are very closely related to each other; they are generated in the same manner and are propagated through the same medium, but they differ from each other in their wave length, and as a consequence in their mode of action on bodies.

In sound the particles of air vibrate to and fro in the direction of propagation; in light and radiant heat the particles of ether vibrate to and fro in a direction perpendicular to that of propagation. In sound the vibrations are *longitudinal*, or in the direction of the rays; in light and radiant heat they are *transversal*, or perpendicular to the rays.

The idea of transversal vibrations may be illustrated by a rope made fast at one end and held by the hand at the other. If the free end be moved rapidly to and fro, at right angles to the rope a succession of waves will run along the rope, whilst the particles of the rope simply vibrate back and forth in perpendiculars to the rope. If a stone be dropped into a pool of still water, a series of waves will be propagated outward, whilst the particles of water simply rise and fall, their motion being perpendicular to the direction of propagation.

Luminous Bodies.—Sources of Light.

260. Bodies that emit light are said to be *luminous;* those that are seen by light derived from others are said to be *illuminated.* Luminous bodies generate light; illuminated bodies reflect and diffuse it. The sun is a luminous body; the moon is illuminated by it.

The principal sources of light are *the sun, the stars, heat, chemical combination, phosphorescence,* and *electricity.*

How is light imparted to the ether? How propagated? *Relation between light and radiant heat. Difference. Illustrate the idea of transversal vibrations.* (**260.**) Define a luminous body. An illuminated body. Illustrate.

The ultimate cause of the sun's light is unknown. The sun is surrounded by a gaseous envelope, called the *photosphere*, which appears to be in a state of intense ignition. The molecular vibrations of this envelope are undoubtedly the immediate sources of solar light and solar heat. The stars are similar to the sun, but on account of their enormous distances from us they send us but a small amount of light and heat.

If a body be heated its molecules are thrown into vibration, and when its temperature reaches 900° or 1000° F., it begins to be luminous in the dark. Beyond that its brightness increases as its temperature rises.

The light developed by chemical combinations is mostly due to the heat that accompanies them. Combustion is an example; the affinity between the oxygen of the air and the carbon of the fuel causes them to rush together under favorable circumstances, thus generating heat and ultimately light itself.

Phosphorescence is the property that some bodies have of giving out light under certain conditions; it is often observed in decaying animal and vegetable matter and in some minerals.

Electricity is the source of a species of light that rivals in intensity that of the sun itself. It will be treated of hereafter.

Media.—Opaque and Transparent Bodies.

261. A MEDIUM is anything that transmits light; thus, free space, air, water, and glass, are *media*.

Media owe their property of transmitting light to the ether which pervades them. This ether exists in the spaces between the particles of all bodies, but not always in such a state as to permit the transmission of light.

A TRANSPARENT BODY is one that permits light to pass through it freely, as glass, diamonds, rock-crystal, and water.

When bodies permit light to pass through them, but not in such quantity as to allow objects to be seen through them,

---

What is Phosphorescence? Illustrate. *What is its cause?* (**261.**) What is a Medium? Examples. What is a Transparent Body?

they are called *translucent*. Thus, scraped horn, ground glass, oiled paper, and thin porcelain are translucent.

An OPAQUE BODY is one that does not permit light to pass through it. Thus, iron, wood, and granite are opaque bodies.

Absorption of Light.

262. No body is perfectly transparent; all intercept or absorb more or less light, but some absorb much more than others. If light be transmitted through great thicknesses of media which in thin layers are transparent, a quantity of light is absorbed, and it often happens that the transmitted light is not of sufficient intensity to produce the sensation of sight.

The atmosphere seems perfectly transparent, but it is a known fact that much of the light of the sun is absorbed in reaching the earth, as is shown by the greater brilliancy of the stars in the higher regions, as on mountain tops. In the high regions of the atmosphere, objects are more clearly seen than nearer the earth; indeed so great is the clearness of vision in these regions, that it becomes exceedingly difficult to judge of distances. Opaque bodies absorb all of the light falling upon them which is not reflected.

The physical cause of absorption of light by bodies is some peculiarity of molecular constitution, which breaks up and neutralizes the waves of light that enter them.

Rays of Light. - Pencils. - Beams.

263. A RAY of Light is a line along which light is propagated. It is normal to the advancing wave front. When the source is very distant the wave fronts are sensibly plane and the rays parallel.

When the ether is uniformly distributed throughout a medium,

A Translucent Body? An Opaque Body? (**262.**) Explain the phenomenon of absorption. *Effect of atmospheric absorption?* Physical cause of absorption? (**263.**) What is a ray of light?

the waves of light are concentric spheres, and the rays of light are straight lines, because a perpendicular to one wave front will be perpendicular to all of the successive stages of that front. Media, in which the ether is uniformly distributed, are, with respect to light, called *homogeneous*. All other media are called *heterogeneous*.

When the waves of light are not concentric spheres, the rays of light are curved. Such, for example, are the rays of light transmitted through the atmosphere.

A PENCIL OF RAYS is a small group of rays meeting in a common point, such as the rays proceeding from a candle or a lamp.

When the rays *proceed from* a common point, they are are said to be *divergent*. When they *proceed towards* a common point, they are said to be *convergent*.

A BEAM OF RAYS is a small group of parallel rays, such as enter a small hole in a shutter, from a distant body, as the sun.

Velocity of Light.

264. It was shown by RŒMER, a Danish astronomer, in 1678, that light occupies nearly $8\frac{1}{4}$ minutes in coming from the sun to the earth, which gives a velocity of 186,000 miles per second.

He ascertained the velocity of light by a succession of observations on the eclipses of Jupiter's first satellite. In Fig. 155, S represents the sun, T, the earth, J, Jupiter, and e, Jupiter's first satellite. The darkened portion of the figure beyond Jupiter represents the shadow of that planet cast by the sun: It is known by computation, that Jupiter's first satellite revolves about that planet once in 42 hours, 28 minutes, and 36 seconds, and by entering the shadow of Jupiter, is eclipsed at each revolution.

What is the direction of a ray in a homogeneous medium? What is a homogeneous medium? A heterogeneous medium? Direction of a ray in such a medium? What is a Pencil of Rays? Example. Convergent? Divergent? What is a Beam of Rays? Example. (**264.**) What is the velocity of light? By whom determined?

ROEMER found that as the earth moved from T, its nearest position to Jupiter, towards t, its most remote position, the interval between the consecutive eclipses of the satellite gradually grew longer, whilst in moving from t back again to T, these intervals grew shorter. The total retardation in passing from T to t, was found to be nearly 16½ minutes, and the total acceleration in the remaining half of the earth's revolution was also found to be 16½ minutes. This was accounted for by the fact that the earth was moving away from Jupiter in the first case, and therefore the

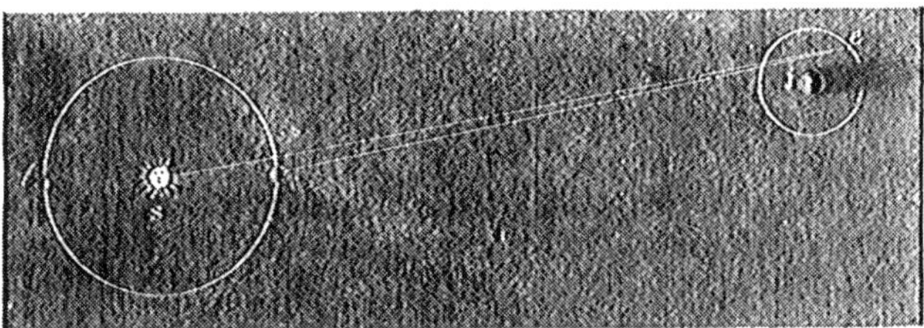

Fig. 155

light had to travel further and further at each eclipse to reach the observer, whilst in the second case, the reverse happened.

ROEMER therefore inferred that it required 16½ minutes for a ray of light to traverse the diameter of the earth's orbit, or 8¼ minutes for it to pass over the radius of that orbit, that is, over a distance equal to that of the earth from the sun.

ROEMER's deduction has been confirmed by observations made on the aberration of light, and also by direct experiment.

Explain the process of ROEMER's discovery. His deduction. Has it been confirmed?

It is difficult to conceive a velocity so great as 186,000 miles per second, a speed that would carry a ray of light around the earth eight times in a single second of time. Some idea, however, may be had of the velocity of light, from the fact that it would require more than two and a half centuries for one of our most rapid express trains of cars to run a distance over which light passes in 8¼ minutes.

It takes light more than four hours to reach us from Neptune, the most distant of the planets of our system, and it is capable of proof that light occupies more than three years in coming to us from the nearest of the fixed stars. Now, if astronomers are right in the inference that the remotest stars visible in our telescopes are more than a thousand times as distant as the nearest ones, then indeed must the light that makes us aware of their existence, have set out on its journey long centuries before the beginning of the Christian era. These conclusions serve to show the vastness of the material universe, and the comparative littleness of our own planet.

Intensity of Light.—Photometry.

265. The INTENSITY OF LIGHT is the amount of disturbance that it imparts to the ether. It can be shown mathematically, for light coming from the same sources, that *the intensity varies inversely as the square of the distance from its source.*

Hence we see that light follows the same law, with regard to its intensity, that is observed for gravity and sound. The law of variation of intensity can be verified, experimentally, by means of an instrument called a *photometer.*

A PHOTOMETER is an instrument for comparing the intensities of different lights.

Several different instruments have been devised for this purpose, one of the simplest being that shown in Fig. 156.

It consists of a vertical screen of ground glass, *A*, and a vertical solid rod, *B*, situated a short distance in front of it.

Give some illustrations of the immense velocity of light. (**265.**) What is the Intensity of Light? How does it vary with the distance? What is a Photometer? Explain the one shown in Fig. 156.

If two equal lights are placed at equal distances from *B*, it is found that the shadows which *B* casts upon *A*, are of the same tint. If one light be placed at any distance, and four

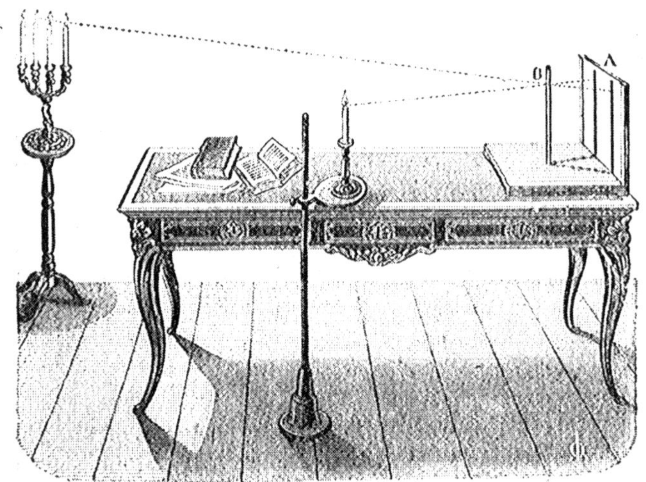

Fig. 156.

equal lights be placed at twice the distance, the shadows will be of the same tint; this is the case shown in the figure. It will require nine equal lights at three times the distance, sixteen at four times the distance, and so on, to produce the same effect. This experiment confirms the law of variation of intensity according to the inverse square of the distance.

To use the photometer to compare the intensities of any two lights, let them be placed, by trial, at such distances from *B*, that the shadows cast on *A* are of exactly the same tint; then will their intensities be to each other inversely as the squares of their distances from the rod, *B*.

How is the photometer used?

II.—REFLECTION OF LIGHT.—MIRRORS.

Reflection of Light.

266. When light passes obliquely from one medium to another, it is separated into two parts, one of which is driven back and remains in the first medium, whilst the other passes on and enters the second medium. The part that is driven back is said to be *reflected*, and the deviating surface is called a *reflector*.

Reflection of light is explained in the same way as reflection of sound. In case of light the wave lengths are so small that the most highly polished surfaces are comparatively rough. Hence, only a part of the reflected light appears to follow the regular laws; the rest is irregularly reflected or diffused. The amount of light reflected, as well as the relation between that which is regularly and that which is irregularly reflected, depends on the obliquity of incidence, the nature of the second medium, and the polish of the deviating surface.

Light that is irregularly reflected enables us to see objects; thus, the light falling on a sheet of paper is scattered or diffused so as to render it visible in all directions. If a reflector were perfectly smooth it would be invisible; we should simply see in it the images of other objects.

It is the diffused light reflected by the clouds, the air, the earth, and objects upon it, that illuminates our rooms and renders objects visible which do not receive the direct rays of the sun.

If we look out from our houses we see objects clearly by means of this diffuse light, because they receive much light, and therefore reflect much; but if we look from without into a house, we see objects with less distinctness, because they receive but little light, and therefore they reflect but little.

It is now proposed to explain the laws of regular reflection.

(**266.**) What is reflection of light? What is a reflector? *How is reflection of light explained? What is diffused light? How are we able to see non-luminous bodies?*

Definitions of Terms.

267. The ray that falls upon a reflecting surface is called *the incident ray;* thus, *CD*, Fig. 157, is an incident ray.

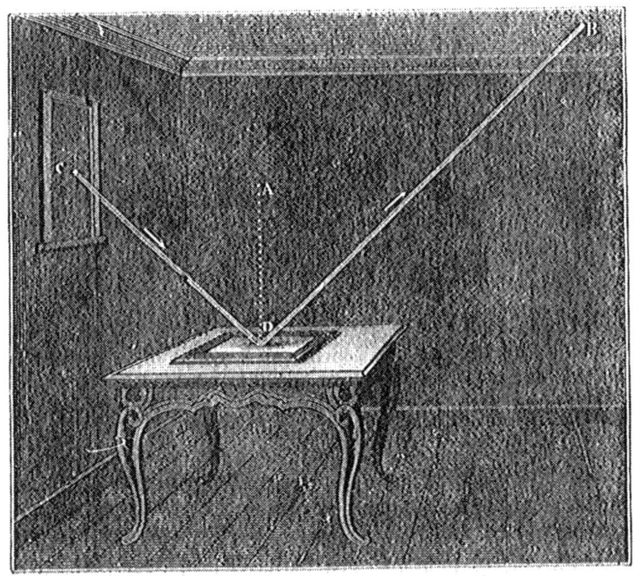

Fig. 157.

The point where the incident ray meets the reflecting surface, is called *the point of incidence;* thus, *D* is a point of incidence.

The angle that the incident ray makes with the normal to the reflecting surface at the point of incidence, is called *the angle of incidence;* thus, *CDA* is an angle of incidence.

(**267.**) What is an incident ray? Example. The point of incidence? Example. The angle of incidence? Example.

The plane that passes through the incident ray and the normal is called *the plane of incidence;* thus, the plane through *CD* and *DA*, is a plane of incidence.

The ray driven off from the reflecting surface is called *the reflected ray;* thus, *DB* is a reflected ray.

The angle that the reflected ray makes with the normal is called *the angle of reflection;* thus, *BDA* is an angle of reflection.

The plane of the reflected ray and the normal is called *the plane of reflection;* thus, the plane of *BD* and *DA* is a plane of reflection.

Laws of Reflection.

268. The following laws are shown by theory, and confirmed by experiment:

1. *The planes of incidence and reflection coincide;* both are normal to the reflecting surface at the point of incidence.

2. *The angles of incidence and reflection are equal;* this is true whatever may be the angle of incidence.

Direction in which objects are seen.

269. Whenever the rays of light proceed directly from an object to the eye, we see the body exactly where it is. When by reflection, or any other cause, the rays are bent from their primitive direction, we no longer see bodies in their proper position. They appear to be in the direction from which the ray enters the eye.

This is illustrated in Fig. 158. *A*, represents a body from which a ray of light, proceeding in the direction *AB*, is deviated or bent at *B*, so as to assume the new direction, *BC*. The eye receives the ray from the direction *BC*,

The plane of incidence? Example. The reflected ray? Example. The angle of reflection? Example. The plane of reflection? Example. (**268.**) What is the first law of reflection? What is the second law of reflection? (**269.**) In what direction do objects appear to the eye? Illustrate.

and in consequence the object, *A*, appears to be situated at

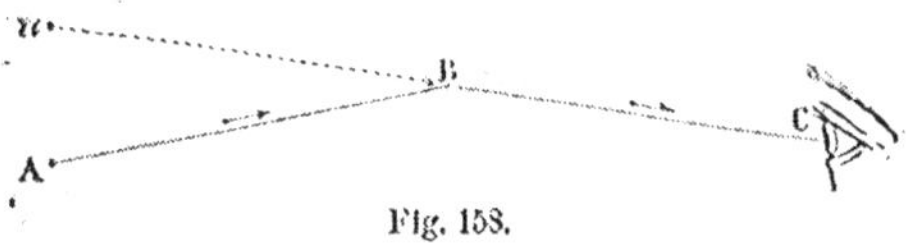

Fig. 158.

some point, *a*. This principle is of importance in explaining certain phenomena produced by reflectors and lenses.

Mirrors.

270. A MIRROR is a body with a polished surface, employed to form images of objects.

The best reflecting surfaces are those of polished metals. Our ordinary looking-glasses are composed of plates of smooth glass, upon the back of which is fastened a thin layer of tin and quicksilver.

This mixture, called an amalgam, offers an excellent reflecting surface, and it is from it that the principal reflection takes place. The glass serves to give the proper smoothness to the amalgam, as well as to protect it from injury and tarnish. There is, however, a reflection from the outer surface of the glass, giving rise to feeble images, which render such reflectors objectionable for optical purposes. Hence it is, that reflectors for telescopes, and the like, are generally made of alloys, or mixtures of hard metals, which admit of a high polish. Such a mirror is called a *speculum*.

Mirrors are of two kinds, *plane* and *curved*.

Plane Mirrors.

271. A PLANE MIRROR is one in which the reflecting surface is plane.

(**270.**) What is a Mirror? What are the best reflecting surfaces? What are looking-glasses? Explain their construction? What is a speculum? How many kinds of mirrors are there? What are they? (**271.**) What is a Plane Mirror?

We have an example of plane mirrors in the ordinary looking-glasses of our houses. The surface of still water, which reflects surrounding objects, and the surface of quicksilver, when at rest, are additional examples. The latter is often used with the sextant in measuring the altitudes of the stars; it is also used in adjusting astronomical instruments.

Images formed by Plane Reflectors.

272. An Image of an object is a picture or representation of that object, formed by a reflector, or by a lens.

Fig. 159.

The manner of forming images by plane reflectors is illustrated in Fig. 159. A pencil of rays coming from a

Give examples. (**272**) What is an Image of an object? Explain the manner of forming the image of a point.

point, is reflected so as to reach the eye. Because the angles of incidence and reflection are equal (Art. 268), each ray will have the same inclination to the mirror after reflection that it had before incidence. Hence the reflected rays, on being produced back, will meet at a point as far behind the reflector as the point of the object is in front of it. Now, because the eye sees objects in the direction from which the rays reach it (Art. 269), the point appears to be as far behind the mirror as it really is in front of it. The representation of the point thus formed, is its image.

What has been said of a single point is true of all points. Hence, if we suppose pencils of rays to proceed from every point of an object, as shown in Fig. 160, each point will

Fig. 160.

Explain the manner of forming the image of an object.

have its own image as far behind the mirror as the point is in front of it. The assemblage of images thus formed makes up the image of the object.

Nature of the Images formed.

273. It will be seen from an inspection of Fig. 160, that the image of the child's right hand is on the left of the image in the glass, and that the image of the child's left foot is on the right of the image in the glass, that is, the image is reversed laterally. This comes from the fact, that the image of each point is as far behind the mirror as the point is in front. Hence we say, that *an object and its image are symmetrically situated with respect to the mirror.*

We see also from what has been said, that *the image is erect, and equal in size with the object.*

The rays that reach the eye *appear* to come from an image which does not in reality exist. The image is only *apparent.* Such images are called *virtual.*

A VIRTUAL IMAGE is an image that appears to exist, and which would be found by producing the deviated pencils of rays backward, till they meet in points.

Multiple Images from Looking-glasses.

274. Metallic mirrors, or *specula,* as they are called, having but one reflecting surface, form but a single image. Glass mirrors have two reflecting surfaces, the front surface of the glass, and the metallic surface at the back of the glass. An image is formed by each of these surfaces, but that formed by the latter is the more striking, because the first surface reflects only a small portion of the light.

This formation of two images by glass mirrors renders them unfit for many optical purposes. The double image, formed by placing a point against the glass, enables us to judge of the thickness of the glass.

(**273.**) How are the object and its image by a plane reflector situated? Is the image real or apparent? Why? What is a Virtual Image? (**274.**) *Why do glass mirrors form two images? What is the objection to this duplication? How do we judge of the thickness of glass?*

Reflection by Transparent Bodies.

275. We have just seen that glass, notwithstanding its transparency, reflects light enough to form an image. The same is the case with other transparent bodies, of which water forms a conspicuous example.

Fig. 161

Fig. 161 represents the phenomenon of reflection from the surface of still water. It shows how the reflected rays produce images of objects above the water, which are symmetrically disposed with respect to the surface of the water. The case is entirely the same as though the images had been formed by a horizontal looking-glass.

Curved Mirrors.

276. A CURVED MIRROR is one in which the reflecting surface is curved. The most important class of curved mirrors, is that in which the reflecting surface is a portion

(**275.**) *Do transparent bodies reflect light? Explain the reflection from water.*
(**276.**) What is a Curved Mirror?

of a sphere. When the reflection takes place from the hollow or concave side, the mirror is called *concave;* when the reflection takes place from the outer or convex side, the mirror is called *convex.*

Concave Mirrors.

277. A CONCAVE MIRROR is one in which the reflection takes place from the concave side of a curved surface.

We shall consider the case in which the reflecting surface is a segment of a sphere.

The following definitions apply equally to concave and convex mirrors:

The middle point of the mirror is called its *vertex.* The centre of the sphere, of which the mirror forms a part, is called the *optical centre.* The indefinite straight line through the optical centre and the vertex, is called the *principal axis,* or sometimes simply *the axis.* Any plane section through the axis is called *a principal section.*

Thus, *MN,* Fig. 162, represents a principal section of a concave mirror, *A* is its vertex, *C* its optical centre, and *AX* is its principal axis.

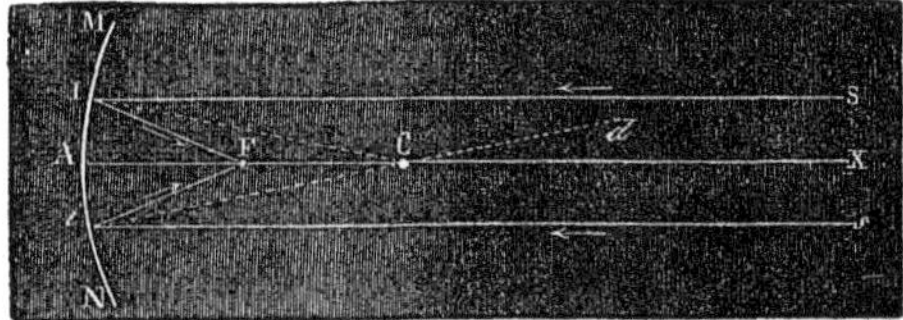

Fig. 162.

It is to be observed, that in practice the surface of a curved mirror is only a very small part of the surface of the sphere of which it forms a part.

Concave? Convex? (**277.**) What is a Concave Mirror? What is the vertex? The optical centre? The principal axis? A principal section? Illustrate.

Principal Focus of a Concave Mirror.

278. A Focus is a point in which deviated rays meet. If the incident rays are parallel to the axis, the focus is called the *Principal Focus.*

In Fig. 162, *SI* and *si*, are two rays parallel to the axis. *CI* and *Ci* are normals at the points of incidence, *I* and *i*. *IF* and *iF* are reflected rays, making the angles of reflection equal to the angles of incidence. When the mirror is small, compared with the whole sphere, all the rays parallel to the axis are reflected to the same point, *F*. Hence, from the definition, *F* is the principal focus. It can be shown that the principal focus is on the axis, and midway between the vertex and optical centre. We shall always designate the principal focus by the letter *F*.

Fig. 163.

(**278.**) What is a Focus? The Principal Focus? Illustrate.

Fig. 163 shows the manner of determining the principal focus by experiment, making use of a beam of light coming from the sun. In this form the concave reflector may be used to collect the rays for the purpose of developing a great amount of heat.

Conjugate Foci.

279. If the rays of light emanate from some point of the axis not infinitely distant from the mirror, they will be brought to a focus at some point of the axis, generally

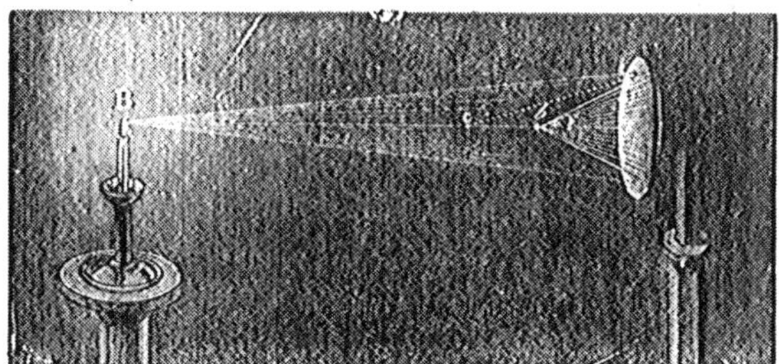

Fig. 164.

different from *F*. Thus, in Fig. 164, the pencil of rays, coming from the point *B*, are brought to a focus at *b*, between *F* and *C*. Had the rays emanated from *b*, they would have been brought to a focus at *B*. These points are so related as to receive the name of *conjugate foci*. Hence we have the following definition:

CONJUGATE FOCI are any two points so related that a pencil of light, emanating from either one, is brought to a focus at the other.

That one from which the light actually proceeds is called the *radiant*; thus, in Fig. 164, *B* is the radiant.

Explain the manner of determining the principal focus by experiment. (**279.**) What are Conjugate Foci? The radiant?

The following are some properties of conjugate foci of concave mirrors:

If the radiant is on the axis and at an infinite distance from the mirror, the rays will be parallel, and the corresponding focus is at *F*, (Fig, 162).

As the radiant approaches the mirror, the focus recedes from it.

If the radiant is beyond the optical centre, *C*, the focus is between *F* and *C*.

If the radiant is at *C*, the focus is at *C* also.

If the radiant is between *C* and *F*, the focus is beyond *C*, in the direction *CX*.

If the radiant is at *F*, the focus is at an infinite distance, that is, the reflected rays are parallel.

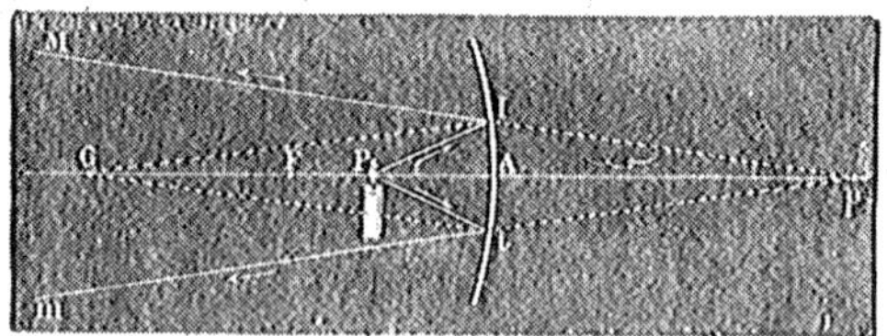

Fig. 165.

If the radiant is between *F* and *A*, as shown in Fig. 165, the rays are reflected so as to diverge, and on being produced backwards, meet at *p*. In this case the focus is behind the mirror, and is said to be *virtual* (Art. 273).

If the radiant is not on the axis, the pencil of rays is oblique, but it is still brought to a focus, and if not far distant from the axis, the radiant and focus enjoy properties entirely analogous to those just explained.

If the radiant is at an infinite distance, where is the conjugate focus? If the radiant approaches the mirror, how does the focus move? Where do they meet? If the radiant is at the principal focus, where is the conjugate focus? When is the focus virtual? Explain the law of an oblique pencil of rays.

Formation of Images by Concave Reflectors.

280. If an object be placed in front of a concave mirror, a pencil of rays will proceed from each point of the object, which after reflection will be brought to a focus either real or virtual. The collection of foci thus formed, make up the *image of the object.*

Real Images.

281. If the object is further from the mirror than the principal focus, the image will be inverted, and real.

Fig. 166.

Fig. 166 represents an inverted image formed by a concave reflector. That the image is real, may be shown by throwing it on a screen.

(**280.**) How is an image of an object formed? (**281.**) When is the image real and inverted?

Fig. 167 shows the course of the rays in forming a real image by means of a concave mirror. In this case the image of a distant church is formed and thrown upon a sheet of paper; the image is a perfect picture, not only in outline but in color; its only defect is that it is inverted.

Fig 167.

When the object is at a greater distance from the mirror than the optical centre, the image is less than the object; when the object is between the optical centre and the principal focus, the image is greater than the object. In this case the reflector may be used as a magnifier.

Explain the course of the rays in forming an image. When is the image smaller than the object? When larger?

Virtual Images.

282. When the object is between the principal focus and the mirror, the image is virtual and erect, as shown in Fig. 168. Furthermore, it is larger than the object, or magnified.

Fig. 168.

Fig. 169 shows the course of the rays in forming a virtual and erect image. The face is between the principal focus, *F*, and the mirror. The pencils of rays from *a* and *b* are reflected so as to diverge from the virtual foci, *A* and *B*. It is easily seen that the image is larger than the object.

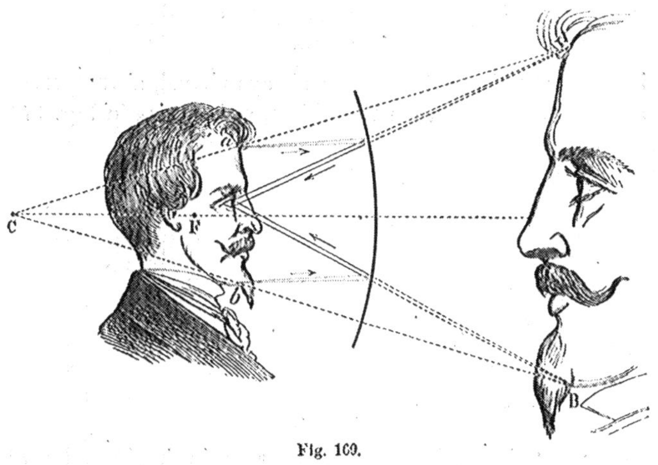

Fig. 169.

(282.) When is the image virtual? Explain the course of the rays in forming a virtual image.

Formation of Images by Convex Reflectors.

283. In convex mirrors the reflection takes place from the outer or convex surface.

From what has been said of concave mirrors, it will readily be seen how images are formed by convex mirrors. The

Fig. 170.

images formed in this case are always virtual, always erect, and always smaller than the object, as is shown in Fig. 170.

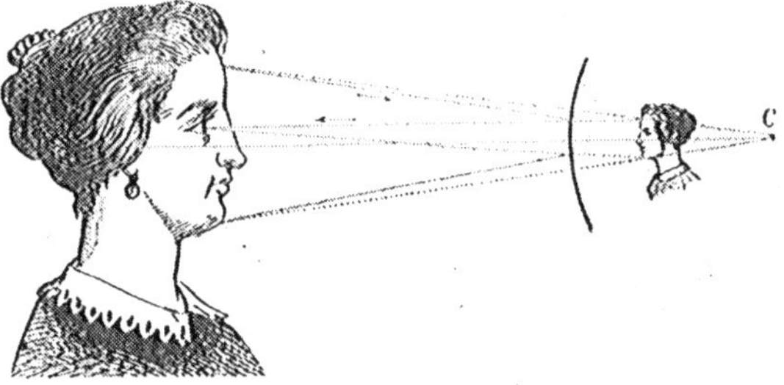

Fig. 171.

Fig. 171 shows the course of the rays in the formation of

(283) Do convex reflectors form erect or inverted images? Are they magnified or diminished? Explain the course of the rays in this reflector.

an image by means of a convex mirror. After what has been said in the preceding article, this figure needs no explanation.

III.—REFRACTION OF LIGHT.—LENSES.

Refraction.

284. If a beam of light fall obliquely on a surface that separates two media, it is divided into two parts; one is reflected and remains in the first medium, the other enters the second medium, and is partially absorbed and partially transmitted. The transmitted rays change direction at the point of incidence. This change of direction is called *refraction.* Its amount depends on the nature of the media, and also on the obliquity of incidence.

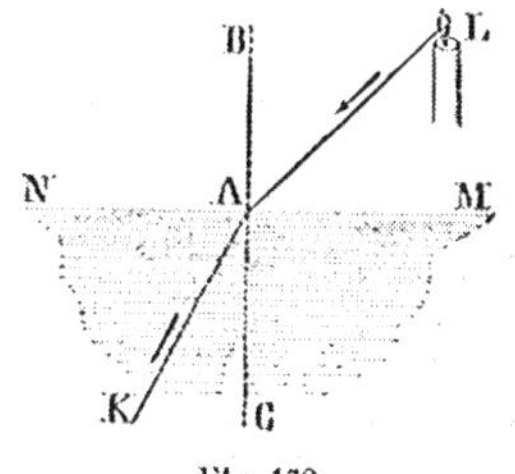

Fig. 172.

The cause of this change of direction is a change in the elasticity and density of the ether in passing from one medium into the other, which causes a change in the velocity of the ray. Thus the density and elasticity of ether in water are different from what they are in the atmosphere, so that light travels considerably faster in the latter medium than in the former. This causes a ray, on passing from air into water to bend towards the normal at the point of incidence, as shown in Fig. 172. Thus, LA, is bent from its course so as to take the direction *AK*. In passing from water to air, the ray is bent away from the normal, just the reverse of what happens when light passes from air into water.

Definitions.

285. The ray before refraction is called the *incident ray;* thus, *LA* (Fig. 172), is an incident ray.

(**284.**) What is Refraction? *What is the cause of refraction? Which way is the ray bent?* (**285.**) What is the incident ray? Illustrate.

The point at which the ray is deviated or bent, is called *the point of incidence;* thus, *A* is a point of incidence.

The ray after deviation is called *the refracted ray;* thus, *AK* is a refracted ray.

The angle that the incident ray makes with the normal at the point of incidence is called *the angle of incidence,* and the plane of this angle is *the plane of incidence.* Thus, *LAB* is an angle of incidence, and the plane, *LAB*, is the plane of incidence.

The angle that the refracted ray makes with the normal at the point of incidence is called *the angle of refraction,* and the plane of this angle is *the plane of refraction;* thus, the angle *KAC* is an angle of refraction, and the plane of this angle is a plane of refraction.

Laws of Refraction.

286. When light passes from any given medium into another, no matter what may be the angle of incidence, it always conforms to the following laws:

1. *The planes of incidence and refraction coincide, both being normal to the surface separating the media, at the point of incidence.*

2. *The sine of the angle of incidence is equal to the sine of the angle of refraction multiplied by a constant quantity.*

This constant multiplier is called the index of refraction, and is equal to the ratio of the velocities of light in the two media.

What is the point of incidence? Illustrate. The refracted ray? Illustrate. The angle and plane of incidence? Illustrate. The angle and plane of refraction? Illustrate. (**286.**) What is the first law of refraction? The second law? What is the index of refraction?

The second law may be illustrated by the figure in the margin. Let A be the point of incidence on a surface separating air from water. With A as a centre, describe a circle, $BmCp$. Let LA be an incident ray, and AK the refracted ray. Draw mn and pq perpendicular to the normal, BC. Then will these lines be the sines of the angles of incidence and refraction, and we shall have in the particular case of air and water, mn equal to pq multiplied by $1\frac{1}{3}$, whatever may be the inclination of LA. Here $1\frac{1}{3}$ is the *index* of refraction. For air and glass the index of refraction is $1\frac{1}{2}$.

Refractive power of Bodies.

287. Different bodies possess different refractive powers. NEWTON observed that, as a general rule, the refractive power was greatest for combustible bodies, or bodies containing combustible elements, such as alcohol, ether, oils, &c., which contain both hydrogen and carbon. He found that the diamond was more highly refractive than any other body, and hence inferred that it was a combustible body, an inference that has since been confirmed. It is to its high refractive power that the diamond owes its brilliancy as a jewel. Gases are not so highly refractive as liquids, but their refractive power may be increased by compression, which augments their density.

Experimental proofs of Refraction.

288. If a beam of light be introduced through a hole in a shutter of a dark room, and allowed to fall upon the surface of water in a glass vessel, as shown in Fig. 173, the bending of the beam as it enters the water may be seen by

How may the second law be illustrated? (**287.**) *Do all bodies refract equally? Explain* NEWTON'S *views?* (**288**) Explain the method of proving refraction experimentally?

the eye. The course of a ray in the air may be rendered more apparent by filling the air with fine dust or smoke, as, for example, the smoke from gunpowder.

Fig. 173.

Let a piece of money be placed at the bottom of an empty vessel, and then take a position such that the coin shall just be hidden by the side of the vessel. Whilst in this position, if water be poured into the vessel, the rays from the coin will be refracted so as to render it visible. The effect of refraction in this and similar cases, is to make the bottom of the vessel appear higher than it is in reality, as shown in Fig. 174.

Explain a second method?

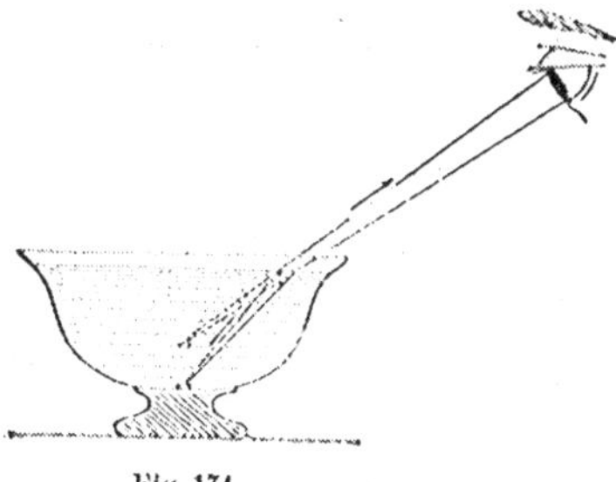

Fig. 174.

Some effects of Refraction.

289. One of the effects of refraction was explained in the last article. The principle has numerous applications. To a person standing on the shore, a fish in the water appears higher than his

Fig. 175.

real position, as is shown in Fig. 175.

(**289.**) *Why does a fish seem higher in the water than he really is?*

If a stick be partially plunged into water, the portion immersed will be thrown up by refraction, and the stick will appear bent, as shown in Fig. 176.

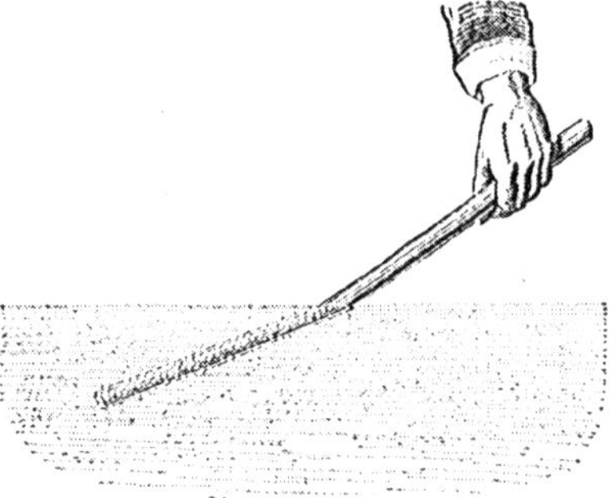

Fig. 176.

Refraction has the effect to make the heavenly bodies appear higher than they are, and thereby causes them to rise earlier and set later than they would do were there no atmosphere.

Total Reflection.

290. If light fall on a surface that separates a medium from one that is less refractive, there is a limit beyond which it will not pass from the first medium into the second, at that limit light is *totally* reflected.

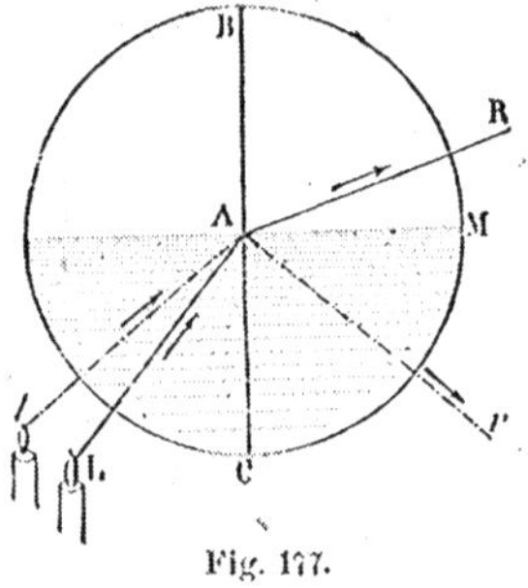

Fig. 177.

Let *BMC* be a glass globe half full of water. The ray, *LA*, being normal to the globe, is not refracted in entering, but if the angle, *CAL*, be small enough, it is refracted at *A*, taking same direction *AR*. If the angle of incidence exceed 41°, the ray can no longer pass through the surface *AM*, but is totally reflected and remains in the first medium, taking same direction *Ar*.

Explain total reflection. *Illustrate.*

Mirage.

291. MIRAGE is an atmospheric phenomenon dependent on extraordinary refraction and total reflection.

Sometimes a layer of atmosphere next the earth becomes a reflector, and in that case portions of the earth appear to the traveller like lakes and ponds; such appearances are frequent in desert countries when the heat is intense. To heighten the illusion, trees are often seen reflected from the surfaces of these apparent ponds. An example of this kind is shown in Fig. 178. The rays coming from the top

Fig. 178.

of the tree on the left of the picture, are totally reflected at *a*, from a layer of the atmosphere, and reach the eye of the observer at the tent. The observer refers the position of the tree top backwards along the direction of the dotted line, which causes the tree to appear inverted. In this case both the tree and its image are seen.

Now if we suppose both to be thrown up by extraordinary refraction, we shall have a phenomenon not unfrequently noticed, in which the object is seen elevated in the air, accompanied by an inverted image.

Explain the phenomena of mirage.

Double Refraction.—Polarization.

291*a*. Certain crystalline substances have the power to separate a transmitted beam into two parts, so that objects seen through them appear double. This phenomenon, called *double refraction*, depends on the molecular arrangement of the body, which causes the contained ether to have different degrees of elasticity in different directions.

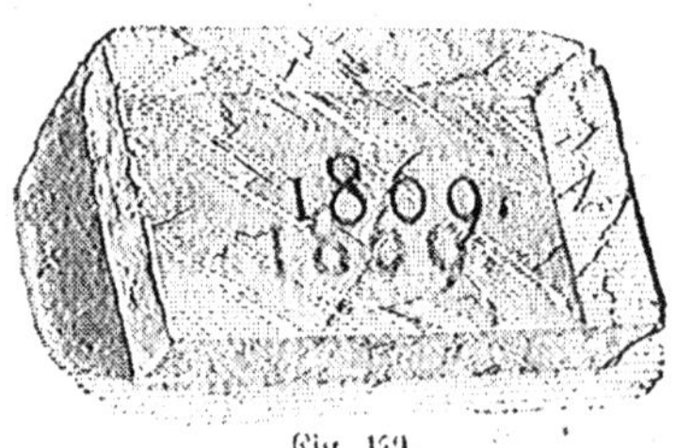

Fig. 179.

Iceland spar is an example of double refracting bodies. Its crystals can be reduced by cleavage to the form of an equilateral rhomb, as shown in the figure. The particles are symmetrically arranged about the shortest diagonal *ab*, and this is called the axis. The contained ether has its maximum density in the direction of the axis, and its minimum density in directions perpendicular to the axis. In consequence of these unequal elasticities the transmitted wave is divided into two, which advance with unequal velocities; hence the phenomena of double refraction.

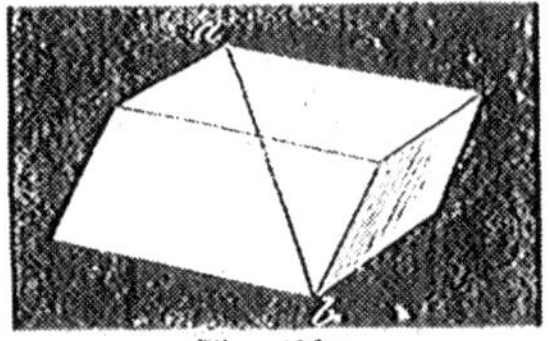

Fig. 179*a*.

The two parts into which a ray is divided do not move according to the same law. One follows both the laws of refraction already explained; it is called the *ordinary ray*. The other does not, as a general thing, follow either of those laws; it is called the *extraordinary ray*. When transmission takes place in the direction of the axis, the two rays coincide; when in a plane perpendicular to the axis, the two rays are most separated.

The class of bodies to which Iceland spar belongs have but one direction in which the refracted rays coincide; these are called *uniaxial*. There are bodies that have two such directions; these are called *biaxial*.

What is double refraction? *What is Iceland spar? Why does it separate rays into two parts? What is the ordinary ray? Extraordinary?*

If a beam of light be transmitted through a crystal of Iceland spar, the parts into which it is divided are of equal intensity. If one of these parts be transmitted through a second crystal, the parts into which it is divided are of unequal intensity, and the degree of inequality depends on the relative positions of the crystals. Hence, light that has been doubly refracted, differs from common light; it is *polarized.*

The vibrations that constitute light are *transversal*, that is, they are perpendicular to the direction of propagation. In *common* light the vibrations take place in every possible direction consistent with this law; in *polarized* light they take place in lines perpendicular to a single plane, called the *plane of polarization.*

Light is best studied by allowing it to fall perpendicularly on a plate of tourmaline, cut parallel to the axis of the crystal. Such a plate allows no vibrations to pass except they be parallel to the axis. Hence the emergent beam is polarized. Let such a beam fall perpendicularly on a second plate, similar to the first. If the axes of these plates are parallel, the entire beam is wholly transmitted; if the axes are perpendicular to each other, the beam is wholly intercepted; if the axes are oblique to each other, the beam is partially transmitted and partially intercepted.

If the rays that have passed through a crystal of Iceland spar be tested by a plate of tourmaline, it is found that they are polarized in planes which are perpendicular to each other.

Light may be polarized by reflection. We have seen, when light falls on a surface separating two media, that it is separated into two parts, one of which is refracted and the other reflected. When these two parts are perpendicular to each other, the reflected ray is polarized in a plane normal to the reflecting surface.

Light may also be polarized by refraction by an ordinary medium. The plane of polarization is then perpendicular to that of the reflected ray.

The mode of vibration in polarized light may be illustrated by a rope pressing between two horizontal slats of a board fence, the first end being fastened on one side of the fence, and the second end being held by the hand on the other side. If the hand be moved rapidly in any direction, the wave motion will be reduced to a single plane after passing the fence.

What is polarized light? The plane of polarization? Action of plate of tourmaline? Use as an analyzer of polarized light? Polarization by reflection?

Media with parallel Faces.

292. When a ray of light, *Lm*, Fig. 180, falls upon a medium bounded by plane faces, as a plate of glass, for example, it is refracted towards the normal and passes through the plate in some direction, *mn;* here it is refracted as much from the normal as it was towards it in the first instance, and the ray emerges in the direction *no*, parallel to *Lm*. The two refractions do not change the direction of the ray, but simply shift it slightly to one side or the other. Hence, in looking through a window, we do not see the direction of objects changed by the intervening glass.

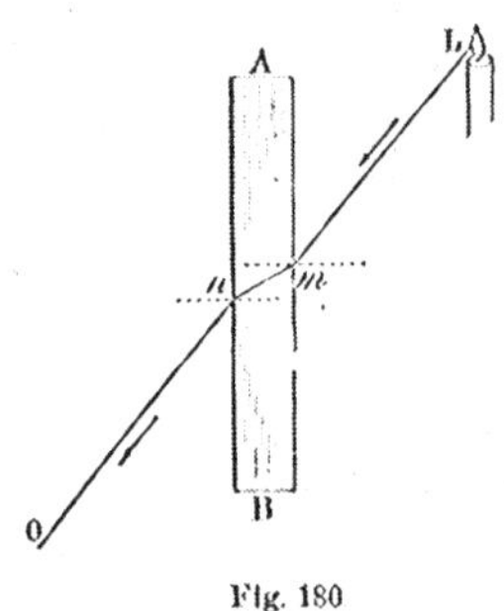

Fig. 180

Prisms.

293. A Prism is a refractive medium bounded by plane faces intersecting each other.

Fig. 181 represents a prism mounted for optical experiments. It consists of a piece of glass with three plane faces,

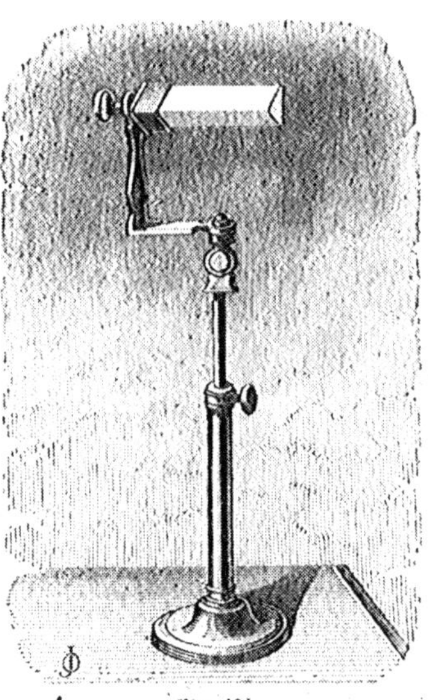
Fig. 181.

(**292.**) Is a ray of light bent from its course in passing through a medium with parallel faces? Explain the phenomenon. (**293.**) What is a Prism? Explain it.

meeting in parallel lines called *edges*. It is placed on a stand so that it can be elevated or depressed, and it also is capable of being turned around an axis parallel to the edges, by means of a button shown on the left.

Prisms produce upon light which traverses them, two remarkable effects: 1st, *a considerable deviation*; 2d, *a decomposition of light* into its elements.

These effects are simultaneous, but we shall at present only consider the first one, leaving the second to be studied hereafter under the name of *dispersion*.

Course of Luminous Rays in a Prism.

294. In order to follow the course of a ray of light in passing through a prism, let *nmo*, Fig. 182, represent a section of a prism made by a plane perpendicular to the edges. A ray of light, *La*, falling upon the face, *nm*, is refracted towards the normal, and passes through the prism in the direction, *ab*; here it falls upon the second face, *mo*, and is again refracted, but this time from the normal, and emerging into the air, takes the direction, *bc*. An eye situated at *c*, refers the object, *L*, backwards along the ray, *cb*, so that it appears to be situated at *r*. The total deviation is the angle between its original direction, *La*, and its final direction, *cr*.

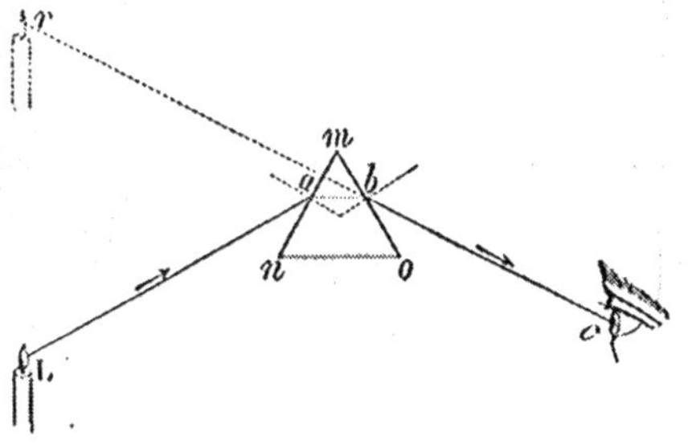

Fig. 182.

We see from the figure that the ray is bent from the

What effect has a prism on light? (294.) Explain the course of a ray through a prism.

edge in which the refracting faces meet, that is, it is bent towards the thick part of the prism; this deviation has the effect to make the object appear as though thrown towards that edge. The angle, *nmo*, is called the refracting angle of the prism.

Fig. 183.

Fig. 183 shows the manner of displacement, caused by viewing an object through a prism. If the prism is vertical, the displacement is towards the right or left, according to the position of the refracting angle.

Lenses.

295. A Lens is a refracting medium, bounded by curved surfaces, or by one curved and one plane surface.

Which way is the ray bent? Explain Fig. 183 (**295.**) What is a Lens?

Lenses are usually made of glass, and are bounded by spherical surfaces, or by one spherical and one plane surface. The surfaces are made spherical, because they are more easily wrought by the glass grinder.

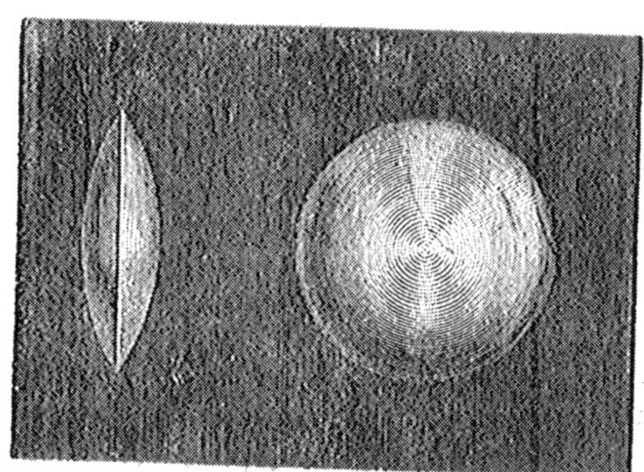

Fig. 184. Fig. 185.

Fig. 184 represents a side view, and Fig. 185 represents a front view of a lens, bounded by two spherical surfaces.

Classification of Lenses.

296. Lenses are divided into six classes, according to the nature and position of the bounding surfaces, sections of which are shown in Figs. 186 and 187.

The first three, represented in Fig. 186, are thicker in the middle than at their edges. These *converge* or collect rays of light, and are called *convergent lenses*.

The last three are thinner in the middle than at their edges. These *diverge* or scatter rays of light, and are called *divergent lenses*.

Of what are lenses made? (**296.**) How many kinds of lenses are there? What are convergent lenses? Divergent lenses?

These different lenses are named and described in the following definitions:

1. The *double convex* lens, *M*, bounded by two convex surfaces.

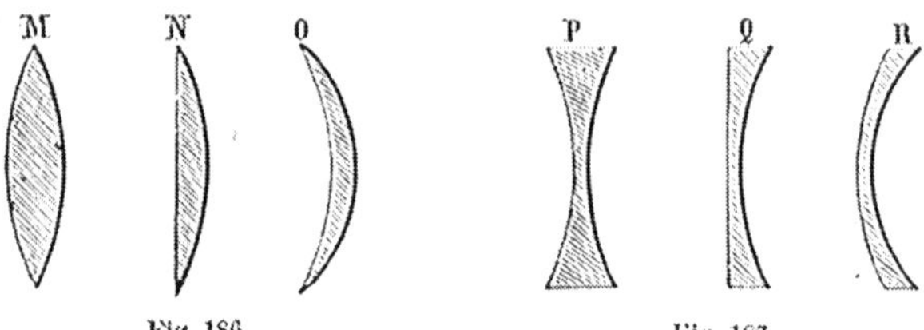

Fig. 186. Fig. 187.

2. The *plano-convex* lens, *N*, bounded by one convex and one plane surface.

3. The *meniscus*, *O*, bounded by one concave and one convex surface, the concave surface being the *least* curved.

4. The *double concave* lens, *P*, bounded by two concave surfaces.

5. The *plano-concave* lens, *Q*, bounded by one concave and one plane surface.

6. The *concavo-convex* lens, *R*, bounded by one concave and one convex surface, the concave surface being the *most* curved.

In studying the effect of these lenses, it will be sufficient to consider the *double convex* and the *double concave* lenses as specimens of the classes to which they belong, the former representing the *convergent*, and the latter the *divergent* classes.

Name and describe the six kinds of lenses separately. What two are taken as specimens?

Definitions of Terms.

297. The centres of the bounding surfaces of a lens are called *Centres of Curvature;* thus, in Fig. 188, *c*, and *C*, are centres of curvature.

In the double convex lens the centre of curvature of each surface is on the opposite side of the lens; in the double concave lens the reverse is the case. In the meniscus and the concavo-convex lens, both centres are on the same side of the lens. In the plano-convex and the plano-concave lens, the centre of curvature of the plane surface is at an infinite distance, and in a perpendicular to the plane surface at its middle point.

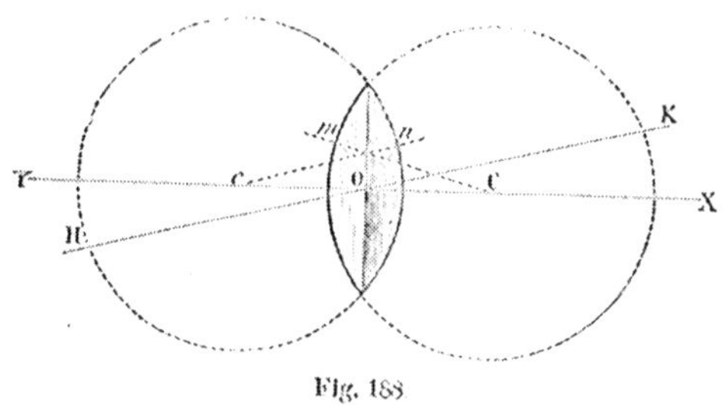

Fig. 188

The straight line through the centres of curvature is called the *axis* of the lens; thus, in Fig. 188, *XY* is the axis.

It is demonstrated in higher optics, that there is always one point on the axis of a lens, such that the rays of light passing through it, are not deviated by the lens. This point is called the *optical centre*, and is of much use in the construction of images.

In practice it is usual to make the surfaces which bound double convex and double concave lenses, equally curved.

(297) What are the Centres of Curvature of a lens? *Where are they in the double convex lens? Double concave? Meniscus? Plano-concave and plano-convex?* What is the axis? What is the optical centre? Its use? In practice, how are the curvatures of the surfaces?

When this is the case, as we shall suppose in what follows, the optical centre is on the axis, and midway between the two surfaces of the lens; thus, in Fig. 188, *O* is the optical centre, and any ray, *HK*, passing through it, is not deviated by the lens.

To find a normal at any point of the surface of a lens, we draw a line from that point to the corresponding centre of curvature; thus, *mC* and *nc*, are normals at the points *m* and *n*.

Action of Convex Lenses on Light.

298. When a ray of light falls upon one surface of a double convex lens, it is refracted towards the normal, passes through the lens, is again incident upon the second surface, and is refracted from the normal. This action is entirely analogous to that of a prism, the deviation being towards the thicker portion in both cases. In fact, if we suppose planes to be drawn tangent to the surfaces at the points of incidence and emergence, they may be regarded as the faces of a prism through which the ray passes.

Principal Focus.

299. If a beam of light, parallel to the axis, falls upon a lens, it will be collected by refraction in a single point. This

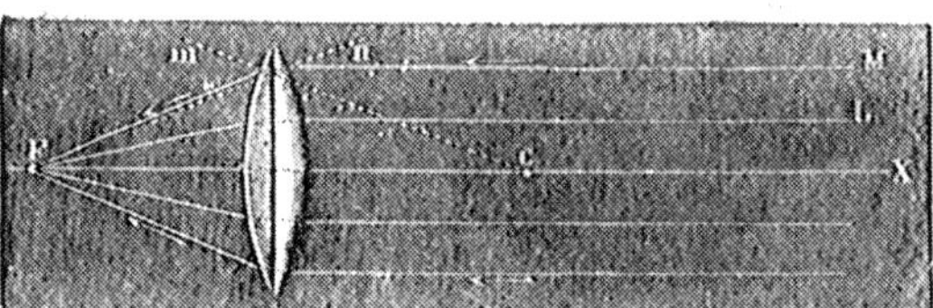

Fig. 189.

Where is the optical centre in this case? How do you find a normal? (**298.**) Explain the action of a convex lens on light. (**299.**) What is the principal focus?

point is called the *principal focus*, and its distance from the lens is called the *principal focal distance*.

The course of the rays is indicated in Fig. 189, in which the rays parallel to CX, are brought to a focus at F. Here, F is the principal focus.

It is to be observed that the rays will not be brought accurately to a focus, except in the case in which the surface of the lens is small, when compared with that of the whole sphere of which it forms part. This scattering of the rays from a focus is called *spherical aberration*. It is remedied in practice by covering up a part of the surface on which light falls, by a paper cover with an *aperture* in its centre.

Had the rays fallen upon the other side of the lens, they would have been brought to a focus as far to the right of the lens, as F is to the left of it.

Conjugate Foci.

300. CONJUGATE FOCI are any two points so situated on the axis of a lens, that a pencil of light coming from one is brought to a focus at the other. That from which the light actually comes is called the *radiant*.

In Fig. 191, a pencil of rays, coming from L, is brought to a focus at l, had the light come from l, it would have been brought to a focus at L; L and l are conjugate foci, and in the case figured, L is the radiant.

When the radiant is at an infinite distance, the rays are parallel, and the corresponding focus is at F; this is the *principal focus*. As we have already seen, there are two such foci, one on each side of the lens. It will be sufficient for our purpose to suppose the light to come from the right, in which case the principal focus is on the left, at F.

Principal focal distance? Explain the course of the rays. *What is spherical aberration? How remedied?* (**300.**) What are Conjugate Foci? What is the radiant? Illustrate. When the radiant is at an infinite distance, where is the conjugate focus?

When the radiant is anywhere on the axis at a greater distance than the principal focal distance, the corresponding focus will also be at a greater distance from the lens than the principal focal distance, as shown in Fig. 190.

Fig. 190.

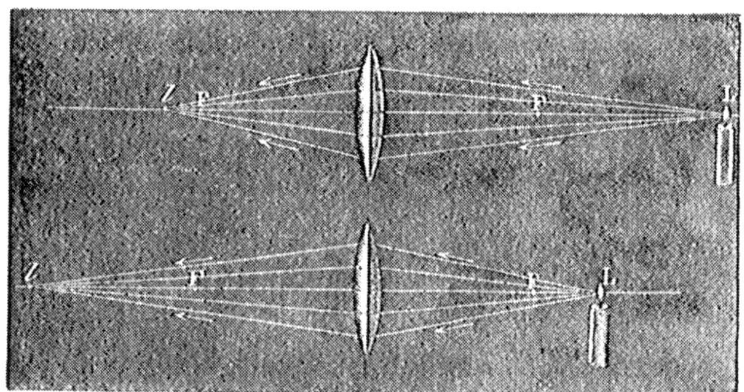

Fig. 191.

If the radiant approach the lens, the corresponding focus will recede from it, as is shown in Fig. 191.

If the radiant is at the principal focal distance, the refracted rays will be parallel, that is, the corresponding focus will be at an infinite distance, as is shown in the upper diagram, (Fig. 192).

If the radiant is still nearer the lens, the rays will diverge after deviation, and will only meet the axis on being produced backwards, in which case the focus is *virtual*, as is shown in the lower diagram, (Fig. 193). In this diagram *L* is the radiant, and *l* the virtual focus.

Thus far we have supposed the radiant to be situated on the principal axis; if it is on any line through the optical centre not much inclined to the axis, the corresponding

When the radiant is at a distance greater than the principal focal distance, where is the conjugate focus? When the radiant approaches the lens? When at the principal focus? If still nearer the lens? Suppose the radiant not on the axis?

focus will be on that line, and the laws which regulate the positions of conjugate foci, already considered, will be applicable.

These principles are of use in the discussion of images formed by lenses.

Fig 192.

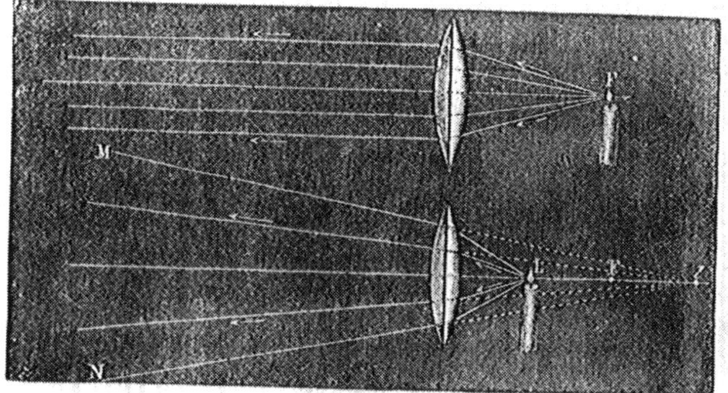

Fig. 193.

Formation of Images by Convex Lenses.

301. If an object be placed in front of a lens, each point of it may be regarded as a radiant sending out a pencil of rays. Each pencil is brought to a focus somewhere behind the lens. The assemblage of these foci makes up a picture of the object, which is called its *image*. When the object is at a greater distance from the lens than the principal focal distance, the image will be real and inverted. The course of the rays is shown in Fig. 194. The image is real, as may be shown by throwing it upon a screen; so long as the image is real, it is inverted, as may be seen by allowing it to fall upon a screen, or it may otherwise be shown from the fact that the axis of each pencil passes through the optical centre; hence the image of each point is on the opposite side of the axis from the point.

(301.) Explain the formation of an image by a lens.

With respect to the size of the image in this case, it may be either greater or smaller than the object. When the object is farther from the lens than twice the principal focal distance, the image is smaller than the object; when the object is at twice the focal distance, the image is of the

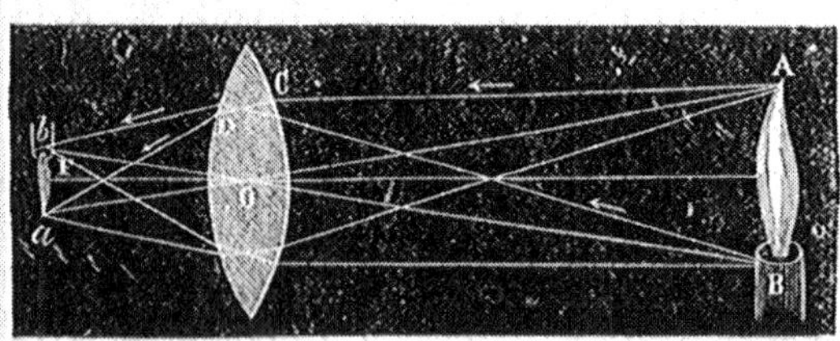

Fig. 194.

same size as the object; when the distance is less than twice the principal focal distance, and greater than the principal focal distance, the image is greater than the object.

These principles may be shown experimentally as follows:

Let a convex lens be placed in a dark room, and suppose its principal focal distance to have been determined by means of a beam of solar rays. Let a candle be placed in front of the lens, and a screen behind it to receive its image, as shown in Fig. 195.

When the distance of the candle from the lens is more than twice the principal focal distance, its image will be less than the object; and the more remote the candle, the less will be its image.

If the candle be moved towards the lens, its image will grow larger, until, at twice the principal focal distance, the size of the image and object will be equal.

If the candle be moved still nearer, the size of the image will be

How does the size of the image compare with that of the object in different cases? *Explain in detail the method of illustrating the foregoing principles by experiment.*

increased, that is, it will become greater than the object, as is shown in Fig. 196.

If the distance of the object does not become smaller than the principal focal distance, the image will be inverted, as is shown in Figs. 195 and 196.

Fig. 195.

If the object approach still nearer the lens, that is, if its distance becomes less than the principal focal distance, the image will increase, it will become erect, and furthermore it will be virtual. The course of the rays in this case is shown in Fig. 197. Here, *ab* is the object, and *AB* is its image, which can only be seen by looking through the lens.

In this case the lens becomes what is called a *single microscope*.

When the object is at the principal focal distance from the lens, the image is infinite; that is, it disappears.

What is a single microscope? Illustrate.

Fig. 196.

The phenomena just described may be observed by looking through a convex lens at the letters on a printed page. When the letters are at a short distance from the lens, they are magnified and erect;

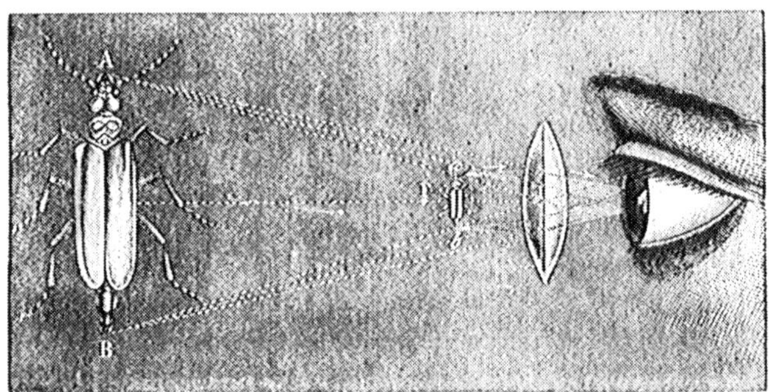

Fig. 197.

on removing the lens, they disappear at the principal focal distance, and finally reappear inverted and diminished in size.

Formation of Images by Concave Lenses.

302. Concave lenses being thinner in the middle than at the edges, have the effect to diverge parallel rays. If the rays are already divergent, these lenses make them still more so.

This is shown in Fig. 198, in which a pencil of rays, coming from the radiant, *L*, are made to diverge, as though they proceeded from a point, *l*, nearer the lens. This point, *l*, is the virtual focus, corresponding to the radiant, *L*. To an eye situated on the left of the lens, the light, *L*, appears to be situated at *l*.

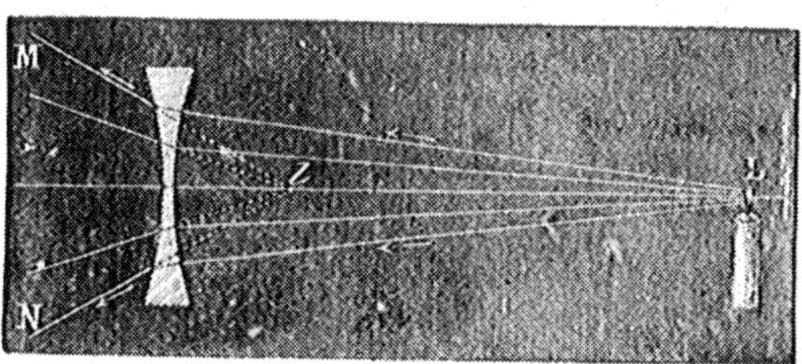

Fig. 198.

From what has been said, it is plain that the images formed by concave lenses are virtual. They are also erect, as in Fig. 198.

The course of the rays, in forming an image in the case of a concave lens, is shown in Fig. 199. In that figure, *AB* represents the object. A pencil of rays, coming from *A*, is deviated so as to appear to come from *a*, situated on a line drawn from *A* to the optical centre of the lens *O*. A pencil, coming from *B*, is deviated so as to appear to come from *b*,

(**302**) What is the effect of a divergent lens upon light? *Explain Fig.* 198. What kind of images are formed by concave lenses? Explain the course of the rays in a concave lens.

on the line Bo. Hence, ab is the image of the object AB,

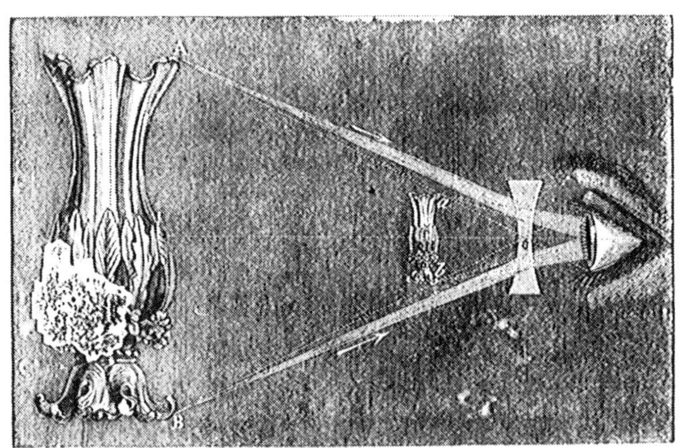

Fig 199.

and is, as we see, smaller than the object, being nearer the optical centre, and furthermore it is erect.

Burning-glasses.

303. Rays of heat are subject to the same laws of reflection and refraction as rays of light. When a beam of solar light falls upon a convex lens, there is not only a concentration of light at the focus, but of heat also.

The heat concentrated is so great as to inflame combustible bodies, such as paper, cloth, wood, and the like. In the case of large lenses, the heat becomes sufficiently powerful to fuse metals. This property of lenses has been used to procure fire: the lens in this case is called *a burning-glass.* Lenses carelessly exposed may sometimes cause dangerous results, by setting fire to inflammable materials. This effect may result from spherical vessels of glass filled with water, which possess all the properties of lenses.

Do concave lenses magnify or diminish objects? (**303.**) How are rays of heat affected by lenses? *What is a burning-glass? Explain its action.*

A curious application of this principle is shown in Fig. 200. A lens is arranged with its axis in the meridian, so that its principal focus shall fall upon the vent of a small cannon. When the sun

Fig. 200.

crosses the meridian, the rays are concentrated upon the vent, and if the gun has been loaded and primed beforehand, it will be discharged at midday.

Light-houses.

304. LIGHT-HOUSES are towers, erected along the coast, upon the tops of which are lanterns. These lanterns are lighted at night as guides to mariners.

One of the most famous light-houses of antiquity was that on the little Island of *Pharos*, near Alexandria, in Egypt. From the location of this light-house the French derive the name *pharo*, which they apply to all light-houses. In former times light-houses were illuminated by fires built with wood, coal, or some bituminous substances.

Explain Fig 200. (**304.**) What is a light-house? *Give an account of the ancient light-houses.*

These methods of illumination were afterwards replaced by oil lamps placed in the foci of concave reflectors, which served to concentrate the rays, and thus to heighten their illuminating effect. But the reflectors, being made of metal, were soon tarnished, and the light afforded became feeble.

In 1822, FRESNEL, already distinguished by his discoveries in optics, and by his researches on the wave theory of light, invented a new system of illumination, which is now being adopted in all civilized countries.

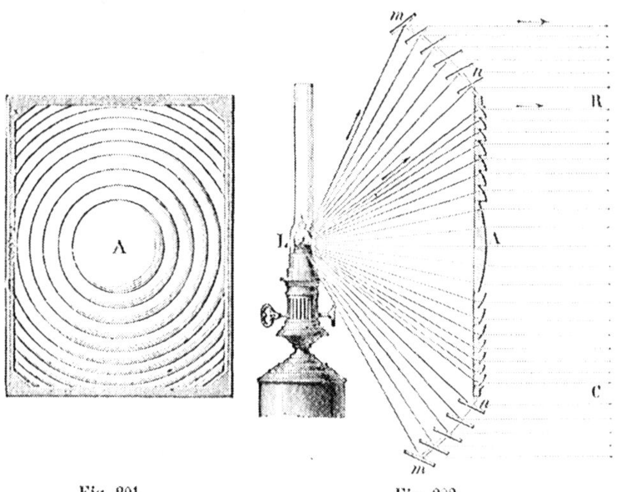

Fig. 201. Fig. 202.

Abandoning the reflectors, which became tarnished by the influence of sea fogs, he substituted for them plano-convex lenses, in the principal foci of which he placed powerful lamps with four concentric wicks, each of which, for the quantity of oil consumed, and the amount of light given out, was found to be equivalent to seventeen carcel-lamps. The difficulty of constructing large plano-convex lenses, together with their great absorption of light, led finally to the adoption of a particular system of lenses, known as *échelon lenses*.

These lenses will be understood by examining Figs. 201 and 202;

Explain the principle of reflectors. What modification did FRESNEL introduce? *Explain the échelon lens.*

Fig. 201 shows a front view, and Fig. 202 a section or profile of an échelon lens.

A lens of this kind consists of a plano-convex lens, *A*, about a foot in diameter, around which are disposed several annular lenses, which are also plano-convex, and whose curvature is so calculated that each one shall have the same principal focus as the central lens, *A*.

A lamp, *L*, being placed at the principal focus of this refracting system, as shown in Fig 202, the light emanating from it is refracted into an immense beam, *RC*, of parallel rays.

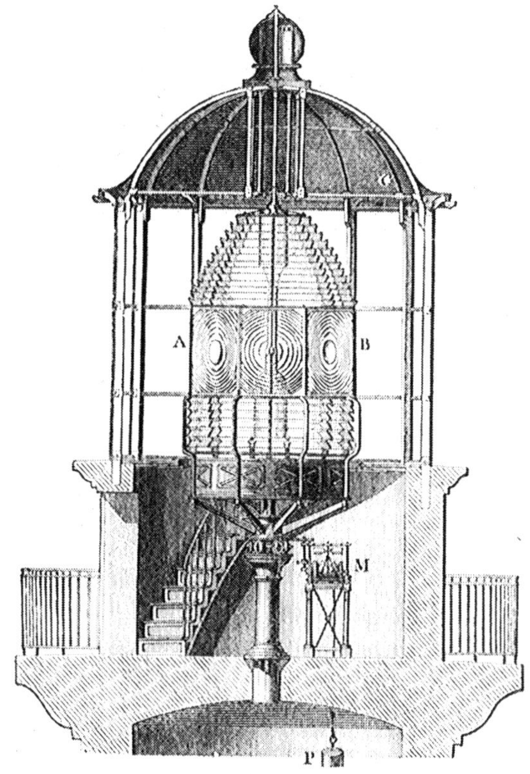

Fig. 203.

Explain the reflectors used by FRESNEL.

Besides this refracting system, several ranges of reflectors, *mn*, are so disposed as to reflect such light as would otherwise be lost, to increase the beam of light formed by refraction.

By this double combination, an immense beam of light is afforded, which renders the light visible for fifteen or twenty leagues; but this beam is only visible in a single direction. To remedy this defect, FRESNEL united eight systems similar to that just described, which combination presents the appearance of a pyramid of glass, nine or ten feet in height.

Fig. 203 represents a section of the lantern of a light-house of the first order, which was actually constructed by M. SAUTTER, and exhibited at the great "Universal Exposition" of France, in 1855.

In order to illuminate all points of the horizon, the system is made to revolve on a vertical axis by clock-work. The clock-work is shown at *M* in the figure, and the weight at *P*. To prevent friction the system turns upon six wheels, or rollers, shown in the figure to the left of *M*.

In consequence of this rotation an observer at any point will see eight flashes of light during one revolution, which are followed by as many intervals of darkness, called *eclipses*. By suitably regulating the number of revolutions in any given time, different light-houses may be distinguished from each other.

IV.—DECOMPOSITION OF LIGHT.—COLORS OF BODIES.

Solar Spectrum.

305. If a beam of sunlight pass through a prism, it is bent from its course and at the same time is spread out into a brilliantly colored band, called the *solar spectrum*. The spreading of the rays is called *dispersion;* it is caused by unequal refrangibility of the different colored rays. The angular dispersion of rays is different for different media.

How far is a FRESNEL light visible? How are all points of the horizon illuminated? Explain flashes and eclipses. (**305.**) What is the solar spectrum? What is dispersion.

The method of forming a spectrum is shown in the figure. The beam of light that enters a hole in the shutter falls on a prism whose refracting edge is turned downward; the whole beam is bent upward, and at the same time its elements are dispersed so as to form the elongated spectrum seen on the screen.

Fig. 205.

This spectrum consists of almost an infinite number of rays of different tints, but it is customary to consider only seven principal colors. These, in the order of refrangibility, are as follows: 1°, *red*, at *r*; 2°, *orange*, at *o*; 3°, *yellow*, at *y*; 4°, *green*, at *g*; 5°, *blue*, at *b*; 6°, *indigo*, at *i*; and 7°, *violet*, at *v*.

Besides the colored portions of the solar spectrum, there is an invisible portion below the red, where the heat is greater than anywhere else, and another portion above the violet, where the chemical effect is greater than anywhere else. The corresponding rays are called *heat rays* and *actinic* or *chemical rays*.

Name the colors in order of refrangibility. What are heat rays? *Chemical or actinic rays?*

If a colored ray of the spectrum pass through a hole in a screen, and then fall on a second prism, it is deviated as before, but there is no further change of color; hence, the colors of the spectrum are said to be *simple*.

The wave lengths corresponding to different colored rays have been measured, and it is found that for red rays it is about *the forty thousandth part of an inch*, and for violet rays it is no more than *the sixty thousandth part of an inch*. For the invisible rays of heat the wave length is greater than it is for red light; for the invisible actinic rays the wave length is less than it is for violet light. The phenomena of dispersion indicate that shorter waves are more retarded than longer ones in passing through a medium; hence, the rays near the heat end of the spectrum are *least* refracted, and those near the actinic end are *most* refracted.

Color in light corresponds to *pitch* in sound. The colors near the heat end of the spectrum correspond to the graver sounds, and those near the actinic end to the more acute sounds. The *range* of visible colors is greater than that of audible sounds. The range of visible colors is scarcely *one* octave, that of audible sounds is more than *ten* octaves.

Fraunhofer's Lines.—The Spectroscope.

306. The solar spectrum is not continuous; rays corresponding to certain degrees of refrangibility are wanting; hence, it is crossed at intervals by dark lines. These are seen to best advantage in a spectrum formed by passing a beam of sunlight through a narrow slit, and then decomposing it by a prism whose edges are parallel to the slit. The prism should be of flint glass and free from flaws.

The dark lines of the solar spectrum were noticed by WOLLASTON as early as 1802, but they were first studied and mapped by Fraunhofer in 1814; from that fact they have been called *Fraunhofer's lines.*

Fraunhofer's chart contains between five and six hundred lines irregularly distributed. In it the most prominent lines are designated

What is the wave length of red? Of violet? What relation is there between color and pitch? (**306.**) What are Fraunhofer's lines?

by letters, and these serve as points of comparison to which others may be referred. The line marked *A* is at the beginning, and *B* is near the middle of the red space; *C* is a well-marked line near the boundary of the red and orange; *D* consists of two strong lines close together in the orange; *E* consists of broad lines in the green, the middle one being the strongest; *F*, *G*, and *H*, are well-marked lines, *F* being in the blue, *G* in the indigo, and *H* in the violet. Between *A* and *B* is a band of lines named *a*, and between *E* and *F* are three strong lines called *b*, the two farthest from *E* being close together.

Fraunhofer counted *nine* lines between *B* and *C*; *thirty* between *C* and *D*; *eighty-four* between *D* and *E*; *seventy-five* between *E* and *F*; *one hundred and eighty-five* between *F* and *G*; and *one hundred and ninety* between *G* and *H*. Recent observations have increased the number of dark lines till they are now counted by thousands.

Fraunhofer found the spectra of the fixed stars to be crossed by dark lines, but the lines are differently arranged in the different stars, and in none are they arranged as in the solar spectrum. Recently the range of observation has been vastly increased, and on the results of these examinations a new branch of science has been founded, called *spectrum analysis*.

The instrument used for forming and examining the spectra of bodies is called a *spectroscope*. It usually consists of three parts: a *collimator*, a *train of prisms*, and a *telescope*.

The *collimator* is used to form a thin beam of parallel rays, and consists of a narrow slit and a double convex lens; the slit is formed by two jaws of metal that can be moved to and from each other, so as to give as narrow an opening as may be desired; the lens is behind the slit, and at a distance from it equal to its principal focal distance; hence, it renders the rays that pass through it parallel to each other. The *train of prisms* serves to disperse the light; it consists of any number of prisms having their edges parallel to the slit, and so placed that the light shall pass through them all in succession. The *telescope* is used to view the spectrum formed, and is usually provided with a micrometer for measuring the distances between the lines of the spectrum; it admits of a certain amount of angular motion, so that it can be made to embrace in succession every part of the spectrum.

Describe the positions of the principal lines of the solar spectrum. What is a spectroscope? Of how many parts is it composed? *Describe the collimator. The train of prisms. The telescope.*

Spectrum Analysis.—Explanation of Fraunhofer's Lines.

307. Metals and their compounds impart characteristic colors to flames; thus, sodium and its compounds impart a yellow color to a Bunsen burner; the compounds of copper render it green, the compounds of zinc make it purple, and the compounds of strontian give it a red color. These colors are due to the vapors of the corresponding substances, and are peculiar to those vapors. If these or any other incandescent vapors be examined with the spectroscope, their spectra are found to consist of bright bands, each corresponding to a definite degree of refrangibility. The number, color, and position of the bands in every case are perfectly characteristic, and always serve to identify the body producing the spectrum. This mode of determining the presence of bodies is called spectrum analysis.

If two or more metals be vaporized in the flame at the same time, the bands peculiar to each are formed as though the others did not exist. If a mineral substance containing many different metals be volatilized the spectrum will show the bands characteristic of each. Bunsen and Kirchoff discovered the new metals Rubidium and Caesium, by means of bands shown by the spectroscope, which differed from those of all the metals previously known, and in like manner Mr. Crookes discovered the new metal Thallium.

The method of spectrum analysis is exceedingly delicate; the presence of the minutest portion of any substance in the form of incandescent vapor is instantly made manifest by its characteristic lines in the spectrum.

It has been shown that an incandescent solid or liquid gives a continuous spectrum. If light from such a source be transmitted through the vapors of any substances, and then examined with the spectroscope, the resulting spec-

(**307.**) What is spectrum analysis? Character of the spectrum of an incandescent vapor. Explain the action of sodium, copper, zinc, and strontian on the flame of a Bunsen burner.

trum will be crossed by dark lines having the same position as the bright lines belonging to the spectra of the vapors. Hence, it appears that every body in a state of vapor is opaque to the class of rays that it emits when rendered incandescent.

The principle just elucidated has been applied to explain the dark lines of the solar spectrum. It is supposed that the body of the sun is an incandescent solid, or perhaps a glowing liquid, and consequently that it emits white light. It is further supposed that the body of the sun is surrounded by a layer of gaseous matter containing vapors of various substances, including many of the known metals. This envelope, called the photosphere, being at a lower temperature than the nucleus, is in a condition to absorb the very rays that it would itself emit, if it were incandescent. The absorbed or missing rays form the dark lines of the spectrum. Were the central nucleus abolished, the solar spectrum would be transformed into a system of brilliant bands. These would correspond to the bands of a spectrum given by a flame charged by metallic vapors. They would constitute the spectrum of the solar photosphere.

The following metals have been shown to exist in the photosphere of the sun: viz., *sodium*, *calcium*, *barium*, *magnesium*, *iron*, *chromium*, *nickel*, *copper*, *zinc*, *strontium*, *cadium*, *cobalt*, *hydrogen*, *manganese*, *aluminum*, and *titanium*—sixteen in all.

The spectra of the fixed stars indicate that those bodies are similar in constitution to our sun, but the number and position of the dark lines show that their photospheres do not contain the same elements that are found in our own luminary.

It has long been known that the sun is surrounded during the time of a total eclipse by a great number of irregular rose-colored protuberances. These have been shown by spectrum analysis to consist, for the most part, of incandescent hydrogen; with it are mixed vapors of sodium and magnesium. The protuberances form part of an irregular envelope surrounding the entire body of the sun, and lying outside of its photosphere. This layer constitutes what has been named the chromosphere, and within a few years a method has

---

What is the character of the spectrum of an incandescent solid? Action on light transmitted through a vapor. *Explanation of Fraunhofer's lines. What is the constitution of the sun? What metals have been detected in the sun? What is the constitution of the fixed stars? Describe the chromosphere.*

been discovered for observing its spectrum without the necessity of waiting for a total eclipse.

Heat Rays and Actinic Rays.

307*a*. The seven rays enumerated differ in illuminating power, the middle rays being those which possess the greatest illuminating power. That is, the most powerfully illuminating rays lie midway between the heat rays and the actinic rays.

If a thermometer be held for a time in the different rays, beginning at the violet, it will show an increase of heat till it comes outside of the red rays, where it is greatest.

The actinic rays are those that produce chemical changes. If a strip of paper, prepared with nitrate of silver, be placed in the spectrum, it will be least changed in the red, and in passing towards the violet end, this change will increase till it becomes the greatest beyond the violet.

Recomposition of Light.

308. The colors of the spectrum may be reunited so as to produce white light.

1. If it be acted on by a second prism exactly like the first, with its refracting edge turned in the opposite direction, it will be recomposed and will emerge as white light.

This amounts to nothing more than passing light through a medium bounded by parallel plane faces.

2. If it be received on a double convex lens, as shown in Figure 206, it will be recomposed and an image will be formed free from color.

The manner of performing this experiment is shown in Fig. 206.

(**307*a*.**) Which are the most illuminating rays? How is their heating power? What are actinic rays? Which produce the greatest chemical effect? How shown? (**308.**) May the rays of the spectrum be reunited? First method. Second method.

Fig. 206

3. If the decomposed light be received upon a concave mirror, it will in like manner be recomposed and a colorless image produced.

4. If a circular disk of card-board be painted as shown in Fig. 207, in sectors, the colors being distributed according to intensity and tint, as in the spectrum, it will be found on rotating the disk rapidly by a piece of mechanism shown in Fig. 208, that the separate colors blend into a single one, which is a grayish white.

The color from any sector produces upon the eye an impression

---

Third method? Fourth method?

Fig. 207. Fig. 208.

that lasts for an appreciable length of time. In the experiment, the rotation is so rapid that the impressions from all of the colors coexist at the same instant, and the effect is the same as though the colors were mixed.

That the impression produced by light lasts for an appreciable length of time, may be shown by whirling a lighted stick round in a circle; it will present the appearance of a continuous circle of fire.

Color of Opaque Bodies.

309. The color of a body may be *temporary* or *permanent*. Temporary colors arise from some modification of light, of a transient character.

How does it appear that the impression of color lasts for a short time? How may it be shown? (**309.**) From what does the color of a body arise?

Thus, by refraction, certain drops of water in the air are colored, producing the rainbow; the color of these drops is due to their position with respect to the eye and the sun. The colors of soap-bubbles are dependent upon interference, a principle not yet ex plained, and are transitory.

The colors of finely-grooved surfaces are due to interference. These colors are independent of the physical constitution of the body, and depend solely upon the fineness and shape of the grooves.

The play of colors upon mother-of-pearl is due to fine grooves or striæ, as may be shown by taking an impression of a piece of it in white wax; the colors of the wax, thus prepared, are entirely analogous with those of the mother-of-pearl, from which the impression was taken.

With respect to the permanent colors of bodies, various opinions have been held. NEWTON conceived that bodies had the power of absorbing some of the rays of the spectrum and reflecting the remainder. According to this theory, the color of a body would be that arising from a mixture of the reflected rays. Thus, vermilion was supposed to have the power of reflecting the red rays only, whilst all of the others were absorbed. All bodies placed in a red light appear red, in a blue light, blue, and so on for other colors.

ARAGO was of the opinion that the colors of bodies arose from light admitted into the body and then emitted again, undergoing certain modifications. Color would, according to this theory, depend upon the molecular condition of the body. According to this view, color is a modification of light, entirely analogous to that modification of sound which we call the *tone*.

ARAGO's theory was based upon a difference of property between reflected and refracted light. On examining the colors of opaque bodies, he found that the light agreed with that which had been refracted, rather than with that which had been reflected.

Explain temporary colors in case of rain-drops. Of grooved surfaces. Of mother-of-pearl. What is NEWTON's theory of colors of bodies? What is ARAGO's theory? *On what was* ARAGO's *theory based?*

Colors of Transparent Bodies.

310. All transparent bodies absorb more or less of the light which enters them, and if sufficiently thick, must appear colored. Their color is due to that part of the light which is transmitted.

If, for example, all of the solar rays except the red ones are absorbed by a medium, it will appear red by transmitted light. Water when seen in masses by transmitted light, appears of a greenish hue. Air appears blue; hence the color of the sky. As we ascend, the mass above us becomes smaller and loses its blue tint. It is probable that the bluish tint of the heavens is also in a measure due to reflection from the aerial molecules. At sunrise and sunset, the rays of the sun have to traverse a great body of the atmosphere, which absorbs most of the rays except the red ones. Hence it is, that the sun appears red at sunrise and sunset.

Complementary Colors.

311. NEWTON calls two colors *complementary*, when by their mixture they produce white.

If all the rays of the spectrum except the red ones be recomposed by a convex lens, a bluish-green color will result; hence, red and green are complementary. In like manner, it may be shown that blue and orange are complementary, as are also violet and yellow.

Accidental Images.—Accidental Fringes.

312. A curious effect of color upon the eye is manifest in the production of what are called accidental images.

If a wafer upon a black ground be viewed intently for some time, until the nerve of the eye becomes fatigued, and the eye be then directed to a sheet of white paper, an image of the wafer will be seen upon the paper, whose color is complementary to that of the

---

(**310.**) To what is the color of transparent bodies due? *Illustrate by examples.* (**311.**) What are complementary colors? What is the complement of red? Of green? Of blue? Of orange? (**312.**) *What is an accidental image? Illustrate. What is the cause of accidental images?*

wafer. Thus, if the wafer is red, the image will be green; if the wafer is orange, the image will be blue, and so on. These images are called *accidental.*

If the setting sun, which is red, be viewed for some time, and then the eyes be directed to a white wall, a green image of the sun will be seen, which will last for some instants, when a red image will appear; a second green image succeeds it, and so on till the effect entirely ceases.

If we look for some time at a colored object on a white ground, we shall finally observe the object surrounded by a fringe, whose color is complementary to that of the body; thus, if a red wafer be placed upon a sheet of white paper, the fringe will be green. Such fringes are called *accidental.*

Shadows cast upon a wall by the rising or setting sun, are tinged green, the tint of the sun being red at that time.

If we examine several pieces of cloth of the same color, the eye becomes wearied, and in consequence of the accidental complementary color being formed, the last pieces examined appear of a different shade from those first viewed.

Many of the phenomena of color may be explained by the principle of interference of light.

If a molecule of ether be acted on by a wave of light, it will take up a vibratory motion at right angles to the direction of propagation. If it be acted on simultaneously by two waves, its motion will be the resultant of the motions it would receive from each acting separately. If, therefore, the waves are in the same phase (Art. 147), the molecule will have its amplitude of vibration doubled; if they are in opposite phases, they will neutralize each other and the molecule will remain at rest. This action of one system of waves on another is called *interference.*

The brilliant colors of a soap bubble are due to the interference of the two sets of rays that are reflected from the outer and inner surfaces of the film that constitutes the bubble.

The colors of thin plates, like the film of oil on water, the splendid colors of the skimmings of melted lead, the iridescent displays of fractured crystals, and the like, are all due to interference of light.

Explain the effect of gazing at the setting sun. At a colored object on a white ground. Explain the phenomenon of interference of light. What are some of the instances of color from interference?

The Rainbow.

313. The RAINBOW is a brilliantly colored arc, seen after a shower opposite the sun.

The colors being disposed in the same order as in the solar spectrum, would indicate that the bow is due to refraction. Such is shown to be the case. Figure 209 shows the course of the rays in the formation of a rainbow. The rays of light coming from the sun, *S*, fall upon the spherical rain drops, enter them, undergoing refraction, are

Fig. 209.

internally reflected, and then emerge, undergoing a second refraction. The result is that the emergent light is resolved into the seven prismatic colors, which, reaching the eye from different drops, give rise to the colors that are observed.

The ray which enters the drop, *a*, for example, after emergence sends to the eye a red ray, whilst that which enters the drop *c*, sends to the eye a violet-colored ray;

(313.) What is a Rainbow? To what is the bow due? Explain the course of the rays, Fig. 209.

intermediate drops send intermediate colors. Each drop sends a different color to the eye.

Analysis shows that it is only at certain angles that the refracted rays emerge with sufficient intensity to affect the eye with color. Hence it is, that the colored drops are arranged symmetrically about a line drawn through the sun and the eye of the observer. The centre of the bow is in this line; hence, as the sun declines towards the horizon, the bow rises, and at sunset it becomes a semicircle. In looking down into spray with the back turned towards the sun, a complete circular bow may be seen.

The bow that we have described is called the *primary* bow, and the colors in it are arranged in the order of the prismatic colors, the red being on the outside.

Another bow is generally seen, concentric with the primary *bow*, which is called the *secondary* bow. This bow is formed by light which enters the drops is first refracted, then twice internally reflected, and then emerges, being again refracted. The result of this deviation is a bow similar to the first, but having its colors arranged in a reverse order, the red being on the inside.

The inversion of colors arises from the additional reflection that the light experiences. It is observed that the colors of the secondary bow are not so brilliant as in the primary; this is due to the loss of a portion of light, which passes out of the drop at each incidence.

From the nature of the rainbow, and the principle of its formation, it is plain that every observer sees a different bow.

Chromatic Aberration.

314. The light that falls on a lens is decomposed into colored rays of different degrees of refrangibility. These

---

How is the bow formed? Where is its centre? Why does the bow enlarge as the sun declines? How may a complete circular bow be seen? What is a primary bow? A secondary bow? How is it formed? *Why are the colors in the secondary bow reversed in order? How do the colors in the two bows compare in brilliancy? Does each observer see the same bow? Why not?* (**314.**) What is chromatic aberration?

rays are brought to different foci along the axis, giving rise to a multitude of partial images of different colors, which by superposition produce a single image slightly indistinct, and fringed with all the colors of the spectrum. This scattering of the colored rays to different foci, is called *chromatic aberration*.

Fig. 210 shows the phenomenon of chromatic aberration. The red rays being less deviated than the others, are brought to a focus beyond them at *r*, whilst the violet rays being more refrangible than the others, are brought to a focus within them at *v*. Between *v* and *r*, the intermediate colors are also brought to foci.

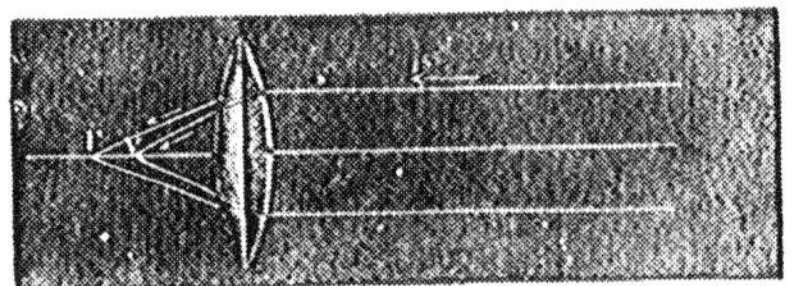

Fig. 210.

Achromatic Combinations.

315. An ACHROMATIC COMBINATION consists of two or more lenses of different kinds of glass, so constructed as to neutralize the effect of dispersion.

The combination usually consists of two lenses: a convex lens made of crown-glass, and a concave lens made of flint-glass, as shown in Fig. 211. Flint-glass disperses light more than crown glass. The combination, having its thickest part at the middle, is convergent. The dispersion of the rays by one of the lenses is exactly neutralized by a dispersion of them in an opposite way, so that the image is nearly colorless.

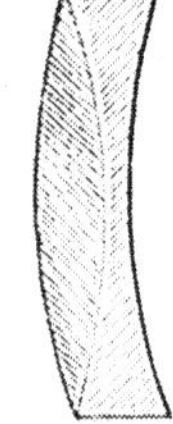

Fig. 211.

Such combinations of lenses are called *achromatic*, and are the ones used in the construction of telescopes.

Illustrate. (315.) What is an Achromatic Combination? Illustrate. What lenses are usually combined? Explain their action.

V.—THEORY AND CONSTRUCTION OF OPTICAL INSTRUMENTS.

Optical Instruments.

316. The properties of mirrors and lenses have led to the construction of a great variety of instruments, which by increasing the limits of vision, have opened to our senses two new worlds, that had else remained unknown to us, the one on account of its minuteness, and the other on account of its immensity.

Of the optical instruments, the most useful and interesting are *microscopes*, so called because used to investigate minute objects, and *telescopes*, so called because they are employed to examine distant objects.

Besides these a great variety of other instruments have been devised, such as the *magic lantern*, the *phantasmagoria*, the *solar microscope*, the *camera obscura*, and the *stereoscope*.

Telescopes.

317. A TELESCOPE is an optical instrument for viewing objects at a distance.

Telescopes may be divided into two classes, *refracting* telescopes, and *reflecting* telescopes.

In the first class a lens, called the *object lens*, is employed to form an image; in the second class a mirror or speculum is employed for the same purpose; in both the image formed is viewed by a lens, or combination of lenses, called the eye-piece. The manner of arranging these component parts, together with the nature of the auxiliary pieces employed, determines the particular kind of telescope.

(**316**) What are some of the most useful optical instruments? Mention some other instruments. (**317**.) What is a Telescope? How many classes of telescopes are there? What is the difference between the two classes? What determines the kind of telescope?

A great variety of devices have been employed to obviate the defects of spherical and chromatic aberration, and at the same time to obtain a sufficiency of illumination to render vision distinct. Hence the variety of telescopes is very great. Only a few of the most important will be described in these pages.

The Galilean Telescope.

318. The GALILEAN TELESCOPE, named from its illustrious discoverer, GALILEO, consists essentially of a *convex object-glass*, which collects the rays from an object, and a *concave eye-piece*, by means of which the rays from each point of the object are rendered parallel, and capable of producing distinct vision.

Fig. 212 shows the course of the rays in the Galilean telescope. Pencils of rays from points of the object, *AB*, falling upon the object lens, *O*, are converged by it, and tend to form an image beyond the eye-piece, *o*. The concave eye-piece is placed so as to intercept the rays coming from

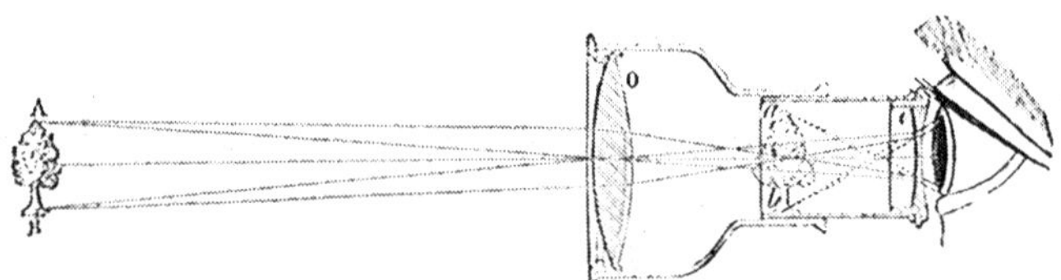

Fig. 212.

the object-glass, being at a distance from the image equal to its principal focal distance. In consequence of this arrangement, the pencil of light coming from *A*, is converged by the object-glass, and falling upon the eye-piece, is diverged and refracted so as to appear to the eye to come from *a*. In like manner the pencil from *B*, appears to the eye to come from *b*.

What are the special objects to be attained in making a telescope? (**318.**) What is a Galilean Telescope? Describe it. Explain the course of the rays in a Galilean telescope.

The image is erect and virtual, and because the visual angle under which the image is seen, is greater than that under which the object would be seen without the telescope, it appears magnified.

Opera-glasses are simply Galilean telescopes. They possess the advantage of showing objects in their proper position, of being short and portable, and of being well illuminated.

The Galilean telescope is not adapted to astronomical observation, because the image formed is virtual; nevertheless it was with such an instrument that GALILEO discovered the satellites of Jupiter.

The Astronomical Telescope.

319. The ASTRONOMICAL TELESCOPE consists essentially of two convex lenses, the one, *o*, being the object-lens, and the other, *O*, the eye-piece. The object-glass forms an inverted image of the object, which is viewed by the eye-piece.

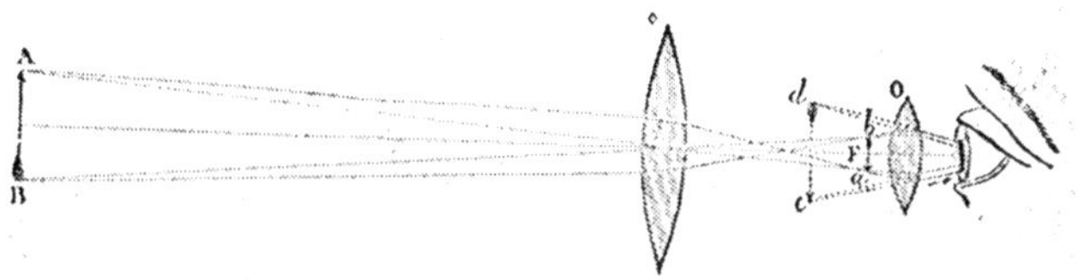

Fig. 213.

Fig. 213 represents the course of the rays in this instrument. A pencil of rays coming from *A*, is converged by *o*, to a focus *a*, whilst a pencil from *B*, is brought to the focus *b*. In this manner the lens *o*, forms an image, *ab*, of an object, *AB*, which image is real and inverted. The eye-piece, *O*, is placed at a distance from *ab* equal to its principal focal distance. The pencil coming from the points, *a* and *b*, of the image, are refracted so as to appear to come from the points, *c* and *d*. The visual angle, that is, the

How is the image? Give an example of a Galilean telescope. Their advantages? Is the Galilean telescope adapted to astronomical purposes? (319.) What is the Astronomical Telescope? Explain the course of the rays in it.

angle formed by the extreme rays which enter the eye, is greater than it would be in viewing the object without the telescope, and consequently the object appears to be magnified.

In this, as in all other telescopes, the eye-piece is capable of being pushed in, or drawn out, to enable the observer to accommodate it to near as well as distant objects.

The object-glass is made as large as possible, and should be

Fig. 214.

How is the eye-piece adjusted?

achromatic (Art. 315). The eye-glass is made quite convex, so as to magnify the image formed by the object-glass.

Fig. 214 represents an astronomical telescope mounted for use. It rests upon a cast-iron stand, with three feet, called a *tripod*. The tripod supports a vertical axis, capable of turning around in its supports; the telescope is attached to the top of this axis by a hinge joint. These arrangements enable the observer to direct the telescope to any point of the heavens. The telescope may be raised or depressed by means of a rack, worked by toothed wheels, set in motion by a crank, as shown at the bottom of the figure.

A smaller telescope with a larger field of view is attached to it, to aid the observer in fixing the instrument on any object. This telescope is called *the seeker*.

The Terrestrial Telescope.

320. The TERRESTRIAL TELESCOPE differs from the astronomical telescope, in having two additional lenses, which together constitute what is called an *erecting piece*. The object of the erecting piece is to invert the image formed by the object-lens, so that objects may appear erect when viewed through the telescope.

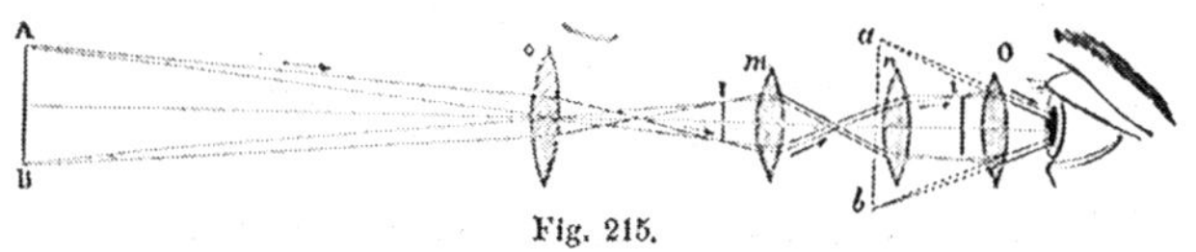

Fig. 215.

Fig. 215 shows the course of the rays in a terrestrial telescope. *AB* is the *object*, *o* is the *object-lens*, *m* and *n*, two convex lenses, constitute the erecting piece, and *O* is the *eye-piece*.

The erecting piece is so placed that the distance of the image, *I*, shall be at a distance from *m*, equal to its principal focal distance.

What is said of the object-glass and of the eye-piece? (320.) In what respect does the Terrestrial differ from the Astronomical Telescope? What is the object of the erecting piece?

A pencil of rays from *A*, falling upon the object-lens, is converged to a focus at the lower end of the image, *I*; the pencil proceeding from *I*, is converted into a beam by the lens, *m*, directed obliquely upwards, which beam is converged to a focus at *i*. In this manner an erect image, *i*, is formed, which is then viewed by the eye-piece, *O*. The eye-piece refracts the pencils coming from the image *i*, so as to make them appear to come from *ab*.

The angle, under which *ab* is seen, is the ***visual angle***, and being greater than the angle under which *AB* would be seen without the telescope, the object is magnified.

The number of times which the visual angle of the image contains the visual angle of the object, is the *magnifying power* of the telescope.

The terrestrial telescope is used at sea and on land for viewing objects at a distance. It may, for convenience, be mounted in the same way as the astronomical telescope shown in Fig. 214.

Reflecting Telescopes.

321. A REFLECTING TELESCOPE is one in which the image of a distant object is formed by means of a reflector or speculum, which image is then viewed by an eye-piece. The eye-piece is either a single lens or a combination of lenses.

One of the first telescopes of this description was constructed by NEWTON, and this is the only one of the kind which we shall describe in detail.

Newtonian Telescope.

322. Fig. 216 shows a NEWTONIAN TELESCOPE, as constructed by M. FROMENT, of Paris, with improvements introduced by that distinguished physicist.

Describe the course of the rays in the terrestrial telescope. What is the magnifying power? *What is the use of the terrestrial telescope?* (**321.**) What is a reflecting telescope? (**322.**) Describe the Newtonian Telescope.

Fig. 216.

Fig. 217 shows the same telescope in section, and indicates the course of the rays of light.

M is a concave mirror placed at the bottom of a long tube. This reflector tends to form a small image of an object at the other end of the tube. But before the rays reach the image, they are intercepted by a prism of glass, *mn*, so arranged that the rays enter its first face without deviation, and strike its second face so as to be totally reflected, which causes the image to be formed at *ab*. The image thus formed is viewed by an eye-piece through the

side of the telescope. The eye-piece in this telescope is made of two plano-convex lenses, as shown in the figure, the combined effect of which is to cause the image to appear in the position *BA*, giving a great power to the telescope.

Fig. 216 shows the manner of viewing the image. It also shows a small *seeker* attached to the tube of the main instrument, which is used in directing the telescope to any required object.

Herschel's Telescope.

323. Sir William Herschel, of London, modified the Newtonian telescope by inclining the mirror, *M*, so as to throw the image to one side of the tube, where it could be viewed by a magnifying eye-piece, the observer's back being turned towards the object.

The large telescope made by this eminent astronomer was forty feet in length, and the speculum had a diameter of about five feet. It was with this gigantic instrument that he made some of his most brilliant discoveries.

Lord Ross's Telescope.

324. Lord Ross, of Ireland, has recently constructed a reflect-

Explain Fig. 216. (**323.**) What modification did Herschel make in the Newtonian telescope? *Describe* Herschel's *telescope.* (**324.**) Describe Lord Ross's telescope.

ing telescope still larger than HERSCHEL'S. The tube is 56 feet in length, and the diameter of the reflector is more than 6 feet. The speculum weighs over 4 tons, and the entire instrument more than 18 tons. This telescope is supported by two walls of masonry 48 feet high, 72 feet long, and 24 feet distant from each other. The instrument is said to have cost the owner $60,000.

Microscopes.

325. A MICROSCOPE is a modification of the telescope, for viewing near objects.

Microscopes, like telescopes, may be composed of a combination of lenses alone, or they may be composed of a combination of reflectors and lenses. Reflecting microscopes are but little used. We shall only describe the refracting microscope, of which there are two kinds, the *simple* and the *compound*.

The Simple Microscope.

326. The SIMPLE MICROSCOPE consists of a double convex lens of short focal distance. It is usually set in a frame of metal or of horn, and held in the hand.

Fig. 218.

Fig. 218 shows the manner of using it. It is held at a distance from the object to be viewed, a little less than its principal focal distance. In this case, each pencil of light falling upon it will be deviated so

(**325.**) What is a Microscope? *How may a microscope be constructed?*
(**326**) What is a Simple Microscope? Explain Fig 218.

as to form a beam, whose axis passes through the point from which the pencil proceeds, and the optical centre.

The object appears of the same size that it would if the eye were placed at the optical centre of the lens. Since the least limit of distinct vision is about eight inches, it follows that a single microscope whose focal distance is one inch, would magnify an object

Fig. 219.

How is the magnifying power determined?

eight times. If the principal length were only one quarter of an inch, it would magnify thirty-two times.

The Compound Microscope.

327. The COMPOUND MICROSCOPE consists essentially of a double convex lens called the *object-lens*, and a second double convex lens called the *eye-piece*.

Fig. 219 represents a compound microscope and the method of using it. Fig. 220 shows the same instrument in section, and makes known the course of the rays. The letters correspond to the same parts in both diagrams.

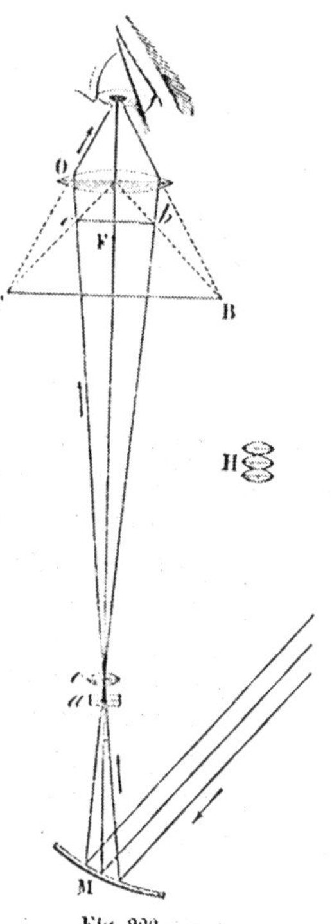

Fig. 220.

The object to be observed is placed at *a*, between two plates of glass upon a support. Over this is a tube, *OAo*, in which are disposed the two lenses, the object-lens, *o*, being at its lower, and the eye-piece, *O*, at its upper extremity. The object, *a*, being placed a little beyond the principal focus of the object-glass, this lens produces a real image, *bc*, which is inverted. The object-glass, *O*, is so placed that its principal focus is a little beyond the image, *bc*. This lens then acts as a simple microscope, and magnifies the image as though it were at *BC*.

(**327.**) What is a Compound Microscope? Explain its construction, and the method of using it.

The magnifying power depends upon the object-lens. This power is increased by combining two or three lenses, as shown at *H*, on the right of Fig. 220. A second lens is often added to the eye-piece, as shown in the Newtonian telescope, Fig. 217, for the purpose of remedying the defect arising from spherical aberration. Moreover, all of the lenses are made achromatic.

Microscopes of this kind are constructed whose magnifying power is 1800; but what is gained in power is often lost in distinctness. A good magnifying power is 600 in length and breadth, which gives 360,000 in surface.

The object, when transparent, is illuminated by a mirror, *M*, which concentrates the light upon it. When the object is opaque, it is illuminated by a lens, *L*, which concentrates the rays upon it.

The microscope is used in the study of botany to discover the laws of the vegetable world; in entomology to study the habits of minute insects; in anatomy and medicine to study the laws of animal physiology; in the arts, to discover the composition of mixtures; in commerce to detect the nature of stuffs, and so on. Its use is almost universal, either as an instrument of research or of curiosity.

The Magic Lantern.

328. The MAGIC LANTERN is an apparatus for forming upon a screen enlarged images of objects painted on glass. It was invented about two hundred years ago, by Father KIRCHER, a German Jesuit.

Fig. 221 represents a magic lantern in use, whilst a section of the same instrument is shown in Fig. 222.

It is composed of a box, in which a lamp is placed before a reflector, *M*; the light is reflected upon a lens, *L*, and is converged so as to illuminate strongly the plate of glass, *ab*, upon which the picture is painted. Finally, a combination of two lenses, *m*, acting as a single convex lens, is placed so

---

Upon what does the magnifying power depend? Why is a second lens added to the eye-piece? How great may the magnifying power be made? How is the object illuminated? What are some of the uses of the microscope? (**328**.) What is a Magic Lantern? By whom invented? Describe the construction and method of using the magic lantern.

Fig. 221.

that the plate, *ab*, shall be a little beyond its principal focus. At this distance the lenses produce (as shown in Fig. 196) a magnified and inverted image of the picture painted on

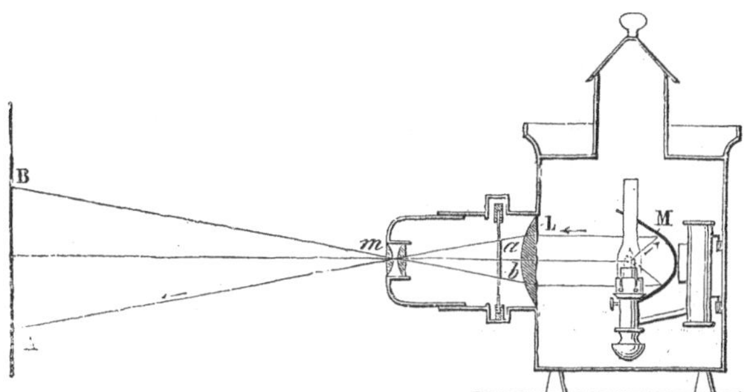

Fig. 222.

the glass. The picture on the glass should be inverted, in order that its image may appear erect.

The image on the screen will be the more magnified, as the plate, *ab*, approaches the principal focus of the compound lens *m*. It will also be the more magnified as the compound lens increases in power.

The Phantasmagoria.

329. The PHANTASMAGORIA differs from the magic lantern only in having an arrangement by which the size of the image on the screen may be increased or diminished at pleasure.

The Polyrama and Dissolving Views.

330. The POLYRAMA consists of a double magic lantern, with two cut-off screens. DISSOLVING VIEWS are obtained by using both lanterns. Thus, if a picture of a daylight scene be painted on one of the slides, and of the same scene by moonlight be painted on the other, the first picture is thrown upon the screen, strongly illuminated, the other one being entirely excluded by a screen that cuts off the second lens. By an arrangement operated by the exhibitor, the light is gradually cut off from the first picture and admitted upon the second, the first fading away insensibly, whilst the second as gradually grows brighter. In this way all the effects intermediate between full daylight and full moonlight may be obtained in succession.

A volcano, calm, and only surmounted by a light cloud of smoke, may be followed by a picture of the same volcano sending forth volumes of flame and smoke. A storm may be made to succeed a smiling landscape, and so on; the illusion is complete.

The Photo-Electric Microscope.

331. The PHOTO-ELECTRIC MICROSCOPE is constructed on the same optical principles as the magic lantern, except

(**329.**) How does the Phantasmagoria differ from the Magic Lantern? (**330.**) What is the Polyrama? Explain the method of producing the Dissolving Views. *Illustrate.* (**331.**) What is the Photo-Electric Microscope?

that the light employed is obtained by passing an electric current between two charcoal points. The pictures on the shades are also made smaller than in the magic lantern, which requires a greater illumination.

Fig. 224.

Fig. 224 represents in detail the arrangement of this instrument. At the foot of the apparatus is a battery for generating electricity, which will be described hereafter. The electricity is conveyed to the charcoal points in the box, *B*, by means of two copper wires, one going to the

---

Explain the arrangement of parts.

upper, and the other to the lower point. The points being slightly separated, the circuit is completed only by the electricity passing across the interval, which gives rise to a light of extreme brilliancy.

In the figure, I represents a parabolic reflector for concentrating the light upon the slide, X, through a lens, C. D is a lens which forms a magnified image of the minute object on a screen. The tube in which the lens, D, is placed, may be drawn out or pushed in to vary the magnifying power of the apparatus.

The magnifying power of this instrument may be made extremely great, and by suitable management it serves to show to a large company the wonders of the microscopic world.

One of the most remarkable experiments made with it, is to show the circulation of the blood. Instead of a picture on the slide, let the tail of a tadpole be placed between two plates of glass and introduced. There will appear upon the screen what seems an illuminated map, all of whose streams flow with a rapid current. It is but the blood circulating with great velocity through the arteries and veins.

The phenomena of crystallization are exceedingly beautiful when seen by this microscope. If a drop of a solution of sal ammoniac, for example, be poured upon a plate of glass, and then introduced into the instrument, the heat will cause the water to evaporate, producing one of the most beautiful examples of crystallization that can be exhibited.

The minute animalculæ of solutions and stagnant water can be shown by this microscope.

When the light of the sun is used instead of the electric light, the apparatus is called the *solar microscope*.

The Diorama.

332. The DIORAMA consists of two pictures, one on each side of a transparent muslin screen, these pictures, as in

How is the magnifying power varied? *What are its advantages? How is the circulation of the blood shown? The phenomena of crystallization? Animalculæ? What is a solar microscope?* (**332.**) What is the Diorama?

the polyrama, being different effects of the same scene. One of these pictures is seen directly, and the other by transmitted light, and the illusion arises from the light being

Fig. 225.

managed so as to produce either of these effects at pleasure.

Fig. 225 explains the manner of exhibiting this kind of

From what does the illusion arise?

picture. The two views are painted on opposite sides of a vertical screen. The first effect is painted upon the front of the screen, and is seen by light that enters a window, *M*, and falling upon a movable mirror, *E*, is thrown so as to illuminate the front of the screen. The room behind the screen being dark, no part of the picture on the other side of the screen is seen.

If, now, the mirror *E*, be lowered gently, the shutters, *NN*, being at the same time slowly opened, the picture on the front of the screen will fade away, to be replaced by that on the other side, now seen by transmitted light. When the mirror is let completely down, and the shutters, *NN*, are completely opened, the only effect that will be seen will be that from behind.

The diorama was invented and perfected by DAGUERRE, the celebrated discoverer of the daguerreotype. Many of his pictures of this kind had a high reputation, among which may be mentioned his *Midnight Mass*, and his *Valley of Goldeau*.

The Camera Obscura.

333. The CAMERA OBSCURA is an instrument used for forming a clear picture of objects upon a screen of ground glass or paper.

It consists, Fig. 226, of a closed box mounted on a stand, having a small hole on one side and a screen for receiving the image on the opposite side. The hole may be of any dimensions, if a concave lens be placed in it capable of filling it, and of such power as to bring the rays to a focus on the opposite screen.

Fig. 226 shows how the image is formed in the camera obscura. The pencil of rays coming from the soldier's cap goes to form an image at the bottom of the box, whilst that coming from his feet goes to form an image at the top of

Explain the method of exhibiting. *Who invented the diorama?* (**333.**) What is the Camera Obscura? Describe it. Explain the course of the rays.

Fig 226

the screen. The image is inverted and reversed in a horizontal direction, but in every other respect, including color, it is a perfect representation of the object pictured.

The camera obscura affords aid in sketching the outlines of a landscape or building, but its principal importance at present consists in its application to the various branches of Photography. It may also be used as a source of amusement.

The images formed by a camera obscura possess the remarkable peculiarity of being entirely independent of the shape of the opening in the box, provided it be quite small. The shape of the images is the same, whether the opening be square, round, triangular, or oblong.

To show this, let us consider the case of a beam of solar light entering a dark room through a hole in a shutter, Fig. 227. With respect to the sun, the hole in the shutter is but a point, hence the group of rays which enter it form in reality a cone whose base is the sun. The prolongation of these rays into the room makes up another

For what is the camera used? What remarkable property do the images possess? How is this illustrated?

Fig 227.

Fig. 228.

cone similar in shape to the first, and if this cone be intercepted by a screen perpendicular to the line joining the hole with the centre of the sun, the image formed will be a circle. If the rays are intercepted by an oblique plane, as in the figure, the image is elliptical, but it never takes the form of the hole when that is small.

In accordance with this principle, we find the illuminated patches of earth formed by light passing between the leaves in a forest of a circular or elliptical shape. This is illustrated in Fig. 228. In an eclipse of the sun, when the visible portion of the sun is of crescent shape, the patches of light all assume the crescent form; that is,

Fig. 229.

Explain the peculiar rounded form of patches of light in the shadow of a forest What form do they take in an eclipse of the sun?

they are images of the visible part of the sun. The reason of this curious phenomenon is evident.

Manner of rendering the Image erect.

334. The manner of producing erect images of external objects in a *camera obscura*, or *dark room*, is shown in Fig. 229. A little above the hole a plane mirror is so placed as to reflect the rays which enter it upon a convex lens fixed at the extremity of a tube. This reflection inverts the beam of light and makes the image erect, which may then be thrown upon a suitable screen for observation.

Such images are perfect representations of the external objects which they represent, being perfectly faithful, not only in form and color, but in motion also. When images of street scenes, with all their life and motion, are thus formed, they are very striking as well as interesting.

Fig. 230.

(334.) How are the images made erect?

Portable Camera for Artists.

335. For taking views, the camera obscura should be light and portable. The best form is that shown in Fig. 230. It consists of a sort of portable tent of black cloth, within which is a table for receiving the image, and at the top of which is a tube bearing a prismatic lens, that produces the combined effect of the mirror and lens, as shown in Fig. 229. The figure projected upon the table may be traced out with a pencil on a sheet of white paper.

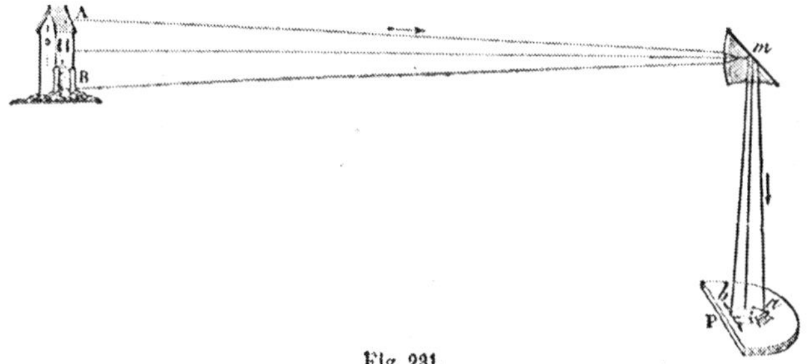

Fig. 231.

Fig. 231 shows the course of the rays in forming the image. The rays coming from the object, *AB*, fall upon the convex face of the lens and are converged, and in this state they reach the plane surface, *m*, which is inclined to the horizon. Being totally reflected from the surface, *m*, they emerge through the slightly concave surface below, and go to form an image, *ab*, on the table, *P*. A sheet of paper is spread on *P*, to receive the image, and on it the outlines may be traced.

The Daguerreotype.

336. One of the most important applications of the camera obscura, is in forming pictures upon plates of pre-

(**335.**) Explain the construction of the Portable Camera for Artists. Explain the course of the rays. (**336.**) What is the most important application of the camera?

pared metal or paper, by the *actinic* or *chemical* action of the light.

The discovery of the daguerreotyping process, like many other discoveries of magnitude, was preceded by many partially successful efforts. One of the most important of them was, perhaps, that of TALBOT, who succeeded in fixing images on prepared paper by means of solar light. The main discovery is, however, due to M. DAGUERRE, who in 1839 announced that he could, by a process occupying but a few minutes, fix the image of a camera upon a metallic plate.

During the few years that have elapsed, improvements have followed each other in rapid succession, until the process of daguerreotyping, in all its various branches, gives remunerative employment to thousands. It is not only one of the most interesting discoveries of modern times, but it has become of immense utility.

Process of Daguerre.

337. The process of Daguerre begins by receiving the image of the camera upon a proper plate, covered with a thin layer of silver, whose surface has been carefully polished and rendered sensitive to light. The polished plate is rendered sensitive by means of iodine. Iodine is solid at ordinary temperatures, but is easily converted into vapor by a slight degree of heat. The plate is held over the vapor of iodine for about two minutes, during which time a thin layer of the silver unites with the iodine, forming a coating of *iodide of silver*, which is exceedingly sensitive to light. The plate thus prepared is placed in the camera, so as to receive the image to be copied, and is acted upon by the rays forming the image. The plate is next exposed for a few minutes to the vapor of mercury. The mercury unites with the silver where it has been acted upon by the light, forming a white amalgam, giving the lights of the picture, whilst the other parts remain dark.

This process was imperfect; the plates required ten or twelve minutes' exposure to light, in order to fix an impression, which rendered the method unsuitable for portraits; the pictures formed were indistinct and easily effaced, and finally, the reflected light

Give a sketch of the history of the Daguerreotype. (337.) *Explain the process of* DAGUERRE.

from the plates diminished the distinctness of vision. All of these defects were remedied by a single man, M. FIZEAU

By using bromine with iodine in preparing the plates, he rendered them so sensitive, that from six to thirty seconds formed a sufficient exposure. He fixed the images and prevented excessive reflection, by using chloride of gold and hyposulphite of soda with gentle heat. This process not only had the effects named, but it also increased the brightness of the picture. Since these, other improvements have been made, till at last in skillful hands it has reached a state of great perfection.

Fig. 232.

Fig. 232 represents the form of camera used in the process of daguerreotyping. It consists of a rectangular wooden box, to one

Explain the modifications of FIZEAU. *Explain the construction and method of using the camera for daguerreotyping.*

face of which is attached a tube, bearing a lens, which forms the image. The opposite face of the box consists of a sliding drawer, holding a plate of ground glass, upon which the image is thrown, and by drawing it out, or sliding it in, the picture may be rendered distinct upon the glass. When the image is clearly defined, the plate of glass is removed, and the prepared silver plate introduced, and the process above described is performd.

Photography.

338. PHOTOGRAPHY is the art of fixing the picture of the camera on paper or glass.

There are two kinds of photographic pictures, *positive* and *negative*. *Positive* pictures are those that have their lights and shades in their proper relative position; *negative* pictures are those in which the lights and shades are reversed in position. A negative picture taken on glass is used to produce a positive one on paper.

To produce a negative, a plate of glass is carefully cleaned and coated with a layer of collodion impregnated with iodide of potassium; the plate is then immersed for about a minute in a bath of nitrate of silver, containing thirty grains of the nitrate to an ounce of water. The double decomposition that ensues gives rise to a layer of iodide of silver, evenly spread on the plate. This operation should be performed in a dark room. The plate is next drained, and when nearly dry, it is inserted in a closed frame and exposed to the action of the camera. The plate is then removed to a dark room, and the picture is brought out, or *developed*, by pouring over it a solution of protosulphate of iron or pyrogallic acid. This brings out, or *develops*, the invisible picture formed by the action of light on the iodide of silver. When the picture is sufficiently brought out, water is poured over the plate, which stops the further development. The parts not acted on by light still contain iodide of silver, which, if not removed, would be affected if exposed to the action of light. This is removed by washing the plate with hyposulphite of soda, which dissolves the iodide

(**338**.) What is Photography? What are positive and negative pictures? *How is a negative on glass produced?*

but does not affect the picture. The plate thus prepared is dried, and then coated with a thin layer of transparent varnish to protect it from injury.

The negative thus produced is used for obtaining positive prints on paper, as follows:—Paper is impregnated with chloride of silver by first immersing it in a solution of nitrate of silver and then in one of chloride of sodium; chloride of silver is thus formed on the paper by double decomposition. The negative is placed over a sheet of this prepared paper and exposed to the action of sunlight for a certain time. The chloride of silver is acted on by the light shining through the negative, being most affected under the light parts of the negative and least affected under the dark parts. The copy thus formed is a positive picture. In order to fix it, the paper is thoroughly washed in a solution of hyposulphite of sodium, which dissolves out the unaltered chloride of silver and prevents the further action of light. The picture is next immersed in a bath of chloride of gold to give it the proper tone.

To obtain a positive picture on glass, prepare the plate as before. After exposure to the action of the camera, develop by a solution of protosulphate of iron; this gives a negative; then pour over the plate a solution of cyanide of potassium; this rapidly changes the negative into a positive. This completed, the picture is washed, dried, and a coating of varnish is poured over its surface.

Besides the methods given above, there are many others in use, producing particular varieties of picture, but all depend on the same fundamental principle, that is, the extreme sensitiveness of salts of silver to the action of light.

Structure of the Eye.

339. The Eye is a collection of refractive media, by means of which we are made acquainted with the external world through the sense of sight.

As an optical instrument the eye is inimitably perfect; it has not the faults either of spherical or chromatic aberration, and withal, it possesses the remarkable property of self-adaptation to great as well as small distances.

The shape of the eye is spherical, with a slight protuber-

How are negatives on glass used to obtain positive prints? How are positives on glass obtained? (**339.**) What is the eye?

ance in front; the average diameter of the human eye is a little less than nine tenths of an inch. Fig. 233 represents a section of an eye, with some of the coverings thrown back so as to show the position of the parts.

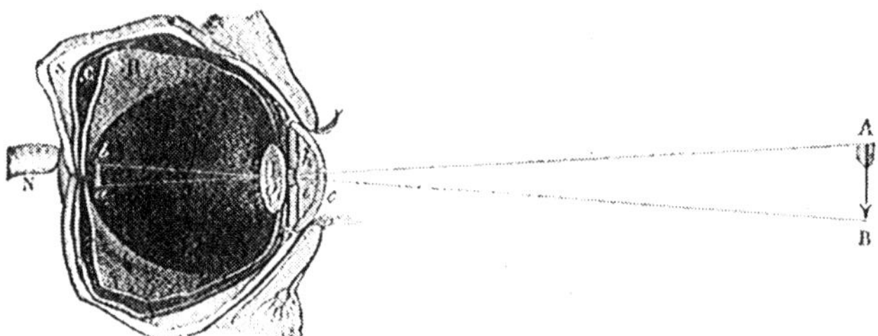

Fig. 233.

The anterior part of the eye is limited by a perfectly transparent membrane, *c*, called the *cornea*. The remainder of the exterior coating is an opaque white membrane, called the *sclerotic coat*. The cornea is set in the sclerotic coat, as a watch-glass is set in its frame.

Immediately behind the cornea is a transparent fluid, limpid as water, called the *aqueous humor*. In this floats a circular curtain, *hi*, attached by its outer edge to the sclerotic coat, and having a small circular opening at its middle. The curtain is called the *iris*, and the hole in its centre is called the *pupil*. The iris gives color to the eye, being black, blue, gray, &c.; it is muscular, and by the contraction and expansion of the fibres the pupil may be enlarged or diminished; it is through the pupil that rays of light enter the eye.

Behind the iris is a double convex lens, *o*, called the *crystalline lens*; it is of the consistence of gristle, perfectly transparent, more curved behind than in front, and is denser towards its middle than at the edges. This lens serves to

Its size? What is the character and position of the cornea? Of the sclerotic coat? Of the aqueous humor? The iris? The pupil? The crystalline lens?

converge the rays to foci behind it. Immediately behind the crystalline lens is a medium *nearly* filling the remainder of the cavity of the eye, called the *vitreous humor;* it is of the consistence of jelly, and perfectly transparent, permitting the rays to pass through it.

Immediately behind the vitreous humor is a thin white expansion of the optic nerve, lining nearly all of the sclerotic coat; this is called the *retina*, and is the seat of vision. Behind the retina, and between it and the sclerotic coat, is a fine velvety coating called the *choroid coat*, covered with a black pigment, which absorbs the rays that pass the retina, preventing internal reflection. The sensation of sight is conveyed to the brain by the optic nerve, which goes to the brain.

The Mechanism of Vision.

340. The action of the eye is similar to that of the camera obscura, except more perfect; the *pupil* corresponds to the hole in the shutter, the *crystalline lens* forms the image, and the retina is the screen on which the image falls. The image formed is of course inverted, as shown in Fig. 233, but the mind refers objects along the rays which produce the sensation of sight, hence points appear in their proper position; that is, we see objects *erect*.

Limit of Distinct Vision.—Defects of Sight.

341. When an object is placed very near the eye, the lens has not sufficient power to bring the rays to foci on the retina, and an indistinctness of vision is the consequence. The least distance at which an object can be seen distinctly is very different in different individuals. It may, on an average, be put down at six inches. Sometimes this limit is not the same for both eyes in the same individual.

The vitreous humor? The retina? The choroid coat? The optic nerve? (**340.**) Describe the mechanism of vision. (**341.**) What is the average limit of distinct vision?

When the limit of distinct vision is much less than six inches, the individual is said to be *short-sighted;* when it is much greater than six inches, he is said to be *long-sighted.*

SHORT-SIGHTEDNESS comes from too great convexity of the cornea, or crystalline lens, or both. The effect is to bring the rays to foci before reaching the retina, giving an indistinctness to vision. This defect is remedied by using spectacles with concave lenses, which diverge the rays before falling upon the cornea, and thus enable the media of the eye to bring them to foci upon the retina. If the eyes are unlike, the lenses should be of different power.

LONG-SIGHTEDNESS is a defect just the reverse of short-sightedness. It arises from too great flatness in the cornea, or crystalline lens, so that rays of light are brought to foci behind the retina. This defect is remedied by using spectacles with convex lenses.

Short-sightedness is a defect of youth, and is gradually removed as the individual advances in years; long-sightedness is a defect of advanced age, and once commenced, it gradually increases with years, probably because the organs which secrete the media of the eye become feeble as life advances.

The best form of convex glasses for spectacles is the meniscus, *O*, Fig. 186, and the best form of concave glasses is the concavo-convex, *R*, Fig. 187. These glasses are called *periscopic*, because they permit a wider range of vision than other forms of lenses.

Vision with two Eyes.

342. An image of every object viewed is formed in each eye, yet vision is not double, but single. This is regarded

When is a person short-sighted? When long-sighted? What is the cause of short-sightedness? How is it remedied? What is the cause of long-sightedness? How is it remedied? What are periscopic glasses? (**342.**) How are we enabled to see clearly with two eyes?

by some as a matter of habit; others refer it to the fact that each nervous filament coming from the brain to the eye is divided into two parts, one going to each eye.

Simultaneous vision with two eyes is supposed to give us the idea of *relief*, or form of objects, a view which receives confirmation from the action of the *stereoscope*.

The Stereoscope.

343. The STEREOSCOPE is an apparatus employed to give to flat pictures the appearance of relief; that is, the appearance of having three dimensions.

It was invented by WHEATSTONE and improved by BREWSTER. At the present day it is offered for sale in a great variety of forms, and constitutes an instructive and amusing instrument.

When we look at an object with both eyes, each eye sees a slightly different portion of it. Thus, if we look at a small cube, as a *die*, for example, first with one eye and then with the other, the head remaining fast, we shall observe that the perspective of the cube is different in the two cases. This will be the more apparent the nearer the body.

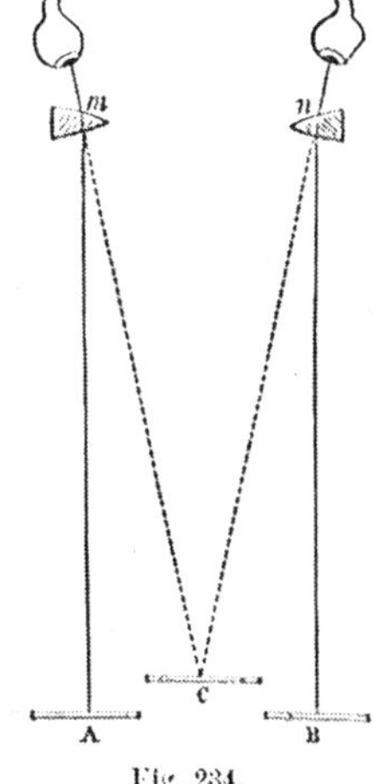

Fig. 234.

If the cube has one face directly in front of the observer, and the right eye is closed, the other eye will see the front face and also the left hand face, but not the right; if, however, the left eye is closed, the other eye will see the front face and also the right hand face, but not the left.

Whence do we derive our notion of relief in bodies? (**343.**) What is the Stereoscope? *By whom invented?* Explain the theory and construction of the stereoscope in detail.

Hence we know that the two images formed by the two eyes are not absolutely alike. It is this difference of images which gives the idea of relief in looking at a solid body.

If, now, we suppose two pictures to be made of an object, the one as it would appear to the right eye and the other as it would appear to the left eye, and then look at them with both eyes through lenses that cause the pictures to coincide, the impression is precisely the same as though the object itself were before the eyes. The illusion is so complete, that it is almost impossible to believe that we are simply viewing pictures on a flat surface.

Such is the theory of the stereoscope. Fig. 234 shows the course of the rays in this instrument as just described. *A* represents a picture of the object as it would be seen by the right eye alone; *B*, a picture of the same object as it would be seen by the left eye alone; *m* and *n* are lenses which deviate the rays so as to make the pictures appear to be coincident in *C*.

The lenses, *m* and *n*, ought to be perfectly symmetrical, and BREWSTER attained this result by cutting a double convex lens in

Fig. 235.

Explain the course of the rays in the stereoscope.

two, and placing the right hand half before the left eye, and the other half before the right eye. The pictures must be perfectly executed, which can be done only by means of the daguerreotype or photographic process. The pictures are made by using two cameras inclined to each other in the proper angle.

Fig. 235 represents two stereoscopic pictures of FRANKLIN, taken from a statue. We see the left hand one more in front, the right hand one more in profile. On placing them in the stereoscope we see a single image in relief. This image stands out in relief, presenting all the appearance of the statue from which the pictures are taken.

The best form of the stereoscope is that of DUBOSCQ. The lenses are large, and touch each other, so that they are adapted to eyes which are at any distance apart, which is not the case in the instrument shown in Fig. 234. In that instrument the eyes must be at a certain distance apart, which does not permit the same instrument to be used by both children and adults.

Explain BREWSTER'S *form.*

CHAPTER VII.

MAGNETISM.

I.—GENERAL PROPERTIES OF MAGNETS.

Definition of Magnetism.

344. MAGNETISM, as a science, is that branch of Physics which treats of the properties of magnets, and of their action upon each other.

Magnets.

345. A MAGNET is a body which exercises a particular power of attraction upon iron and a few other metals.

Magnets are either *natural* or *artificial.*

Natural magnets are certain ores of iron, and are generally known under the name of loadstones.

The magnet is so called from the town of Magnesia, in Lydia, where it was first noticed by the Greeks. In its natural form it consists of a mixture of two oxides of iron, with a small proportion of quartz and alumina. It is now found in considerable quantities in Sweden and Norway, as well as in many other countries.

The magnet possesses the remarkable power, when freely suspended, of directing itself towards a particular point of the horizon, and it is to this property that its importance is

(344.) What is Magnetism as a science? What is a Magnet? How many kinds of magnets are there? What are natural magnets? *Whence the name? What is the constitution of a natural magnet?* What remarkable property does the magnet possess?

chiefly due. It may be suspended by a thread, or by balancing it on a pivot. In practice the latter method is the one most usually adopted.

Artificial magnets are bars of tempered steel, to which the property of the natural magnet has been imparted. The artificial magnet is far more valuable than the natural magnet, and is generally used in practice.

Steel is a mixture of iron with a small quantity of carbon, and when heated and then plunged into water, it becomes exceedingly hard, and capable of retaining the magnetism that may be imparted to it.

Artificial magnets for experiment are made of oblong bars, from twelve to fifteen inches in length, as represented in Figs. 245 and 246. They are sometimes made in the form of a horse-shoe, as shown in Fig. 247. Sometimes they are made in the form of a thin long needle, as shown in Fig. 239. This is the form in which they are constructed for pointing out the direction of the magnetic meridian, as in compasses. In this form they are also used in many magnetic experiments.

Magnets may be made of soft iron or untempered steel, but they do not retain their magnetism when the exciting cause is removed. Such magnets are called *temporary magnets.*

Distribution of Force in Magnets.

346. The force with which a magnet attracts iron, is not the same in all of its parts. The attraction is strongest at its extremities, from which it decreases towards its middle, where it is nothing.

This may be shown by plunging one end of a magnetized bar into iron filings; on withdrawing it, the filings will be seen adhering to it in long filaments, as shown in Fig. 237.

If the entire bar be rolled in the filings, it will be found that they adhere to both ends, but not to the middle.

What is an artificial magnet? What is steel? *Describe an artificial magnet. What are temporary magnets?* (**346.**) Where is the attraction strongest? *How shown?*

The two ends, where the attraction is strongest, are called *poles*, and the central part, where the attraction is nothing, is called the *equator*, or the *neutral line*.

Fig. 237.

Every magnet has two poles and one neutral line, whether the magnet be natural or artificial. Sometimes, besides the two principal poles, there are other minor poles, called *secondary poles*. In artificial magnets these arise from inequality of temper in the steel bars, or from want of proper care in magnetizing them. We shall suppose each magnet to have but two poles.

The action of a magnet upon iron takes place through intermediate bodies. If a magnetized bar be covered with a sheet of paper, and

What are poles? Equator? *What are secondary poles? Their origin? How is it shown that magnetism is exerted through intermediate bodies?*

then fine iron filings be sifted uniformly over the paper, they will be seen arranging themselves in regular curves around each pole, as shown in Fig. 238. No action is observed about the neutral line, the filings falling there as on any other surface.

Fig. 238.

Hypothesis of two Magnetic Fluids.

347. If we compare the action of the two poles upon soft iron, we observe the same phenomena at both. It is not so, however, when we compare the action of two magnets upon each other. If to the same pole of a magnetic needle, *ab*, balanced on a pivot (Fig. 239), we present in succession the two poles of a magnetized bar, held in the hand, we observe the curious phenomena, that if the pole, *a*, of the needle is attracted by the pole, *B*, of the bar, the pole, *b*, will be repelled by it; if the pole, *a*, is repelled, the pole, *b*, will be attracted.

To explain these phenomena, it has been supposed that there are two *magnetic fluids*, that is to say, two kinds of

(347.) What is the action of one magnet upon another? What is the theory of two fluids?

subtile matter surrounding the molecules of the magnet, each fluid repelling its own kind, and attracting the other kind.

According to this hypothesis, a body is magnetized when these fluids are separated and driven to its opposite extremities. The difference of the two poles arises from the nature

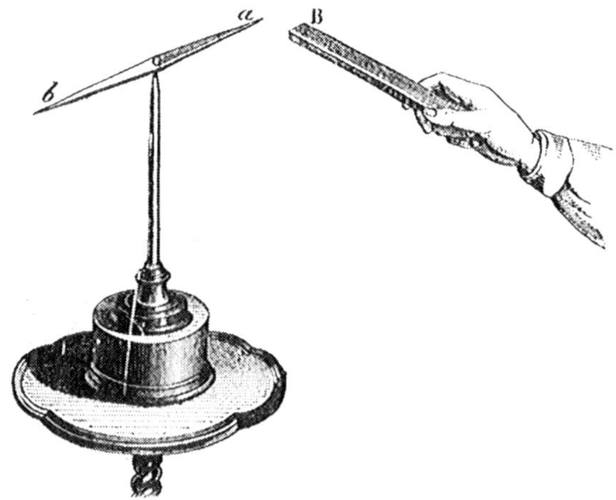

Fig. 239.

of the fluids which predominate in them; the poles which contain the same kind of fluid, *repel*, those which contain opposite kinds, *attract each other*. The attraction and re pulsion are mutual.

Another theory supposes but one kind of magnetic fluid, and explains the phenomena by supposing this to exist in excess at one pole, and in defect at the opposite pole. Either theory explains the phenomena, but that of two fluids is the most easily applied, and for that reason, *solely*, it is adopted.

The earth, as we shall see hereafter, resembles a huge magnet, acting upon magnetic needles in the same way that magnetized bars

When is a body magnetized according to this theory? *What other theory is there? Describe the magnetic action of the earth.*

do. Its magnetic poles are near the geographic poles of the earth, and the neutral line coincides very nearly with the equator. Consequently the fluid which is supposed to predominate near the north pole of the earth is called the *boreal fluid*, and that which is supposed to predominate near the south pole of the earth is called the *austral fluid*.

Because dissimilar poles attract and similar ones repel, it follows that the pole of a balanced magnetic needle which turns towards the north must contain the *austral fluid*, whilst the one which turns towards the south must contain the *boreal fluid*.

Laws of Attraction and Repulsion.

348. The following laws have been suggested by theory and confirmed by experiment:

1. *Magnetic poles of contrary names attract, and those of the same name repel each other.*

2. *The forces of attraction and repulsion both vary inversely as the square of the distance between the attracting and repelling poles.*

Magnetic and Magnetized Bodies.

349. A MAGNETIC BODY is one which contains the two magnetic fluids, but in a state of equilibrium, that is, balancing each other; thus, iron, steel, nickel, and cobalt, are such bodies.

MAGNETIZED BODIES also contain the two fluids, but the difference between them and magnetic bodies is, that in the former the two fluids are separated, each producing an opposite effect, whilst in the latter the fluids are combined and produce no effect. In a word, magnetic bodies are

What is the boreal fluid? The austral fluid? Which turns towards the north? Why? (**348.**) What is the first law of magnetic attraction and repulsion? The second law? (**349.**) What is a Magnetic Body? Examples. What are Magnetized Bodies?

capable of being magnetized, but are not yet magnets; they present neither poles nor neutral line.

When a magnetic substance is brought into contact with one of the poles of the magnet, as the *boreal* pole, for example, the latter, acting by its attraction upon the *austral* fluid, and by its repulsion upon the *boreal* fluid, separates them, giving rise to poles, producing a real magnet.

If a magnetized bar be presented to a magnetic body, as an iron ring, it converts it into a magnet in the manner just described. If a second ring be presented to the first, it is in like manner converted into a magnet, and so on for a third, fourth, &c. The magnets thus formed adhere to each other, as shown in Fig. 240. If the bar be removed, the rings cease to be magnets, the chain falls to pieces, and the rings separate. This mode of exciting magnetic phenomena is called magnetizing by *induction*. According to the theory of two fluids, it is in consequence of this action that a magnet is capable of attracting magnetic bodies. It first acts by induction to convert them into magnets, and then it attracts them according to the laws laid down in the last article.

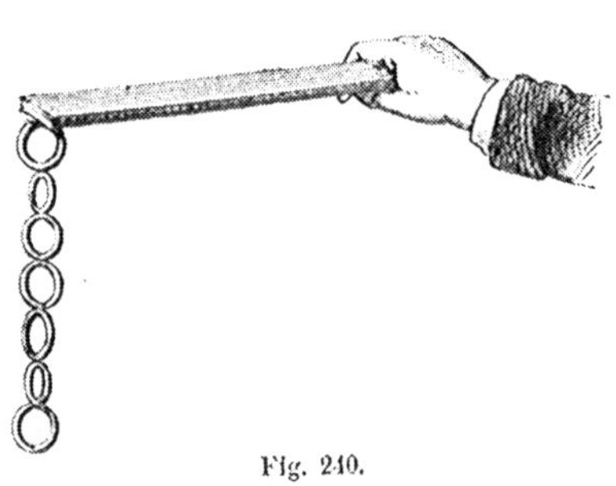

Fig. 240.

Fig. 241 represents a common child's toy. A small swan made of glass has a piece of iron in its head, and on presenting to it a magnet, the swan approaches it, swimming along the surface of the water upon which it is placed. The magnet may be concealed in a

---

How are magnets produced? Illustrate. What is magnetic induction? Explain it on the two fluid theory. *Explain the magnetic swan.*

Fig. 241

piece of bread, in which case the swan seems desirous of feeding upon the bread.

The Coercive Force.

350. The force required to separate the two fluids in a magnetic body is called the Coercive Force.

The fluids are not separable with equal ease in all bodies. In some, as, for example, in soft iron, they yield easily and separate at once; in others, as in hardened steel, for example, the fluids yield with difficulty, and a powerful magnet is required to effect the separation, and it is effected only after a greater or shorter length of time. The harder and better tempered the steel, the more difficult it becomes to separate the two fluids.

(**350.**) What is the Coercive Force? How is it in different bodies?

In soft iron the coercive force required to separate the fluids is very small, in hardened steel it is very great. Soft iron brought in contact with a bar magnet becomes a magnet instantly, and on being removed returns to its neutral condition, ceasing to be a magnet. With hardened steel the reverse is the case; it takes considerable force and some time to render it a magnet, and on being removed from the bar it continues to be a magnet. The force which resisted the separation of the fluids in the first instance, now acts to prevent their reunion, so that the steel magnet retains its magnetism for a long time.

II.—TERRESTRIAL MAGNETISM.—COMPASSES.

Directive Force of Magnets.

351. When a permanent magnet is balanced so that it can turn freely in a horizontal direction, it assumes, after a few oscillations, a determinate direction, which is very nearly north and south.

Fig. 242 shows the manner of balancing a needle, and indicates the north and south direction which it assumes. In this figure, as in all others illustrating the subject of magnetism, the pole which contains the *austral* fluid is designated by the letter *A*, whilst that which contains the *boreal* fluid is designated by the letter *B*.

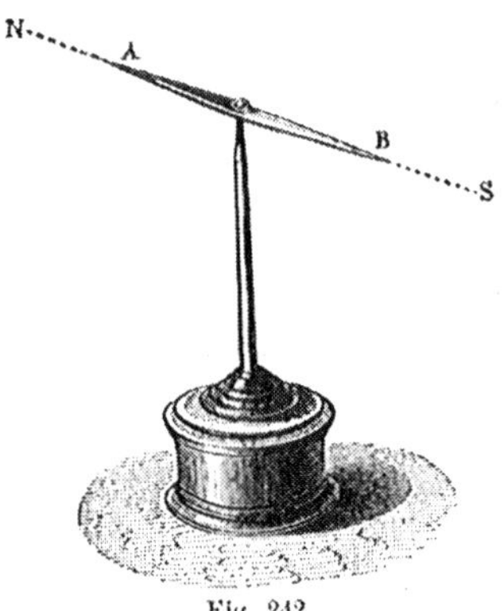

Fig. 242.

It will be noticed that it is the austral pole which turns towards the north, and the

Illustrate. (351.) What direction does a free magnet take? How is a needle balanced? In what other way may it be balanced?

boreal pole which turns towards the south, the reason of which will be seen hereafter.

If, instead of mounting the needle on a pivot, it be attached to a piece of cork and placed in a vessel of water, so that the needle may float in a horizontal position, it will turn itself slowly around and come to rest in the same general direction as though it were balanced on a pivot. In this experiment it will be found that the needle once in the meridian, does not advance either towards the north or south. Hence we infer that the force exerted upon the needle is simply a *directive* one.

The force which causes a movable magnet to direct itself north and south is called the *directive force.*

Since the phenomenon described takes place at all points of the earth's surface, the earth has been regarded as an immense magnet, having its boreal and austral poles near the north and south poles of the earth, and a neutral line near the equator. This immense magnet acting upon the smaller magnets described, would produce all of the effects observed. When we come to explain the action of electric currents, it will be seen that there is another explanation of the *directive* power of the earth.

Magnetic Meridian. — Declination. — Variations.

352. When a balanced magnetic needle comes to a state of rest, it points out the line of magnetic north and south. If a plane be passed through the needle in this position and the centre of the earth, it is called the plane of the *magnetic meridian*, or simply the magnetic meridian.

This does not, in general, coincide with the plane of the true meridian, which is determined by a plane passing through the place and the axis of the earth. The angle which the magnetic meridian at any place makes with the

---

How is it shown that the magnetic force is simply directive? What is the directive force? *Why has the earth been regarded as a magnet? Where are its poles?* (352) What is the magnetic meridian? What is the declination of the needle?

true meridian of the same place is called the *declination of the needle.* In short, the declination of the needle is its variation from true north and south. This is different at different places on the earth, and even at the same place at different times.

When the north end of the needle points to the east of true north, the declination is said to be *to the east;* when to the west of true north the declination is said to be *to the west.*

There is a line running from near Cleveland, Ohio, to Charleston, S. C., along which the needle points to the true north; this is called a line of *no declination.*

The line of no declination is travelling slowly to the westward at a rate which would carry it around the globe in about 1000 years. For all points of the United States east of the line of no declination, the declination of the needle is to the west; for all points to the west of it, the declination is to the east; that is, the north end of the needle in all cases is inclined towards the line of *no declination.*

For all points in the United States to the east of the line of no declination, the declination is slowly increasing, whilst for all points to the west of it, the declination is slowly decreasing.

Besides this slow change in declination, the needle undergoes slight changes, some of which are pretty regular and others very irregular. In our latitude the north end of the needle moves towards the west during the early part of every day, through an angle of 10 or 15 minutes, and moves back again during the latter part of the day. This is called the *diurnal variation.* In the southern hemisphere this motion is reversed. There is also a small change of similar character which takes place every year, called the *annual* variation.

When is it to the east? To the west? What is the line of no declination? *How does this line move? At what rate? Where is the declination to the west? To the east? How does the declination vary in the United States?* What is the diurnal variation? The annual variation?

Irregular changes are called *perturbations*. They usually take place during thunder storms, during the appearance of the aurora borealis, and in general, when there is any sudden change in the electrical condition of the atmosphere.

The Compass.

353. The property possessed by magnets of arranging themselves in the magnetic meridian has been utilized in the construction of COMPASSES.

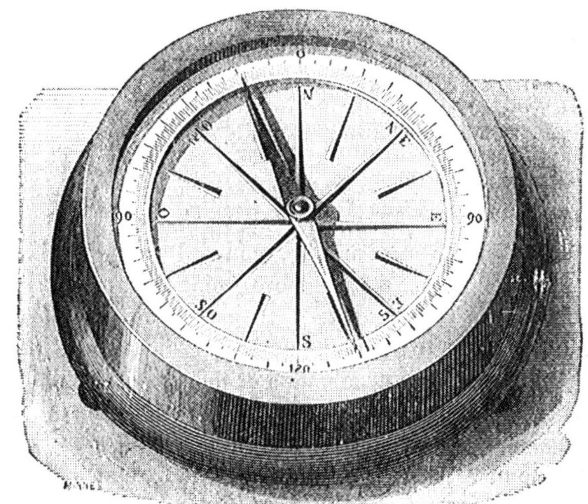

Fig. 243.

Fig. 243 represents a compass. It consists of a compass-box, having a pivot at its centre, on which is poised a delicate magnetic needle. Around the rim of the box is a graduated circle, whose diameter is somewhat less than the length of the needle, and of which the pin is the centre. The pin is of hard steel, carefully pointed; a piece of hard stone is let

What are perturbations? Illustrate. (**353.**) What is a Compass? Describe it.

into the needle, in which is a conical hole to rest upon the pivot, to diminish the friction between the needle and its support. In addition to the graduation on the circle, the bottom of the box is divided into sixteen equal parts, indicating the *points of the compass.*

This instrument under various forms is used for a great variety of purposes. It is used in navigation, in surveying, and is of importance to the traveller and explorer, to say nothing of its use in mining.

The magnetic declination at any place may easily be found when the true meridian is known. Let the compass be so placed that the line, *NS*, coincides with the true meridian, then when the needle comes to rest, the reading under the head of the needle will be the declination required. In the figure, if we suppose *NS* to be in the true meridian, the declination is 19° west.

The Dipping Needle.

354. When a steel needle, mounted as shown in Fig. 242, is carefully balanced before being magnetized, it is found, after being magnetized, to incline downwards or to *dip*. This dip is towards the north in our latitude, that is, the north end of the needle dips or inclines. The defect of dipping in the compass is remedied by making the other end of the needle a little heavier, by adding a movable weight, as a piece of wire wound round the needle, and capable of sliding along it.

To show the dip and to measure it, the needle is mounted in the way indicated in Fig. 244. The needle is suspended on a horizontal axis, so that it can move up and down freely, and the amount of the dip is indicated by a graduated circle or quadrant. The dip indicated in the figure is 54°, which is the angle made by the needle with

What is its use? How is the magnetic declination found at any place? (354.) What is a dipping needle? How is the compass needle prevented from dipping? How is the dip shown and measured?

the horizon. At any place the dip will be the greatest possible when the needle vibrates in the plane of the magnetic meridian.

Fig. 244.

The dip varies in passing from place to place, increasing as we approach the magnetic poles of the earth, where the dip is 90°; that is, the needle is perpendicular to the horizon.

The dip is subject to irregularities corresponding to those of the declination. The amount of the dip is an important element in forming a correct notion of the laws of terrestrial magnetism, and for this reason many observations have been made and are still making, to determine it at different places, and at different times at the same place.

III.—METHODS OF IMPARTING MAGNETISM.

Magnetizing by Terrestrial Induction.

355. To MAGNETIZE a body is to impart to it the properties of a magnet; that is, to impart to it the property of attracting magnetic bodies.

The only substances that can be permanently magnetized, are steel and the compound oxide of iron, which constitutes the loadstone. A body capable of being magnetized may be converted into a magnet by the inductive influence of

---

How does the dip vary? Is it subject to irregularities? (**355.**) What is meant by magnetizing a body? What substances can be permanently magnetized?

the earth, or more rapidly by being rubbed by another magnet, or finally, by the action of electricity, in which case the operation is instantaneous.

The magnetic ores of iron may exist as magnets in the natural state, or they may possess no trace of magnetic action. But they are highly susceptible to magnetic influence, and once magnetized, they retain their magnetic action by virtue of their strong coercive force.

Natural magnets owe their magnetism to the slow action of the earth, which separates the two fluids in them. The magnetic action of the earth is so great as to be used successfully in forming artificial magnets.

To use this principle, we place a thin bar of iron in the magnetic meridian and incline it to the horizon by an angle equal to the dip. In this position the earth acts upon it by induction, driving the austral fluid to the lower end (in our latitude), and the boreal fluid to the upper end.

The magnetism thus induced is only temporary, for if the bar be moved from its position, the two fluids return to a state of equilibrium. If, however, when the bar is in position, it be struck smartly by a hammer, or if it be violently twisted, sufficient coercive force may be developed to retain the induced magnetism for a time.

Magnetizing by Friction.

356. Bars of steel, and needles for compasses, are usually magnetized by rubbing them with other magnets. The three methods are called the methods by *single touch*, by *separate touch*, and by *double touch*.

To magnetize a steel bar by *single touch*, we hold the

Are the magnetic ores of iron always magnets? To what is the natural magnetization of these ores due? How are bars magnetized by this principle? (356.) How may bars of steel be magnetized? Explain the method of single touch.

body to be magnetized in one hand, and with the other we pass over it a powerful bar magnet, as shown in Fig. 245. After several repetitions of this process, the steel is found to possess all the properties of a magnet. These properties

Fig. 245.

are the more durable in proportion to the hardness of the steel.

To magnetize a steel bar by *separate touch*, we rub it in one direction with one pole of a magnetized bar, and in the opposite direction with the opposite pole.

To magnetize a body by *double touch*, we make use of two magnetized bars, which are placed with their opposite poles in contact with the bar at its middle point, being only separated by a small interval, as shown in Fig. 246; the combined bars are then moved alternately in opposite directions to the two ends of the bar, and the operation is repeated several times. Care must be taken to apply the same number of touches to each end of the bar.

Of separate touch. Of double touch.

Fig. 246.

The method of magnetizing by electricity will be treated of under the head of electrical currents.

Bundles of Magnets.—Armatures.

357. A Bundle of Magnets consists of a group of magnetized bars united, so that their poles of the same name may be coincident.

Sometimes these bundles are composed of straight bars, like that shown in Fig. 245, and sometimes they are curved in the shape of a horse-shoe, as shown in Fig. 247.

Magnets, if abandoned to themselves, would lose in a short time much of their power; hence it is, that *armatures* are employed.

An Armature is a piece of soft iron, placed in contact with the poles of a magnet. Thus, *ab*, in Fig. 247, is an *armature*.

(357.) What is a Bundle of Magnets? What is an Armature?

The poles, acting by induction upon the armature, convert *a* into an austral, and *b* into a boreal pole. These two poles reacting upon the poles of the magnet, *AB*, prevent the recomposition of the two fluids, and thus preserve its magnetism. The armature is sometimes called *a keeper*.

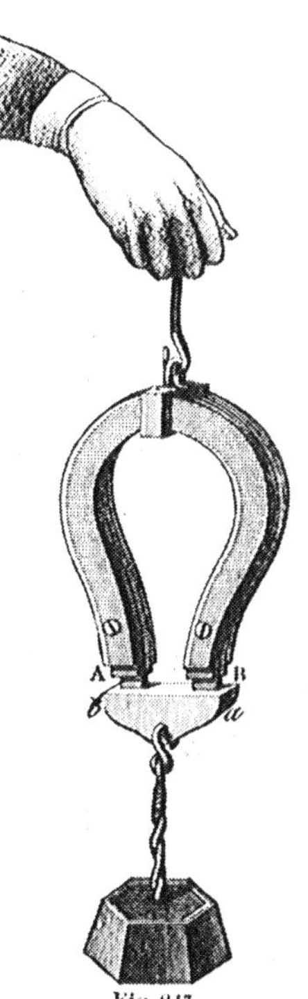

Fig. 247.

If weights be attached to the keeper till it separates from the magnet, we can, from the number of pounds applied, judge of the power of the magnet.

For many kinds of magnetic experiment the horse-shoe form is preferable. It is also the form best adapted to the application of an armature or keeper.

The most powerful horse-shoe magnets are formed by means of electrical currents. Magnets of this kind have been constructed by Prof. Henry, of the Smithsonian Institution, capable of sustaining a weight of more than a ton and a quarter.

A keeper? *How can we judge of the power of a magnet? What are the advantages of the horse-shoe magnet?*

CHAPTER VIII.

STATICAL ELECTRICITY.

I.—FUNDAMENTAL PRINCIPLES.

Definition of Electricity.

358. ELECTRICITY, as a science, is that branch of Physics which treats of the laws of attraction and repulsion exhibited by bodies under certain circumstances. Such phenomena are called *electrical phenomena*. The name electricity is derived from the Greek *elektron*, which means amber.

Discovery of Electrical Properties.

359. Six hundred years before the commencement of the Christian era, THALES, of Miletus, knew that when yellow amber was vigorously rubbed with wool, it acquired the property of attracting light bodies, such as small pieces of paper, barbs of quills, straws and the like. Comparing this action to suction, the ancients said that amber had a power of suction, and sucked light bodies towards it. In consequence of the rarity of amber, whose origin is even in our day unknown, they went so far as to say, that it was formed from the tears of an Indian bird, grieved at the death of King MELEAGER.

Six centuries later, PLINY, an eminent Roman naturalist, writes: "When the friction of the fingers imparts heat and life to yellow amber, it attracts straws, just as the magnet attracts iron." This

(**358.**) Define Electricity as a science. What are electrical phenomena? Whence the name? (**359.**) *Give an outline of the history of electrical discoveries.*

was all of the knowledge had on the subject until the end of the sixteenth century, when WILLIAM GILBERT, an Englishman, called anew the attention of scientific men to the properties of amber, and showed that a great number of other substances, such as glass, resin, silk, sulphur, and the like, acquired the power of attracting light bodies, on being rubbed with woolen cloth or cat's skin.

To repeat these experiments, rub a tube of glass or a stick of sealing-wax with a piece of woolen cloth, then present them to light bodies, as shreds of gold leaf, barbs of quills, or fragments of paper, and the latter will be seen to approach and adhere to the excited glass or sealing-wax. The manner of making these experiments is indicated in Fig. 248.

It will be seen hereafter, that resin and other substances named above, not only develop forces of attraction when rubbed, but also they become luminous, emit sparks, and display a number of other properties, all of which are known as electrical phenomena.

Fig. 248.

Since the beginning of the seventeenth century the progress of discovery in electricity has been rapid, and a multitude of new facts have been developed, which have been so well studied as to form a very extensive branch of natural science.

Sources of Electricity.

360. The sources of electricity may be divided into three classes: *Mechanical*, *Physical*, and *Chemical*.

The mechanical sources are: *friction*, *pressure*, and *separation of the molecules of bodies*. When a piece of sugar is broken suddenly in a dark room, a feeble light is observable,

---

Explain GILBERT'S *experiments, and the manner of making them.* (**360.**) What are the principal sources of electricity? What are the mechanical sources?

which is due to the development of electricity at the moment of separating the molecules.

The physical sources are *variations of temperature*, and the like. Some minerals, particularly tourmaline and topaz, manifest electrical phenomena on being heated or cooled.

The chemical sources are *chemical compositions* and *decompositions of bodies*. Metals, like zinc, iron, and copper, when plunged into acids, are attacked by them, forming compounds known as salts. During these combinations considerable quantities of electricity are developed.

The most powerful of the causes of electricity are *friction* and *chemical action*. These will be studied in their order.

Electroscope.—Electrical Pendulum.

361. An ELECTROSCOPE is an apparatus for showing when a body is electrified.

The most simple electroscope is the ELECTRICAL PENDULUM, which consists of a small ball of elder pith, suspended by a fine silk thread, as shown in Fig. 249. The thread is fastened to the upper end of a stem of copper, which stem has a support of glass.

To ascertain whether a body is electrified or not, the pendulum is presented to it; if it is electrified, the pith ball will be attracted, otherwise not. When the quantity of electricity is too small to produce sensible attraction upon the pith ball, more delicate instruments are sometimes employed, called *electrometers*.

Two kinds of Electricity.

362. That there are two kinds of electricity, may be shown by the action of glass and resinous bodies, after being rubbed, upon pith balls.

What is the chief physical source? The chemical sources? *What is the most powerful cause of electricity?* (**361.**) What is an Electroscope? Describe the Electrical Pendulum. How used? (**362.**) How many kinds of electricity are there?

If a tube of glass be rubbed with a piece of cloth, and then presented to the electrical pendulum, the pith ball will at first be attracted, and after a short time it will be repelled, as shown in Fig. 250. The ball is then charged with the same kind of electricity as that in the glass.

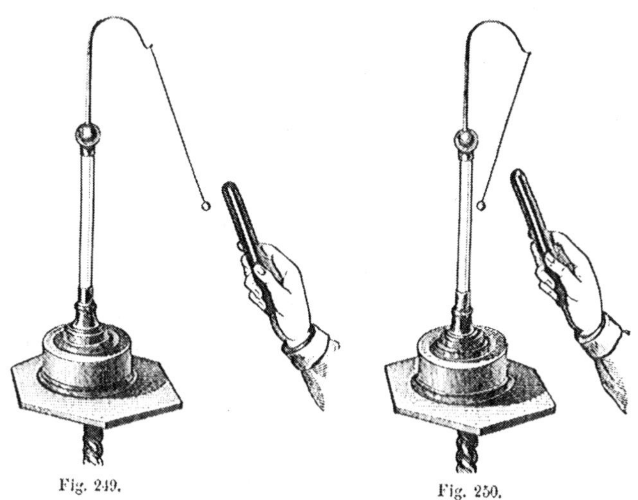

Fig. 249. Fig. 250.

If now a piece of a resinous body, as sealing-wax, be rubbed with cloth and brought near the excited pith ball, the latter is immediately attracted to the former. In like manner, if the sealing-wax be first presented to the pendulum, it will be attracted and then repelled. If then the glass be brought near the pith ball, attraction will be observed. This shows that the action of electricity, as developed in glass and resin, is different, the one repelling when the other attracts. This fact was discovered by DUFAY, in 1734.

The electricity developed in rubbing glass with a piece

How is it shown that there are two kinds?

of silk, has been named *vitreous electricity*, that developed by rubbing resin or sealing-wax with the silk, has been named *resinous electricity*.

Hypothesis of two Electrical Fluids.

363. The discovery of Dufay gave rise to the theory of two electrical fluids, which in unexcited bodies exist in a state of combination, forming what is called a *neutral fluid*. The earth is regarded as a great reservoir of this fluid, which has of itself no obvious properties; hence bodies which only contain it are said to be neutral. If by friction, chemical action, or other cause, the neutral fluid is decomposed, and the two fluids separated, electrical phenomena are at once developed.

These two fluids were at first named *the vitreous*, and *the resinous* fluids, but more recently they have been called *the positive*, and *the negative* fluids; the vitreous being called positive, and the resinous negative. These names were adopted by Franklin the better to express their opposite characters. The positive fluid is often indicated by this sign, +, (*plus*), and the negative fluid by this sign, −, (*minus*.)

The hypothesis of two fluids was first made by Symner, and according to it the development of electricity consists in separating the two fluids. When glass is rubbed with silk, the positive fluid of the two goes to the glass, whilst the negative fluid goes to the silk. When sealing-wax is rubbed with silk, the reverse is the case, the negative fluid goes to the resinous body and the positive fluid to the silk.

It is to be observed that all of the phenomena can be equally well explained by the theory of a single fluid. This is the theory of Franklin, and if we adhere to the hypothesis of two fluids, it is simply because it is more easily applied than that of one fluid.

What are they called? (**363**.) What is the neutral fluid? When are electrical phenomena produced? What other names are given to the two fluids? How are they indicated? *Explain in detail the two fluid hypothesis. Who is the author of the one fluid hypothesis?*

Laws of Electrical Attraction and Repulsion.

364. The following laws have been deduced from theory, and confirmed by experiment:

1. *Fluids of the same name repel each other; fluids of opposite names attract each other.*

2. *The intensities of the attractions and repulsions vary inversely as the square of the distances between them.*

Conductors. — Insulators.

365. CONDUCTORS, or *conducting substances*, are those which permit electricity to pass through them.

INSULATORS, or *non-conducting substances*, are those which do not permit electricity to pass through them.

GRAY observed that electrified bodies returned instantly to a neutral state when brought into contact with the earth, or when placed upon supports of metal, wood, stone, or any moist substance whatever. He also observed that they remained in an electrified condition for a long time when placed upon supports of glass, resin, sulphur, or when suspended by silken cords. From these facts, he concluded that metals, wood, stone, and the like, permitted the electricity to pass freely through them, whilst glass, resin, sulphur, and the like, opposed its passage. He also inferred that the latter class of bodies was not entirely incapable of conducting electricity, but that they were extremely poor conductors. When an electrified body is surrounded by non-conductors it is said to be *insulated*, and any non-conducting support of an electrified body is therefore called an *insulator*.

The best conductors of electricity are the metals; after these come plumbago, well calcined carbon, acid and saline

(364.) What is the first law of attraction and repulsion? The second law? (365.) What are Conductors? Insulators or non-conductors? *What observations were made by* GRAY? *When is a body insulated?* What are the best conductors? Next in order?

solutions, water either in a liquid or vaporous form, the human body or animal tissues, vegetable substances, and in general, all moist or humid substances.

The worst conductors, or best non-conductors are resins, gums, india-rubber, silk, glass, precious stones, spirits of turpentine, oils, air, and gases when perfectly dry.

Methods of Electrifying Bodies.

366. Non-conducting bodies are electrified only by friction, but conductors may be electrified either by friction, by contact, or by induction.

In order to electrify a metal it must be insulated; that is, it must be surrounded by non-conducting bodies, and it must be rubbed by an insulated body.

This may be effected by mounting the metal upon a stand of glass and rubbing it with a non-conductor, such as a piece of silk. Were the metal not insulated, the electricity would flow off to the earth as fast as generated, and were the rubbing body not a non-conductor, the electricity would flow off through the hands and arms of the experimenter.

The method of electrifying by contact depends upon the property of conductibility. If a conductor is brought in contact with an electrified body, a portion of the electricity of the latter at once flows into the former body. If the two bodies are exactly alike, the electricity will be equally distributed over both. If they differ in size or in shape, the electricity will not be equally distributed over both.

The method of electrifying bodies by induction is similar to that of magnetizing bodies by induction, and will be treated of hereafter.

The worst conductors? (366.) How are non-conductors electrified? Can conductors be electrified by friction? *How?* How are bodies electrified by contact?

Accumulation of Electricity on the Surface of Bodies.

367. Experiment shows that when a body is electrified, the electricity all goes to the surface of the body, where it exists in a thin layer, tending continually to escape. It actually does escape as soon as it finds an outlet through a conducting body.

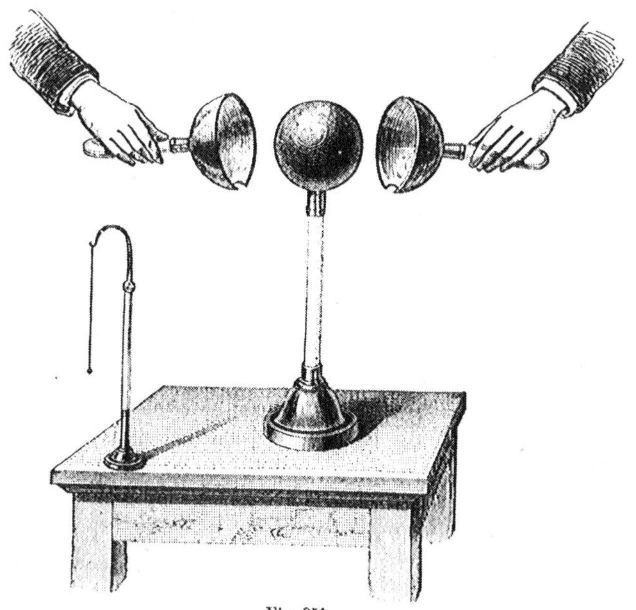

Fig. 251.

Of the various experiments intended to show this fact, we select one that was first performed by COULOMB. He mounted a copper sphere upon an insulating rod of glass, as shown in Fig. 251. He then provided two hollow hemispheres also of copper, which, when put together, exactly

(367.) Where is the electricity of a body found? Explain COULOMB's experiment.

fitted the first sphere, and these he insulated by attaching them to glass handles. Having placed the hemispheres so as to cover the solid sphere, he brought the whole apparatus in contact with an electrified body till it was fully charged.

On removing the apparatus from the electrified body, he separated the two hemispheres abruptly, and applied to each in turn the electrical pendulum, when he found that both were electrified. On testing the solid sphere in like manner, he could discover no trace of electricity; in other words, it was perfectly neutral.

In taking away from the body its outer coating, he had removed every particle of its electricity, which proved that the electricity was entirely upon the surface.

Another fact which indicates the same conclusion is, that a hollow and a solid sphere of the same size and of the same material, will be charged with exactly the same quantity of electricity when made to communicate with the same electrical source.

When the electric fluid is accumulated upon the surface of a body, it tends to escape with a certain force, which is named the *tension.*

The tension augments with the quantity of electricity accumulated. So long as it does not pass a certain limit, it is held by the resistance of the air, but if the tension passes this limit, the electricity escapes with a crackling noise and a brilliant light called the *electric spark.* In moist air the tension is always feeble, because the electricity is slowly conveyed away by the moisture. In a vacuum, there is no resistance to the escape of electricity, and the tension is nothing. The electricity in this case flows off as fast as generated, with a feeble light.

Influence of the Forms of Bodies.—Power of Points.

368. The distribution of electricity over the surfaces of bodies depends upon their form. If a body is spherical, the

What fact confirms COULOMB'S *conclusion?* What is the tension? *What is the electric spark? Why is the tension feeble in moist air? In a vacuum?* (**368.**) What effect has the form of a body?

fluid is equally distributed, as may be shown by an instrument called a *proof-plane.*

The proof-plane consists of a disk of gilt paper attached to the end of a rod of gumlac, which insulates well. Taking the rod in the hand as shown in Fig. 252, it is applied successively at different points of the electrified surface, and after each contact it is presented to the electrical pendulum.

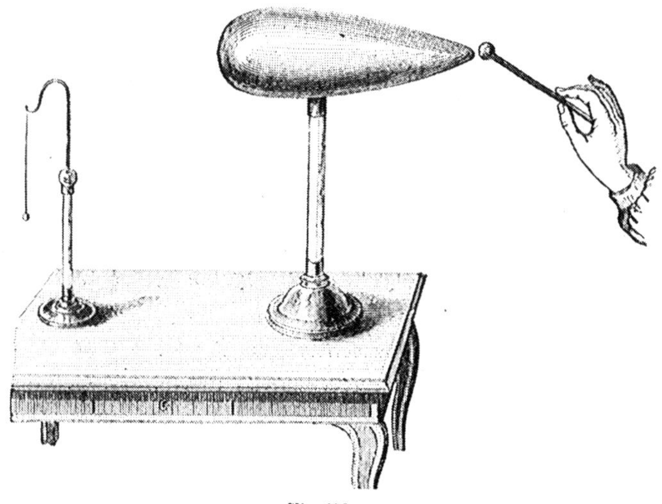

Fig. 252.

If the electrified body is a sphere, the same amount of attraction for the pith ball is shown, wherever the contact may be made; this shows that the proof-plane is equally charged at every point of the sphere, and consequently it is inferred that the distribution is uniform over the whole surface.

When the body is elongated and pointed, as in Fig. 252, different results are obtained. In this case the proof-plane

---

What is a proof-plane? How used?

is more highly charged at the sharp end of the body than at any other point, showing a larger amount of electricity at the point than elsewhere. In general, it may be shown that the greater the curvature of a surface at any part, that is, the nearer it approaches a point, the greater will be the accumulation of electricity there. This shows that electricity tends to accumulate at, or to flow towards the pointed portions of bodies.

The accumulation of electricity at points gives rise to a high tension, which is sufficient to overcome the resistance of the air and to give rise to an escaping current. In fact, metallic bodies of a pointed shape soon lose the electricity imparted to them. and often the escaping current may be felt by placing the hand in front of the point. If the flow takes place in a darkened room, it may be discovered by a feathery jet of faint light.

The property of points, or the power of points, as it is called, was noticed by FRANKLIN and made use of by him in his theory of lightning-rods.

II.—PRINCIPLE OF INDUCTION.—ELECTRICAL MACHINES.

Induction.

369. If an insulated conductor in a neutral state is brought near an electrified body, the fluid of the latter acting upon that of the former, decomposes it, repelling the fluid of the same name, and attracting that of a contrary name. This operation is called INDUCTION, and it may take place not only at considerable distances, but also through non-conducting bodies, such as air, glass, and the like.

The method of electrifying bodies by induction is shown in Fig. 253. On the right of the figure is the prime conductor of an electrical machine, which, as we shall see hereafter, is charged with the positive fluid. On the left is a

What effect has a pointed form? *Discuss the power of points.* (**369.**) What is Induction?

metallic cylinder with spherical ends, and supported by a rod of glass. Attached to its lower surface, at intervals, are pairs of pith ball pendulums, supported by threads of some conducting substance.

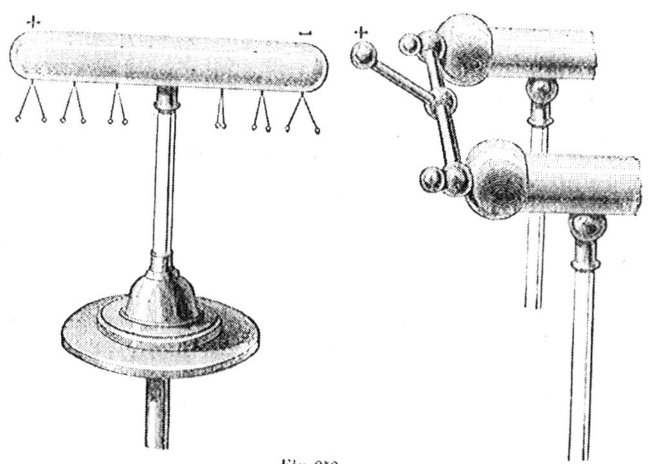

Fig. 253.

When the cylinder is brought slowly towards the electrical machine, we see the pith balls repel each other and diverge. This divergence is unequal at different points, being greatest near the extremities of the cylinder; towards the middle of the cylinder the pith balls remain in contact without repelling each other. We conclude from these facts that the fluids are driven towards the extremities of the cylinder, whilst the central portion remains in a neutral state.

If a stick of resin be rubbed with silk and brought near the pith balls towards the electrical machine, they will be repelled, showing that that end of the cylinder is negatively electrified. If it is brought near the pith balls at the remote

How is an insulated body affected by induction? Explain the phenomena in detail.

extremity of the cylinder, they are attracted, showing that that end of the cylinder is positively electrified. Finally, the electricities in the two ends are equal in quantity, as may be shown by removing the cylinder, when they flow together and neutralize each other.

The positive electricity of the machine, then, simply acts to separate the two fluids, attracting the negative fluid to the end nearest it and repelling the positive fluid to the opposite end of the cylinder. No electricity passes from the machine to the cylinder.

If, whilst the apparatus is in the position shown in the figure, the two electricities being separated, the positive end be touched by a conductor, as the finger, for example, all of the positive fluid will escape, whilst the negative fluid, being held by the attraction of the positive fluid in the machine, remains on the surface of the cylinder. The cylinder is thus charged with negative electricity throughout, as may be shown by applying a rod of electrified glass or resin to the pith balls at the two extremities. Furthermore, it is immaterial whereabouts the cylinder is touched by the conductor, for if touched at any point, the positive fluid escapes, and the cylinder is charged with the negative fluid.

Had the inducing body been charged negatively, the cylinder would in like manner have received a positive charge by induction.

The method of induction is of frequent application in experimental inquiries, and the principle set forth above serves to explain a great variety of electrical phenomena.

The Electrical Machine.

370. The ELECTRICAL MACHINE is a machine by means of which an unlimited amount of electricity may be generated by friction.

How does the positive electricity act? How is the negative fluid drawn off? Had the inducing body been negatively electrified, what would have happened? (**370.**) What is an Electrical Machine?

This machine was invented about two hundred years ago by OTTO VON GUERICKE, the distinguished inventor of the air-pump. The first machine was simply a ball of sulphur fixed upon a wooden axis. On turning the axis, and at the same time pressing one hand

Fig. 254.

against the ball, a quantity of frictional electricity was developed. After various improvements the machine has taken the form shown in Fig. 254. which is the form that is now most generally employed in physical researches.

When invented and by whom?

The principal piece of the machine is a plate of glass, *PP*, three feet or more in diameter. This plate is mounted upon a horizontal axis, and may be turned upon this axis by means of a crank. The wooden frame which supports the axis embraces the plate, and bears four cushions which press against the glass on its opposite faces, two above and two below the axis. The cushions are of leather stuffed with hair; by their friction they give rise to positive electricity in the plate when it is turned.

Two cylinders of brass, *AA*, are mounted on the table which supports the frame-work, and are insulated by glass pillars. These cylinders, called conductors, are united at their remote ends by a third cylinder of brass, as shown in the figure. At their ends nearest the plate they terminate in cylindrical pieces constructed so as to partially embrace the plate but not touch it. These pieces are called combs, from the fact that a great number of projecting teeth are placed on their sides next the plate. Finally, all of the ends of the cylinders in the machine are wrought into spherical forms, to prevent the dissipation of electricity as much as possible. The entire collection of metallic cylinders is called the *prime conductor*.

Use of the Electrical Machine.

371. When the plate is turned rapidly, the friction of the cushions or rubbers develops a great quantity of positive electricity on the glass, whilst the negative fluid goes to the rubbers and is conveyed through the frame to the earth, and thus disappears. The neutral fluid on the conductors is decomposed; the negative fluid flows through the teeth of the combs to the glass plate, tending to neutralize the positive fluid on the plate. The conductors thus

What is the principal piece? How mounted? Describe the cushions. Describe the conductors and the method in which they are mounted. How are they electrified? What is the prime conductor? (**371.**) Explain the operation of the machine.

losing the negative fluid, become charged with positive electricity.

The electrical machine may be arranged to produce negative electricity as follows: The feet of the table are insulated by being placed upon glass supports, and the prime conductor is then connected with the earth by a metallic chain. This chain permits the positive electricity to flow out of the prime conductor, whilst the negative electricity, being unable to escape, accumulates upon the cushions, table, and frame of the instrument.

Measure of the Quantity of Electricity in the Machine.

372. The quantity of electricity in the prime conductor may be shown by an instrument called HENLEY'S electrometer.

This electrometer is represented on the left of the drawing of the machine, and consists of a vertical support of wood, bearing a quadrant divided into degrees. At the centre of the quadrant is attached a small arm of whalebone, turning around an axis and terminating in a pith ball.

When the machine is in operation, the ball rises along the quadrant, and by its divergence from the vertical indicates the quantity of electricity developed.

Precautions in using the Machine.

373. After the prime conductor is electrified, if we cease to turn the plate, and the air is dry, the pith ball will descend slowly, showing a gradual dispersion of the electricity. If the air is damp, the ball descends rapidly, showing a rapid loss of electricity. Electrical experiments seldom succeed in a damp day. In order that they should be successful, the instrument, as well as the surrounding atmosphere, ought to be perfectly dry.

Electricity will be developed more rapidly if the cushions are

---

How may it be arranged to collect negative electricity? (**372.**) How is the quantity of electricity on the prime conductor indicated? Describe the electrometer used. Its action (**373.**) *What effect has dampness on the development of electricity? How is the electricity increased?*

covered with a paste composed of sulphur and tin, or an amalgam of zinc and mercury, or of tin and mercury.

Only a certain amount of electricity can be retained on the prime conductor, after which, if the plate is turned, the tension becomes so great that it escapes through the air or along the glass legs of the conductor, and all that is generated continues thenceforth to be dissipated. The electrometer indicates that the instrument is fully charged, by ceasing to rise, and remaining stationary as the plate is turned.

Finally, in order to attain the best possible results, the machine should not be placed too near the walls, or the furniture of a room, or any thing upon which it can act by induction. In particular all angular objects should be avoided. The prime conductor tends to abstract from surrounding objects their negative electricity, and to return to its neutral condition.

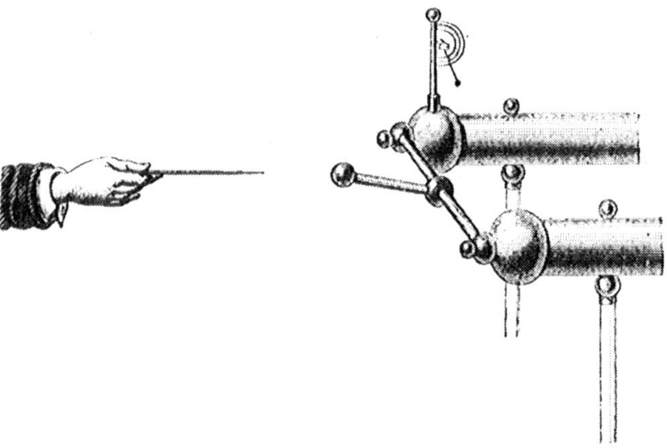

Fig. 255.

The effect of neighboring bodies may be illustrated by bringing a metallic point near a charged prime conductor, as shown in Fig. 255. When the point is at a considerable distance from the conductor, the electrometer begins to fall, showing a loss of electricity.

How do we know when the prime conductor is fully charged? What is the effect of neighboring conductors? How illustrated?

This may be explained by supposing negative electricity to flow from the point to the conductor, in accordance with what has been shown before.

It is sometimes said that the point draws off the electricity from the conductor, but this is not the case; the point abstracts none of the positive electricity, but gives to the conductor negative electricity, which unites with the positive fluid to neutralize it.

Electrophorus.

374. The ELECTROPHORUS is a machine due to VOLTA, by means of which we may obtain considerable quantities of electricity.

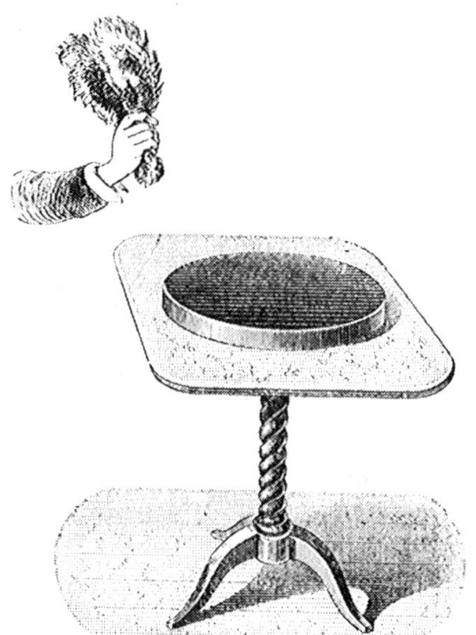

Fig. 256.

It consists of two pieces: one a plate of resin spread on

Effect of the point. (**374.**) What is an Electrophorus? Describe it. How is it used?

a table of wood, and the other a wooden plate, covered with tin foil, and provided with an insulating handle of glass. It is represented in Figs. 256, 257, and 258.

To use this instrument we commence by rubbing the resinous plate vigorously with a cat's skin, as shown in Fig. 256. This develops negative electricity in the resin. We then apply the disk, holding it by its handle. The plate of resin acts upon the disk by induction, drawing the positive fluid to the tin foil on its lower face, and repelling the negative fluid to the foil on the upper face. In this position, if the upper face be touched with the finger, as shown in Fig. 257, the negative fluid will be drawn off into the body, and the disk will be charged with positive electricity.

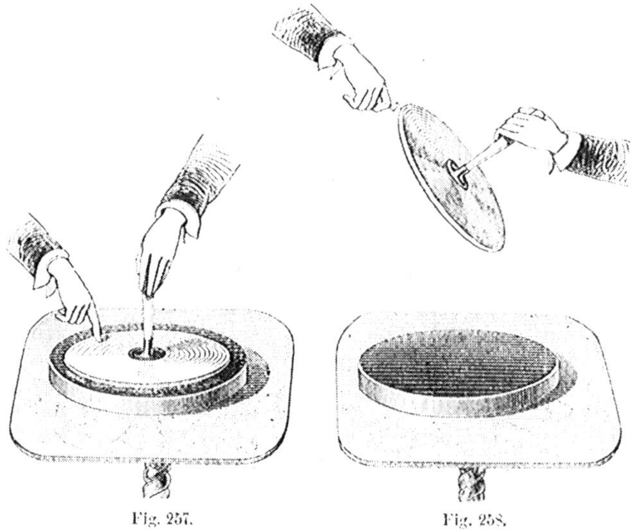

Fig. 257. Fig. 258.

If the disk be raised from the resinous plate by its handle, and touched with the knuckle, as shown in Fig. 258, a spark

Explain its action.

will pass which is due to the negative electricity, passing from the body to the positively electrified plate.

If now we continue to repeat the manipulation, exhibited in Figs. 257 and 258, a succession of sparks may be obtained without the necessity of rubbing the resin again with the cat's skin. If the air is dry, the resin will continue in an electrified state for a very long time.

The principle of continued induction has been applied in constructing electrical machines. That of Holtz, one of the best, is more powerful than the one described in Art. 370.

Gold-leaf Electrometer.

375. The GOLD-LEAF ELECTROMETER is an instrument invented by BENNET, for determining whether a body is

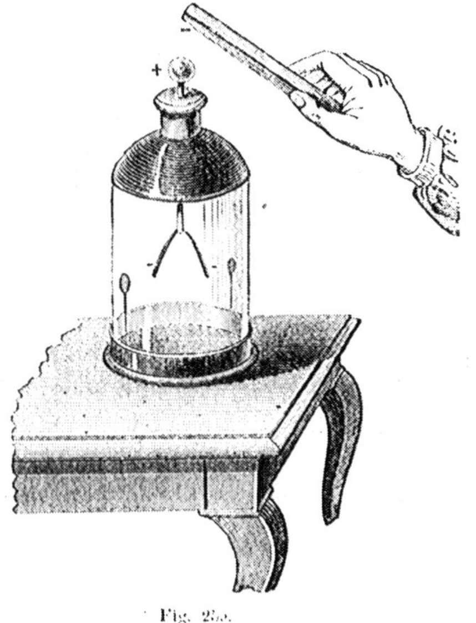

Fig. 259.

How may a succession of sparks be obtained? (**375.**) What is the Gold-leaf Electrometer?

electrified, and to show the kind of electricity with which it is charged.

It consists of a glass bottle, closed at the top by a cork, through which passes a large copper wire. This wire is terminated at its top by a copper ball, and has attached to its lower extremity two slips of gold-leaf. The instrument is represented in Fig. 259.

The cork and the whole top of the bottle are covered with a kind of varnish, made by dissolving sealing-wax in alcohol. The varnish is laid on with a brush, and serves to make the bottle a better non-conductor. This kind of varnish is often used in electrical experiments to render glass non-conducting. Glass in a dry state is a good non-conductor, but it is apt to condense moisture from the air so as to become a conductor. When covered with any resinous varnish, this trouble is removed.

Method of using the Gold-leaf Electrometer.

376. To ascertain whether a body is electrified, we bring the ball of the electrometer near it. If it is electrified, it acts upon the ball and its stem by induction, attracting the fluid of a contrary name into the ball, and repelling that of the same name into the gold leaves, which, being very light and electrified by the same kind of fluid, will diverge. This instrument is very sensitive, showing the slightest amount of electricity.

To test the kind of electricity in a body, bring it near the instrument and touch the ball with the finger. This will draw off the electricity of the same name and leave that of a contrary name to that in the body. Now let a glass rod be rubbed with woollen cloth, so as to excite positive electricity, and then let it touch the ball of the electrometer. If the leaves diverge more, the electricity

Describe it. *Why is the top of the bottle varnished? Describe the varnish.* (376.) How do we ascertain when a body is electrified by this instrument?

in them before was positive, and that of the body in question was consequently negative. If, however, the leaves approach each other, the electricity in them before was negative, and consequently that in the body experimented upon was positive.

This is an exceedingly delicate test, and one of great practical value.

Fig 260.

III.—ELECTRICAL RECREATIONS.

Electrical Spark.—Electrical Shock.

377. An ELECTRICAL SPARK is a brilliant flash of light which passes when a conductor approaches a highly-electrified body.

Is this electrometer of much use? (**377.**) What is an Electrical Spark?

The method of drawing a spark from the prime conductor is shown in Fig. 260. The spark, when received by the human body, is accompanied by a sensation, called an *electrical shock*, which may be very painful and even dangerous.

The spark arises from the combination of the two contrary fluids. The positive fluid acting at a distance by induction, drives the positive fluid of the hand to the earth, and the body of the experimenter becomes negatively electrified. When the tensions of the positive electricity of the machine and the negative electricity of the body, overcome the resistance of the air, they rush together with a sharp crack and a bright light which constitutes the spark. When the electrical machine is powerful, the sparks take a zig-zag course, like lightning from a storm-cloud.

Fig. 261.

What is a shock? *What is the cause of the spark? Explain in detail.*

The Electrical Stool.

378. A spark may be drawn from the human body when properly electrified. For this purpose an ELECTRICAL STOOL, that is, a stool insulated by means of glass legs, is made use of, as shown in Fig. 261. A person standing on the stool, and taking hold of the prime conductor, becomes, when the plate is turned, positively electrified. If a second person now attempts to shake hands with the first, a shock will be experienced, and a spark will pass between them.

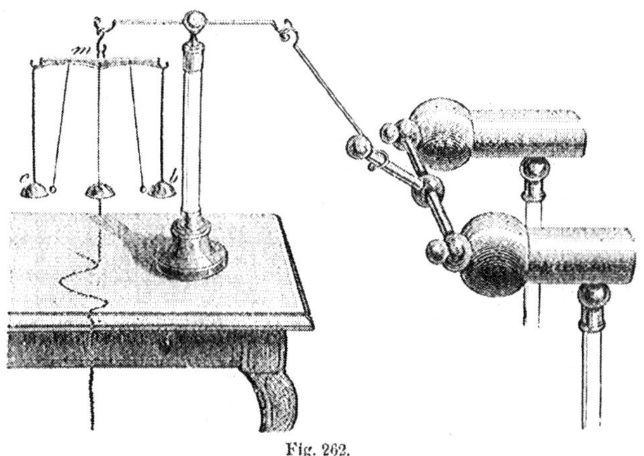

Fig. 262.

The Electrical Chime.

379. The ELECTRICAL CHIME is a collection of bells that are made to ring by means of electrical attractions and repulsions.

It consists, in the case shown in Fig. 262, of three bells suspended from a horizontal bar of wood, *m*. The outer bells, *b* and *c*, are suspended by metallic chains, and the

(**378.**) What is an Electrical Stool? Its use? (**379.**) What is an Electrical Chime? Describe it.

middle one by a silk cord; the middle bell, moreover, is connected with the earth by means of a metallic chain. Between the bells are two balls of metal, suspended from the bar, *m*, by a cord of silk. The entire apparatus is connected with the prime conductor of an electrical machine, as shown in Fig. 262.

When the machine is turned, the outer bells become positively electrified, and attract the balls, which impinge against them, become electrified, and are immediately repelled, striking against the middle bell, where they lose their charge, and are again attracted to the extreme bells, and again repelled. This alternate attraction and repulsion of the balls keep up the ringing as long as the plate is turned.

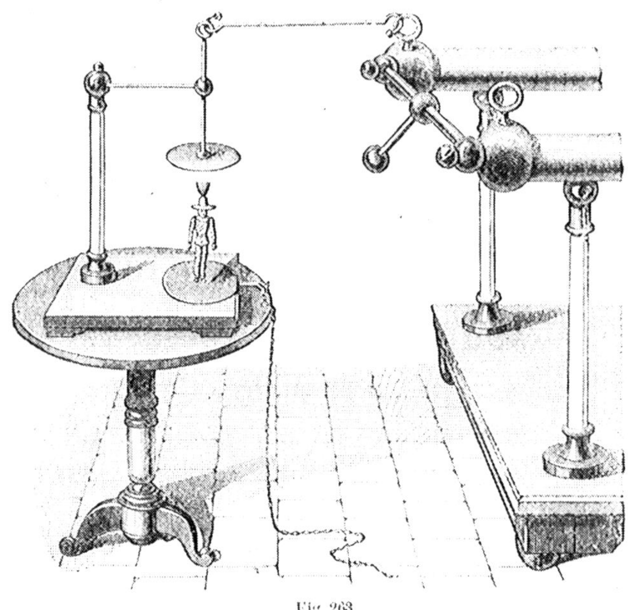

Fig 263.

Explain the action of the electrical chime.

The Electrical Puppet.

380. The ELECTRICAL PUPPET consists of a little figure which is made to dance by means of electrical attraction and repulsion.

It consists of a light image made of elder pith, or some similar substance, placed between two metallic plates, one of which is in connection with the prime conductor of the machine, and the other with the earth by means of a chain, as shown in Fig. 263.

When the machine is turned, the upper plate is electrified, and attracts the image to it. The image is charged and immediately repelled to the lower plate, where it loses its electricity, and is again attracted to the upper plate, and so on, dancing up and down as long as the plate is turned.

The Electrical Wheel.

381. The ELECTRICAL WHEEL consists of four or more arms, bent in the same direction, and attached to a small cap, which is free to rotate about a pivot.

This pivot is attached to the prime conductor, or else to a metallic support, connected with the conductor. Fig. 264 represents such a wheel. It is a reaction wheel, and is made to turn by the es-

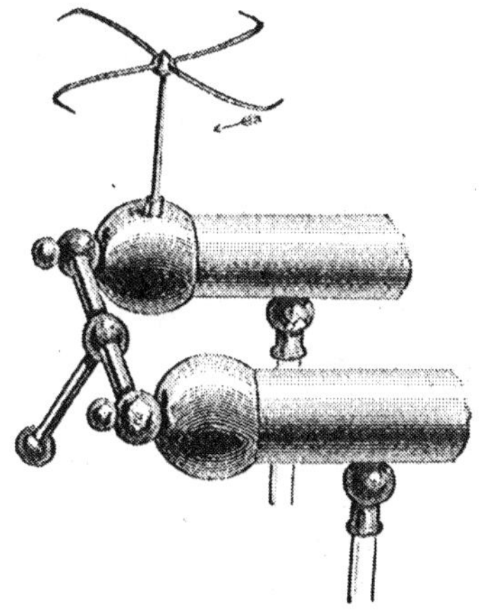

Fig. 264.

(380) What is the Electrical Puppet? Describe it. Explain its action. (381.) What is the Electrical Wheel? Describe it. Explain its action.

cape of electricity from the points. When the machine is turned, the prime conductor and the wheel become electrified; the tension of the electricity at the points becomes very great, and finally escapes with a force that causes the wheel to revolve in a direction indicated by the arrow-head, that is, in a direction contrary to that in which the points are bent. The wheel does not turn in a vacuum, which shows that electricity escapes from points in a vacuum without resistance.

The Electrical Egg.

382. The ELECTRICAL EGG is an egg-shaped light, produced by a flow of electricity through a vacuum.

Fig. 265.

(**382.**) What is the Electrical Egg?

The method of exhibiting this light, and the apparatus employed, are shown in Fig. 265. The apparatus consists of a hollow globe or oval of glass, containing two small metallic spheres of metal at some distance apart. The upper one communicates with the prime conductor, and the lower one with the earth. The globe may be deprived of its internal air by means of the air-pump. Then, if the electrical machine be turned, a flow of electricity will take place from the machine to the earth through the two balls, and because the balls are in a vacuum there will be no obstruction to the flow. If the experiment is made in a darkened room, a beautiful violet-colored light will be seen between the two balls, of the shape shown in the figure.

The Electrical Square.

383. The ELECTRICAL SQUARE consists of a square plate of glass, upon one surface of which a thin strip of tin foil is fastened, running backwards and forwards across the plate, as shown by the black line in Fig. 266. One end of this strip of tin is made to connect with the prime conductor of the electrical machine, and the other end is made to communicate with the earth by a chain. The square is insulated by legs of glass.

When the machine is turned, a current of electricity flows through the strip of tin from the machine to the earth, and no spark is given out. If, however, the tin is broken at any point, there will be a succession of sparks at that point, which will be so close together as to produce a continuous light. If, now, the tin be broken by a penknife so that the points of rupture are arranged in a definite figure, as that of a flower, for instance, a continuous light will be seen at each of these points, and the figure will appear as if traced upon the glass with fire. Any kind of figure may be drawn,

Explain the method of exhibiting it. (**383.**) What is the Electrical Square? Describe it. Explain its action.

Fig. 266

or words may be written on the glass. The experiment is more striking in a darkened room.

The Electrical Cannon.

384. The ELECTRICAL CANNON is a small cannon which is discharged by means of the electrical spark.

This cannon is used not only as an electrical recreation, but it serves also to demonstrate an important scientific fact, viz.: that the

(**384.**) What is the Electrical Cannon? *What is its use?*

electric spark is capable of producing chemical reactions. For example, water is formed of oxygen and hydrogen gases in the proportion of two volumes of the former to one volume of the latter. Now if these two gases be mixed in this proportion, and an electrical spark be passed through the mixture, the gases instantly unite and form water. Moreover, the combination takes place with a brilliant flash of light and a loud report, the report being due to the expansive force of the vapor which is produced at the moment of combination. It is upon these principles that the electrical cannon represented in Fig. 267 is constructed.

Fig. 267.

It consists of a small copper cannon mounted on a stem of glass. In the vent of the cannon is a tube of glass, through which passes a copper wire. This wire terminates externally in a ball, and internally it terminates near the metal of the cannon without touching it. The whole communicates with the earth by a chain.

Illustrate. Describe it.

To use the instrument, it is filled with a mixture of oxygen and hydrogen in the proportions to form water, and the muzzle is then closed by a cork. If a charged electrophorus is brought in contact with the ball, a spark passes between them, and another between the internal extremity of the wire and the metal of the cannon. This spark causes an explosion and drives out the cork.

A similar apparatus, called VOLTA'S pistol, is used for exploding a mixture of oxygen and hydrogen, or of air and hydrogen. It is nothing more than a sheet-iron cylinder closed by a cork. It is exploded by touching its button to the prime conductor of an electrical machine.

IV.—ACCUMULATION OF ELECTRICITY.

Electrical Condenser.

385. An ELECTRICAL CONDENSER is an apparatus employed for the accumulation of electricity.

They are of various forms, but are all essentially composed of two conductors separated by an insulator. The condenser of EPINUS may serve as a type of this class of apparatus.

Condenser of Epinus.

386. The CONDENSER OF EPINUS is composed of two metallic plates, *A* and *B*, Fig. 268, standing upon supports of glass, with an intervening plate of glass, *C*, somewhat larger than either of the metallic plates. These several plates are so mounted that the plates *A* and *B* may be made to approach to, or recede from the plate *C*.

How is it used? *Describe* VOLTA'S *pistol.* (**385.**) What is an Electrical Condenser? (**386.**) Describe the Condenser of EPINUS.

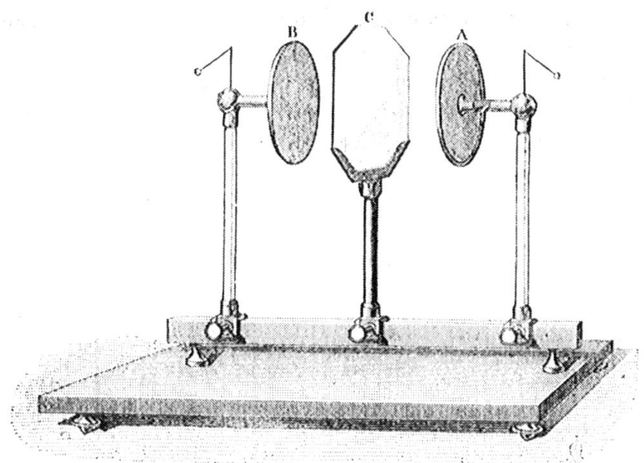

Fig. 268.

Method of using the Condenser.

387. To use the condenser, the plates, *A* and *B*, are moved up to touch the plate, *C*, as shown in Fig. 269. The

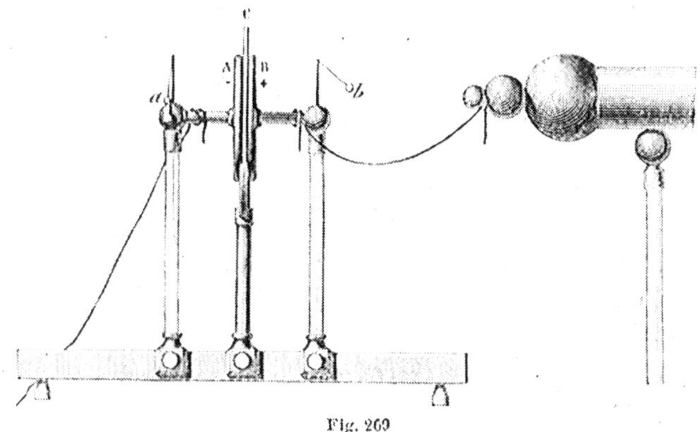

Fig. 269

(**387.**) Describe the method of using it, in detail.

plate, *B*, is made to communicate with the prime conductor of an electrical machine, and the plate, *A*, with the earth. The electrical machine is then put in motion, which charges the plate, *B*, with positive electricity. Were it not for the plate, *A*, the quantity of electricity on each unit of surface of *B* would be the same as on a unit of surface of the prime conductor; but the presence of the plate, *A*, modifies this result.

The plate, *B*, acts by induction upon *A*, and drives its *positive* electricity to the earth, retaining its negative electricity by the force of attraction. The *negative* electricity of *A* now reacts upon *B*, partially neutralizing the effect of its *positive* electricity. The electricity of *B* being partially neutralized, no longer holds that of the prime conductor in equilibrium, and an additional quantity of the positive fluid flows into it, which, acting as before, draws into *A*, from the earth, an additional quantity of the negative fluid, and so on. In this way there is gradually accumulated upon *B* and *A*, large quantities of the positive and negative fluids.

When the apparatus is fully charged, we cut off the communication between *A* and the earth, then that between *B* and the machine, by taking away the chains. In this condition the two electricities on *A* and *B* show no effects, but simply hold each other in equilibrium. There is, however, in consequence of the intervening glass plate, an excess of electricity in *B*, as is shown by the electrical pendulum, *b*, placed in connection with it. A similar pendulum placed in connection with *A*, gives no such indication.

If, now, the plates be separated, as shown in Fig. 268, both electrical pendulums will diverge, as they should do, because the two electricities no longer hold each other in equilibrium. In this condition the electricities of the two plates may be

Explain its theory. How may it be shown that the two plates are differently charged?

tested and shown to be opposite. If a rod of glass be rubbed with silk and brought near the pendulum upon *B*, it will be repelled, indicating positive electricity; if it be brought near the pendulum upon *A*, it will be attracted, indicating negative electricity.

Slow discharge of the Condenser.—Instantaneous discharge.—Discharger.

388. The condenser being charged and placed as in Fig. 269, may be discharged, that is, brought back to its neutral state, in two ways. *First*, by successive contacts, in which case the discharge takes place slowly; or *secondly*, by connecting the plates *A* and *B* by a conductor, in which case the discharge is instantaneous.

If the plate, *A*, is touched, no electricity is drawn off, because all that it contains is held in equilibrium by that in the plate, *B*. If, however, the plate, *B*, is touched, all of its free electricity, that is, all which is not neutralized by that in the plate, *A*, is drawn off. After this, a certain unneutralized portion of electricity will exist upon *A*, which will be indicated by the pendulum. By continuing to touch the plates *alternately*, the whole charge may be drawn off in small quantities.

To obtain an instantaneous discharge, we might touch one plate with one hand and the other plate with the other hand, when the two fluids would flow through the body and neutralize each other; this method produces a shock much more powerful than that produced by the simple spark from the prime conductor.

To avoid this shock we make use of a *discharger*. A discharger consists of a heavy wire bent into an arc, terminated at its two ends by balls, and having a hinge joint in the

(388.) In how many ways may a condenser be discharged? Describe the first method. The second method. Describe the method by successive contacts. How may it be instantaneously discharged? What is a discharger, and what is its use?

middle, so that it can be folded upon itself, as shown in Fig. 271. It is usually provided with a glass handle, by which it is held.

To discharge the condenser, one ball is brought in contact with one plate, and being held there, the discharger is folded so that the other ball will touch the second plate. At the instant of contact a spark is emitted, arising from the combination of the two fluids, which takes place through the discharger. No shock is felt, because the electricity does not pass through the body as in the previous case.

Limit of the Charge in a Condenser.

389. Two circumstances limit the amount of electricity that may be accumulated in a condenser. *First*, the unbalanced electricity in the plate, *B*, goes on augmenting with the charge, until at last its tension becomes equal to that on the prime conductor, after which no more can flow into the condenser from the machine. *Secondly*, the two electricities on the plates, *A* and *B*, tend to unite with an energy which goes on augmenting with the accumulation of electricity on the plates, and may ultimately become so great as to break through the glass and thus cause a union of the two fluids.

The Leyden Jar.

390. The LEYDEN JAR, named from the city where it was invented, is a condenser, differing only in form from that which has been described.

In its improved form it consists of a bottle or jar of thin glass, as shown in Fig. 270, nearly covered on its outside with tin foil, and nearly filled within by loose tin foil, or some other metallic substance in a loose state. A wire passing through the cork extends to the metallic filling within, and terminates externally in a sphere of metal called the *button*.

How is it employed? (**389.**) What circumstances limit the charge in a condenser? (**390.**) What is the Leyden Jar? Describe it.

This is a condenser in which the glass of the bottle serves as an insulator, whilst the metallic substances within and without correspond to the metallic plates in the instrument already described. The interior metal corresponds with the plate, *B*, and may be called the collector, whilst the external metal corresponds with the plate, *A*. What has been said of the condenser holds good for the Leyden jar.

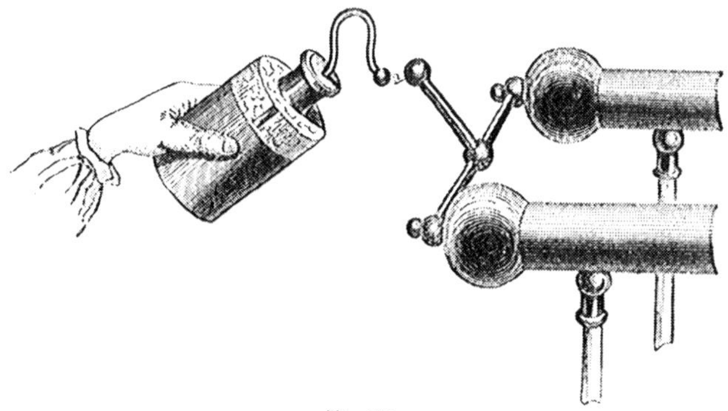

Fig. 270.

The Leyden jar is charged by holding the outer tinned part in the hand, and bringing the button in contact with the prime conductor of an electrical machine, as shown in Fig. 270. The positive fluid is accumulated in the interior, and acts by induction upon the outer coating, which becomes negative, the positive fluid in that coating being conveyed away by the hand through the body. As in the condenser, the two fluids react so as to accumulate a large quantity of positive electricity on the inside of the jar, and of negative electricity on the outside.

After the jar has been charged, if it be held in one hand whilst the other is brought in contact with the button, a shock will be felt through the arms and body, and the jar will return to its neutral state. When it is desirable to discharge the jar without the shock,

How does it resemble a condenser? How is the jar charged? Explain the theory. *How is it discharged?*

the discharger is used, as shown in Fig. 271. One ball of the discharger is made to touch the outer coating, and the other is then

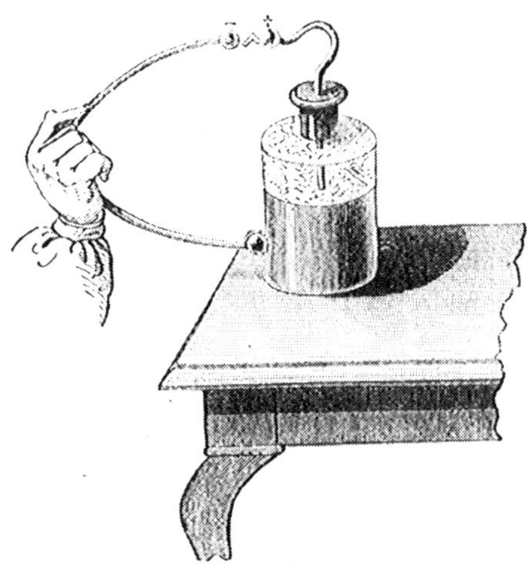

Fig. 271.

brought in contact with the button. In this case there is a spark emitted, and the jar returns to its neutral condition.

Electrical Battery.

391. An Electrical Battery consists of an assemblage of Leyden jars so connected as to act like a single condenser, as shown in Fig. 272.

The jars are placed in a box whose bottom is lined with metal, which serves to connect their outside surfaces. Their inside surfaces are brought into communication by connecting the several buttons with metallic rods.

In batteries the jars are made large, and are covered within and without with tin foil, the interior lining being brought into communication with the button of each jar by a metallic chain. Upon one

(**391.**) What is an Electrical Battery? Describe it. *What kind of jars are used in batteries?*

of the buttons is placed an electrical pendulum, which indicates the excess of the fluid on the inner over that on the outer surface.

The battery is charged in the same manner as the condenser of Epinus (Art. 391). When charged, the chains are removed by a hook with a glass handle.

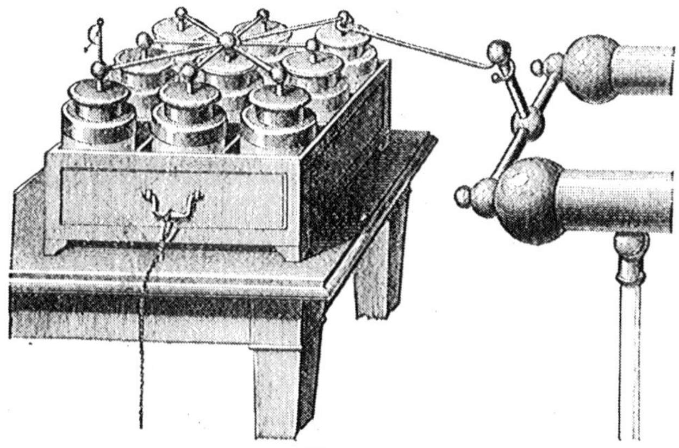

Fig. 272.

In discharging an electrical battery, a discharger is used with two glass handles, as shown in Fig. 276. Care should be taken to touch the external covering before touching the common button with the discharger.

Condensing Electrometer.

392. The gold-leaf electrometer, shown in Fig. 259, is very sensitive, but it may be rendered still more so by the addition of two disks or condensing plates, as shown in Figs. 273 and 274. The inferior disk is attached permanently to the stem, *t*, which supports the strips of gold-leaf. The superior disk has a glass handle by which it can be removed at pleasure. The two disks are of brass, coated

How connected? How charged? How discharged? (**392.**) What is a Condensing Electrometer?

with varnish, which serves as an insulator, taking the place of the glass plates in the condensers already described,

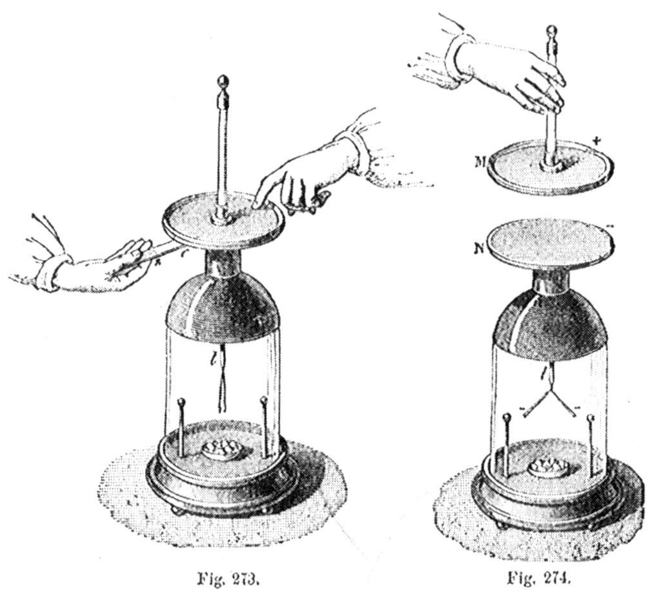

Fig. 273. Fig. 274.

but being very thin, the condensing power is much augmented.

Use of the Condensing Electrometer.

393. To use the condensing electrometer for detecting small quantities of electricity, we place the upper disk upon the lower one, as shown in Fig. 273, then using the lower disk as a collector we bring it into contact with the body to be experimented upon. At the same time we establish a connection between the upper disk and the earth, by touching it with the finger.

(**393.**) How is the condensing electrometer used?

In Fig. 273, the body experimented upon consists of two plates, one of zinc, and the other of copper, fastened together. We shall see hereafter, that by a simple contact of two such plates, the zinc is positively, and the copper negatively electrified. This last metal then being brought into contact with the inferior plate, yields its negative electricity, which acting by induction through the varnish, renders the upper plate positive. When the two electricities have accumulated upon the plates, we first withdraw the finger, and then the plate *cz*. If the upper plate be lifted off, the negative electricity which was before held in equilibrium, becomes free, and the gold leaves diverge, as shown in Fig. 274.

In this manner quantities of electricity may be discovered so small as to be unnoticed by the simple electrometer.

V.—EFFECTS OF ACCUMULATED ELECTRICITY.

Physiological Effects of Electricity.

394. The physiological effects of electricity are the effects which it produces on men and animals. They consist of muscular contractions, accompanied by a greater or less amount of pain, according to the power of the electrical apparatus.

When we receive a simple spark from the prime conductor, we experience only a slight stinging sensation; with a small Leyden jar, the pain is felt extending up the arms to the elbows or shoulders; with a more powerful jar or a battery the shock is felt through the arms and chest, and may be sufficient to produce death.

An electric shock may be given to a great number of persons at the same time. To that end they form a chain by taking each other by the hand, as shown in Fig. 275; then the person at one end takes a Leyden jar in his hand: the circuit is completed by the person at the other end of the chain touching the button of the jar, when the shock is felt simultaneously throughout the ring. NOLLET

Explain Fig. 273. What are the advantages of this instrument? (**394.**) What are the physiological effects of electricity? *Describe the shock. How may the shock be given to a number of persons?*

administered in this manner, in the presence of LOUIS XV., an electrical shock to an entire regiment of soldiers.

Fig. 275.

With a battery, the shock becomes so powerful as to render it dangerous to attempt receiving it. With a battery of only six jars of mean size, it would be hazardous to receive the shock. With more powerful batteries, cats, dogs, and even stronger animals may be killed by a single shock. Fig. 276 represents a dog killed by a shock from a battery of nine jars. The metallic collar of the dog is connected with the exterior coating of the battery, then one ball of the discharger is placed near the posterior part of the dog's spinal column, after which the circuit is completed by touching the button

Examples. What is the effect of a shock from a powerful battery? Explain the experiment.

of the battery with the other ball of the discharger. The animal is killed instantly.

In the Museum at Harlem, in Holland, is a battery whose discharge is capable of killing an ox. There is also a very powerful

Fig. 276.

battery in the Conservatory at Paris, which was given to it by the Physicist CHARLES.

Heating Power of Electricity.

395. The heat developed by electricity is sufficient not only to inflame ether, gunpowder, and the like, but also to melt and volatilize the metals.

Fig. 277 represents the manner of inflaming ether. The ether is poured into a glass vase, through the bottom of

---

The Harlem battery. (**395.**) Is there much heat developed by electricity? Explain the experiment shown in Fig. 277.

which passes a metallic wire, terminating in a button. The wire is connected by a chain with the outer covering of a Leyden jar. When the circuit is completed by touching the button of the apparatus with that of the jar, a spark is given off, and heat enough developed to inflame the ether.

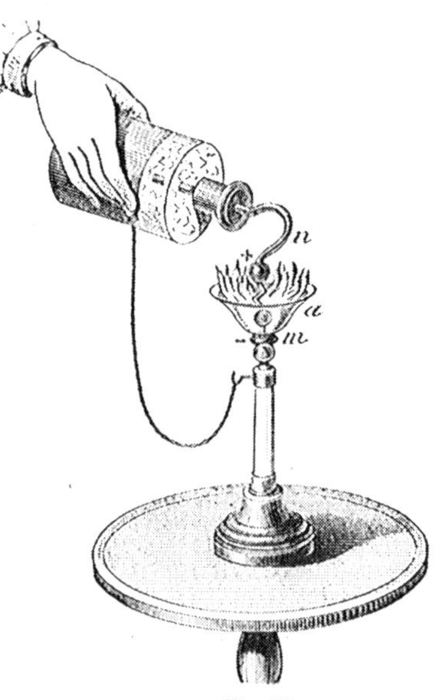

Fig. 277.

This experiment succeeds with a very small jar, or even a simple spark from the prime conductor. The experiment may be made more interesting by standing upon the electrical stool (Fig. 261), and inflaming the ether with the finger. The ether may be inflamed by a spark from a piece of ice held in the hand.

When an electrical battery is discharged through a fine metallic wire, it may be melted or even volatilized, according to the power of the battery.

In performing this experiment it will be best to use the *universal discharger*. This instrument and the manner of using it, are shown in Fig. 278. The discharger consists of two copper wires, *A* and *B*, mounted upon glass supports. The wires can slide freely through the rings that hold them, and can furthermore be turned about hinge

---

What variation may be made in it? How may a wire be melted? *Explain the construction and use of the universal discharger.*

joints, so as to bring their buttons as near as may be desired to any body that is placed upon the stand, *M*.

To melt a wire by electricity, we attach it to the two inner buttons at *i*, then connect one of the wires, *A*, for example, with the exterior coating of the battery, and complete the circuit by connecting *B* with the button of one of the jars of the battery. This is

Fig. 278.

effected in the manner shown in the figure, the connecting chain being managed by means of a hook with a glass handle. At the instant of contact, the wire, if fine enough, is melted into globules, and even volatilized, that is, reduced to vapor, which disappears in the air.

When the wire is a little larger, it simply becomes red hot and gives forth a brilliant light; if still larger, it becomes heated without being luminous. Fine and short wires may be melted under

Explain the experiment of melting a wire in detail.

water in the same manner as in air, but the experiment is more difficult to make.

Mechanical Effects of Electricity.

396. The MECHANICAL EFFECTS OF ELECTRICITY are manifested when large charges of electricity are passed through imperfect conductors. They consist of violent expansions, with tearing, fracturing, and the like.

These effects are generally exhibited by placing the body upon the plate, *M*, of the universal discharger (Fig. 278), and then passing a

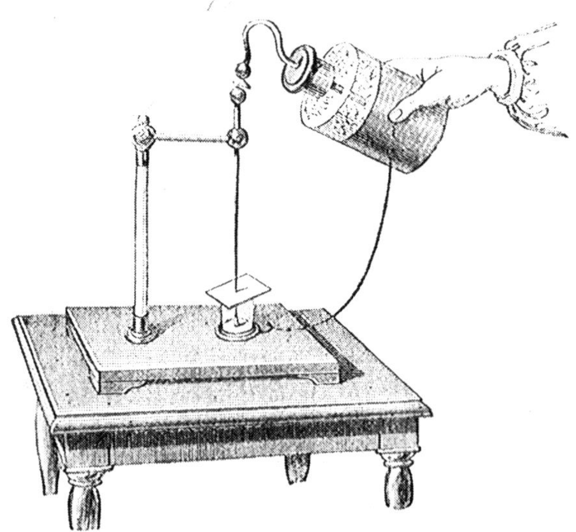

Fig. 279.

powerful charge from a battery through it. In this way a small block of wood may be torn to splinters in an instant.

Fig. 279 represents an apparatus by means of which a hole may be torn in a card by using a single Leyden jar. A card is placed at the top of a glass cylinder, beneath which is a wire projecting from

(**396.**) What are some of the mechanical effects of electricity? *How exhibited? Explain the method of perforating a card by electricity.*

a metallic plate. The plate connects by a chain with the exterior coating of the jar. Above the card is a second wire which is insulated in the manner shown in the figure. When the circuit is completed, by touching the upper wire with the button of the jar, a shock follows, and the card is found to have been pierced as if run through by a needle or pin.

To pierce a plate of glass requires a large battery. The battery belonging to Harlem Museum (Art. 394), is capable of piercing a book of four hundred pages.

A partial account of the chemical effects of electricity has been given in speaking of the electrical cannon. More on this subject will be given when we come to treat of the effects of the Voltaic pile.

VI.—ATMOSPHERIC ELECTRICITY.

Identity of Lightning and the Electrical Spark.

397. FRANKLIN published a memoir in 1749, showing the complete parallelism between the electricity of the clouds and that of the electrical machine. In that memoir he suggested that the electricity of the clouds might be attracted to the earth by means of points, and recommended that the experiment should be made.

In accordance with that suggestion, the experiment was first made by DALIBARD, in May, 1752. He erected in his garden a rod of iron about forty feet high, having its upper extremity terminating in a point. After the passage of a thunder cloud, the rod was found to be electrified, and for the space of fifteen minutes sparks were drawn from it, which were used in charging several Leyden jars.

About a month later, FRANKLIN, without any knowledge of the discovery of DALIBARD, succeeded in attracting electricity from a cloud to the earth. He raised a silken kite, just before a coming thunder storm. The string of

---

Of piercing a plate of glass. (**397.**) Who first showed the identity of lightning and electricity? Explain DALIBARD's experiments. Explain FRANKLIN's experiments.

the kite was of hemp; attached to the lower end of it was a small key, and fastened to the key was a silken cord, by which the kite might be insulated. It was only after the string became damp from the falling rain that the key showed signs of being electrified. He was at last rewarded by obtaining an electric spark. So great was his joy that he could not refrain from bursting into tears.

The complete identity between lightning and the electric spark was thus established, and all, even DALIBARD himself, unite in attributing to FRANKLIN the honor of the discovery.

Atmospheric Electricity.

398. The existence of atmospheric electricity is not confined to clouds alone, for it often exists in the atmosphere when no trace of a cloud is visible. In this case the electricity is positive. It is most abundant in open spaces and at considerable elevations. In houses, in the streets, under trees, and in sheltered localities, no trace of free electricity is discoverable. During storms the electricity of the air is sometimes positive and sometimes negative. All clouds are supposed to be electrified, some positively and some negatively.

The electrical condition of clouds may be determined by metallic rods, by kites, or by small balloons held by a string in the hand.

The electrical state of the atmosphere may be determined in a great variety of ways. M. BECQUEREL employed for this purpose the gold-leaf electrometer shown in Fig. 259. Instead of the button he used a stem of metal, attaching to its upper end a fine and flexible wire. To the second extremity of the wire he fastened an arrow,

(**398.**) What is the nature of the electricity of the air? Where is it most abundant? What is the state of the atmosphere during storms? How is the electrical condition of the clouds determined? *How is the electrical state of the atmosphere determined?*

which, being shot from a bow, ascended into the atmosphere, drawing the wire with it. When the arrow was shot directly upwards, the divergence of the gold leaves indicated the existence of free electricity, and the nature of this electricity was tested as already explained.

Lightning and Thunder.

399. LIGHTNING is nothing else than an elongated electrical spark, which passes between two differently electrified clouds when brought near each other. Sometimes a discharge takes place between a cloud and the earth; this is called a *thunderbolt.*

A flash of lightning is often of great length, and as it takes place along the line of least resistance, it generally follows a zig-zag path, as is often the case with the spark from a Leyden jar. When a flash of lightning is seen in the lower regions of the atmosphere, it has a brilliant white color; but in the higher regions, where the air is rarefied, it assumes a violet hue, similar to that of the electric egg (Art. 382).

THUNDER is the sound which follows a flash of lightning. It is due to vibrations caused by the passage of the spark through the air.

Thunder is not heard till an appreciable time after the flash is perceived. This arises from the fact that light travels with immense velocity, reaching the eye instantaneously, whilst sound travels more slowly, and reaches the ear only after a sensible interval of time. The distance of a clap of thunder may be ascertained by counting the number of seconds between the flash and the report, and allowing five seconds to a mile.

The intensity of the sound diminishes as the distance becomes greater: near by, it is sharp and rattling, like boards falling one

---

(**399.**) What is Lightning? What is a thunderbolt? *Why is the flash often zig-zag in its shape? What is the color of the flash?* What is Thunder? *Why is the thunder only heard after an appreciable time? How may the distance of the flash be determined? What effect has distance on the sound of thunder?*

upon the other; at a greater distance, it is dull and prolonged in a low rumble of varying intensity.

The rattling or rolling of thunder is differently explained. By some it is said to be due to a succession of echoes from the clouds and the earth. Others regard lightning, not as a single spark, but as a succession of sparks, each giving rise to separate explosions that succeed each other so rapidly as to produce a continuous rumbling sound. Others again attribute the rolling of thunder to the zig-zag course of the lightning, the sound from different points of the zig-zag path reaching the ear in times proportional to their distances. In this way the sounds from different points are superposed irregularly, giving rise to irregularity in the resulting sound.

The Thunderbolt.

400. A THUNDERBOLT is a discharge of electricity between a cloud and the earth.

When an electrified cloud passes near the earth, it acts upon it by induction, repelling the fluid of the same name and attracting that of an opposite name. As soon as the tension of the two electricities becomes greater than the resistance of the intervening air, a spark or flash passes, and the thunderbolt is said *to fall*, or the lightning *to strike*. The flash generally passes from the cloud to the earth, but sometimes the reverse is the case. The attraction between the two electricities increases as the distance diminishes. Hence it is that elevated objects are most likely to be struck, such as spires, high trees, lofty buildings, and the like. Good conductors, like metals, moist bodies, trees, and the like, are more likely to be struck than bad conductors. Hence the danger of taking refuge under a tree in a thunder storm.

Effects of the Thunderbolt.

401. The effects of the thunderbolt are extremely various and wonderful. It crushes or fractures bad conductors, inflames com-

How is the rattle or roll of thunder accounted for? (**400.**) What is a Thunderbolt? *Why does lightning strike? Explain the phenomenon. What bodies are most likely to be struck? What least likely?* (**401.**) *Describe the effects of the thunderbolt.*

bustible bodies, melts metals, reverses the poles of magnets, and often kills men and animals. Sometimes it falls slowly in the form of a globe of fire, and then explodes with a noise like a battery of cannon. It is this form of lightning that is most likely to inflame the edifices which it chances to strike.

It is said that a ball of electrical fire fell, in 1718, near Brest, striking a house with such force that the roof sprung up as if a mine had been exploded beneath it, and the stones of the walls were scattered in all directions, some being carried to the distance of a hundred and fifty feet.

The thunderbolt is often accompanied by a peculiar sulphurous odor, which is due to the oxygen of the air becoming electrified, forming a product called *ozone*.

Considering the fearful character of the thunderbolt, but few individuals perish from it. It is estimated that no more than twenty deaths a year occur from this cause throughout the whole of France, which is only one out of two millions of inhabitants.

Means of Safety.

402. It is recommended to those who are fearful of the effects of lightning, that they should wear clothing of silk, or still better, that they should sit in chairs insulated by glass legs or upon a thick plate of this material. They should also keep as far as possible from conductors, particularly the metals. When thus insulated, even if struck, they can experience only a slight shock, which can hardly prove fatal.

In some of the French villages it is customary to ring bells during a storm, with the idea of driving away the cloud, and avoiding the hail which so frequently accompanies thunder storms. This does no good, but simply exposes the bell ringer to additional danger, for high edifices, like church spires, are by far the most likely to be struck, and as the bell ropes are conductors of electricity, the danger to those who hold them is much increased.

Example. What is the cause of the peculiar odor that accompanies lightning?
(**402.**) What are some of the methods of protection from lightning?

The Return Shock.

403. The RETURN SHOCK is a violent, and often fatal shock, felt by men and animals at a great distance from the place where the lightning strikes. (See Fig. 280.)

This phenomenon is due to the inductive influence exerted by an electrified cloud upon bodies beneath it, which are all strongly charged with electricity contrary to that of the cloud. Now if a discharge takes place at any point, the cloud returns to its neutral state, induction ceases instantly, and all of the bodies electrified by induction instantly return to a neutral state. The suddenness of this return is what constitutes the return shock.

Fig. 280.

The return shock may be illustrated on a small scale, by placing a living frog near an electrical machine in motion. Every time that the machine is discharged by placing the finger upon it, the frog experiences a shock, which is nothing else than the return shock above described.

(**403.**) What is the Return Shock? *Explain its cause. How illustrated?*

Lightning-rods.

401. A LIGHTNING-ROD is a rod of metal, placed upon a building or ship, to preserve it from the effect of lightning, as shown in Fig. 281.

Fig. 281.

A lightning-rod should fulfill the following conditions:

1. It should be of sufficient size.

A copper rod of a half inch in diameter, or an iron one of three fourths of an inch in diameter, is large enough to protect any building.

2. If made of more than one piece, the parts should be screwed or welded together, to avoid defective joints.

(**401.**) What is a Lightning-rod? What is the first condition that a lightning-rod should fulfill? *Illustrate.* Second condition?

3. It should terminate above in a single platinum point. The point should be of platinum that it may not be fused. It also prevents the point from rusting.

4. The rod should be carried down into the earth till it meets with a good conducting medium, such as a layer of wet or moist earth.

When no such medium can be reached, a pit should be dug, and after the lower end of the rod has been carried to the bottom, it should be nearly filled with some good conductor, as coke.

The lightning-rod was invented by FRANKLIN, who thought that its protective action consisted in drawing off the electricity from the cloud, and conducting it to the earth. The real explanation of its utility is just the reverse. The cloud acts by induction upon the earth, repelling the electricity of the same name as that in the cloud, and attracting that of an opposite name, which accumulates upon the bodies under the cloud. Now, by arming a body with metallic points communicating with the earth, we permit a flow of electricity from the earth to the cloud. This flow not only prevents the accumulation of electricity upon the body, but it tends gradually to neutralize the electricity of the cloud itself, and thus the rod acts in a double way to prevent the body from being struck.

Electrical Meteors.

405. A METEOR is any atmospheric phenomenon; thus, wind, rain, snow, hail, thunder, and lightning are meteors. Besides thunder and lightning, three other meteors are attributable either wholly or in part to electricity; these are: *hail*, *tornados*, and the *aurora borealis*.

Hail.

406. HAIL consists of globules of ice which fall from the clouds. The globules consist of a coating of ice, disposed

Third condition? Fourth condition? *Who invented the lightning-rod? Explain its mode of action.* (**405.**) What is a Meteor? Mention some of them. (**406.**) What is Hail?

about a central nucleus of compact snow. They are called *hailstones*. Hailstones sometimes are very large, being not infrequently as large as a pigeon's egg, and it is said they sometimes weigh several ounces.

A fall of hail is often preceded by a noise like that of rattling nuts in a bag. This noise is attributed to collisions between the hailstones. A hailstorm is always accompanied by electrical phenomena, and thunder generally precedes or accompanies the fall of hail. From this circumstance it is inferred that hailstorms are in some way due to electrical action. As yet no satisfactory theory has been advanced to account for the formation of hailstones, and especially those enormous ones that are sometimes seen.

VOLTA supposed them to be formed between two clouds oppositely electrified, and that they were alternately repelled from one to the other, like electrical puppets, during which time they were continually increasing in size by congealing the moisture of the clouds upon their surface, till at last they became heavy enough to break through the lower cloud and descend to the earth. This theory is now rejected.

The Tornado.

407. A TORNADO is a violent whirlwind, attended with rain, thunder, and lightning. Tornados often travel considerable distances, overturning buildings and uprooting trees; they are accompanied with a noise like that of heavily-loaded carts driven over a stony road. The flashes of lightning and balls of electrical fire that accompany tornados, indicate their electrical origin.

Two species of tornado are recognized: *terrestrial* and *marine*, according as they take place on land or on water. The latter class present remarkable phenomena. The rotary force of the wind raises the water in the form of a cone, whilst a second cone forms in the cloud, having its apex downwards. These cones move to meet each other, forming a column of water reaching from the ocean to the

---

Describe a hailstone. *Explain the rattling sound preceding a hailstorm. What was* VOLTA'S *theory of the formation of hail?* (**407.**) What is a Tornado? Why is it regarded as of electrical origin? *How many species of tornados?*

cloud. In this form the column of fluid is called a *water-spout*. When a water-spout strikes a ship, it does immense damage.

The Aurora Borealis.

408. The AURORA is a luminous phenomenon, which appears most frequently about the poles of the earth, and more particularly about the *boreal* or northern pole, whence its name.

At the close of twilight, a vague and dim light appears in the horizon in the direction of the magnetic meridian. This light gradually assumes the form of an arch of a pale yellowish color, having its concave side turned towards the earth. From this arch streams of light shoot forth, passing from yellow to pale green, and then to the most brilliant violet purple. These rays or streams of light generally converge to that point of the heavens which is indicated by the dipping needle, and they then appear to form a fragment of an immense cupola, as shown in Fig. 282.

Fig. 282.

What is a water-spout? (**408.**) What is the Aurora? *Describe it.*

Since the aurora is always accompanied by a disturbance of the magnetic needle, and is generally arranged in the direction of the dip, and because the chemical action of electricity is accompanied by precisely analogous phenomena, it is inferred that it is due to electrical action. Such is at present the generally received belief.

Why is it regarded as of electrical origin?

CHAPTER IX.

DYNAMICAL ELECTRICITY.

I.—FUNDAMENTAL PRINCIPLES.

Galvani's Experiment.

409. IT has been observed that chemical combinations are sources of electricity. The form of electricity thus developed is different, but its nature is the same as that produced by friction. The name of GALVANISM has been given to electricity developed by certain chemical combinations, in honor of GALVANI, who first discovered this new way of generating it.

In 1790, GALVANI observed that the body of a frog recently killed, when placed near an electrical machine, manifested signs of excitation whenever sparks were drawn from it. The cause of action was, in fact, the return shock, as has been explained; but GALVANI, ignorant of this fact, began to seek for an explanation of the phenomena. One day he saw a dead frog suspended from a copper hook in a window, and noticed a muscular contraction whenever the wind blew the lower extremities against the iron bars of the window. Here was a case of electrical manifestation which was entirely independent of any electrical machine, and it furnished a clew to one of the most important discoveries in modern science.

This discovery led to an experiment which may be repeated as follows: Having killed a frog and cut off the hinder half of the body, we suspend it by a copper hook, *c*, passed between the back

(409.) What is Galvanism? Why so called? *Explain the method of its discovery. How may* GALVANI'S *experiment be repeated?*

bone and the nerves which run on each side of it, as shown in Fig. 283; then holding a small plate of zinc, z, in the hand, we bring one end of it in contact with the copper stem that holds the hook, and then touch the legs of the frog with the other end. At every contact the muscles contract, reproducing all the motions of life.

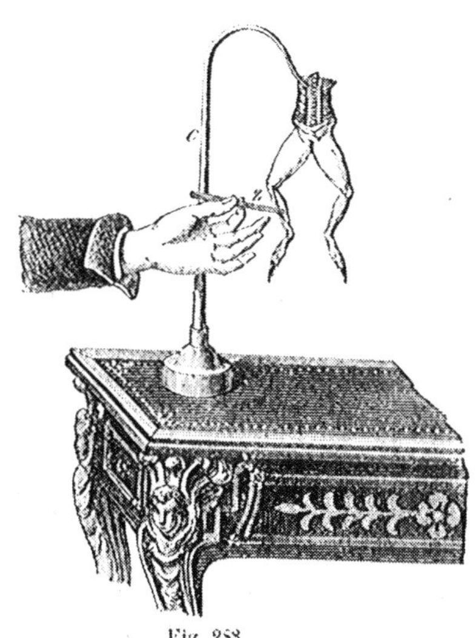

Fig. 283.

GALVANI attributed the phenomena observed, to the electricity existing in animal tissues, which, passing from the nerves to the muscles, through the metals, produced the muscular contractions.

Volta's Theory of Contact.

410. VOLTA repeated the experiment of GALVANI, and after much study, advanced the theory of contact. According to this theory, when two metals or other dissimilar substances are simply brought in contact, there is always a decomposition of the natural electricity of both bodies, the positive fluid going to one and the negative fluid to the other.

In the case of the frog, the electricity was supposed to be developed by the contact of the copper hook and zinc plate, the nerves and muscles serving simply as conductors.

VOLTA called the force which separates the two electricities in cases of contact, the *electro-motive force*, which he supposed to act

To what did GALVANI *attribute the phenomena observed?* (**410.**) What was VOLTA's theory? *What is the electro-motive force?*

like the *coercive force* in magnetism, to prevent a recombination of the separated fluids. He called those bodies which by contact developed much electricity, *good electro-motors*, and those which developed but little, he called *bad electro-motors*. The best electro-motors are zinc and copper soldered together.

In confirmation of his theory, VOLTA performed the experiment explained in speaking of the condensing electrometer, Figs. 273 and 274. This decisive experiment overthrew the theory of GALVANI. The theory of contact has since given way to the chemical theory, which will be explained hereafter.

The Voltaic Pile.

411. In the year 1800, VOLTA invented an apparatus by which he could multiply the number of contacts, and thus produce a more powerful effect. This apparatus is called the *voltaic pile*.

The voltaic pile has received many different forms, but the same principle is applied in all. One of these is shown in Fig. 284. It consists of an assemblage of *couples*, each consisting of a disk of copper and a disk of zinc in contact, and each couple being separated from the next by a layer of cloth moistened with dilute sulphuric acid. The couples are all disposed in the same order, the zinc of each couple being always on the same side of the corresponding disk of copper. When the pile is completed, there will be a disk of zinc at one end and a disk of copper at the other. A connection is made between them by means of the wires, *a* and *b*, one being attached to each of the extreme plates.

In the pile shown in Fig. 284, there are twenty couples, the zinc disk being at the bottom of each couple, and the copper one at the top. The pile is supported by a suitable frame-work.

This apparatus has been much modified, but the name *pile* has been retained for all apparatus of the same kind, and the electricity generated in this way is called *voltaic*, or *galvanic electricity*.

What are good and bad electro-motors? (**411.**) What is the voltaic pile? *Describe the pile figured in the text. What name is given to the electricity of the pile?*

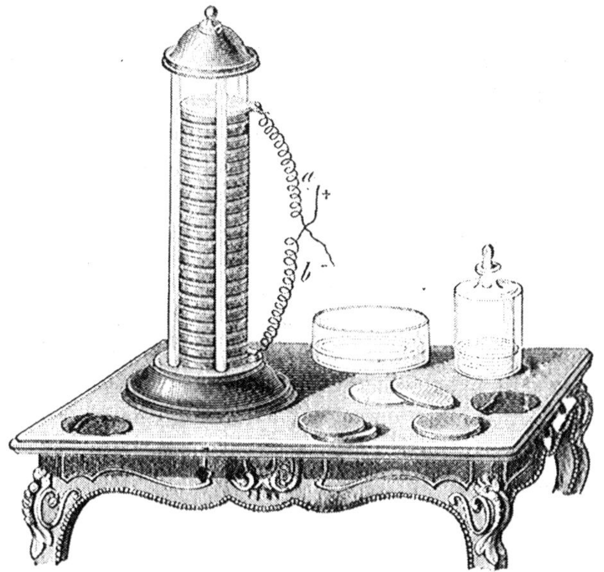

Fig. 284

Electrical Tension in the Pile. — Poles. — Electrodes.

412. In a pile which is insulated, one half is found to be electrified positively and the other half negatively, the middle being neutral. In the zinc and copper pile, that end towards which the zinc plate in each couple is turned, is positive, the other end being negative, as indicated by the signs + and —, in Fig. 284.

The tension of the electricity in either end increases with the number of couples in the pile, but is independent of their size. The tension is greatest at the two extremities; hence these extremities are named *poles;* the one towards the zinc end is the *positive pole,* the one towards the copper end is the *negative pole.*

The wires, *a* and *b* (Fig. 284), which are attached at the two poles for the purpose of completing the circuit, are called *electrodes.*

(**412.**) *How does the tension vary in the pile? Where is it greatest in any pile? What are the poles? How named? What are the electrodes?*

Electrical Currents.

413. So long as the electrodes remain separated, the pile manifests no electrical action, but on being brought near each other, a small spark is seen to pass, which arises from a recombination of the two electricities. The passage of the spark does not discharge the pile, as is the case with the Leyden jar. We see a continual succession of sparks, showing that the process of decomposition is continually kept up in the pile, by which the poles are continually fed with new supplies of the positive and negative fluids.

If the two wires are brought into actual contact, the sparks cease, but the flow of the fluids continues as before, decomposition going on in the pile, and recomposition taking place through the electrodes. This continuous flow of electricity is called *the electric current.* There are, in fact, two currents flowing in opposite directions, according to the two fluid theory, but it is found convenient to consider only one of them, namely, that which flows from the positive to the negative pole. In the figures, hereafter, the direction of this current will be indicated by an arrow, as in Fig. 292.

Chemical Theory of the Pile.

414 FABRONI first suggested that the phenomena of the pile were due to chemical action. In the pile described, the dilute acid in the cloths between the couples, acting upon the zinc, was supposed to be the cause of the development of electricity. This view was adopted by DAVY and WOLLASTON, who made many experiments calculated to sustain it. Finally, DE LA RIVE and BECQUEREL succeeded in demonstrating most conclusively that in every chemical action electricity is developed. They also showed that

(**413.**) What phenomenon is observed when the circuit is completed? What is the electric current? Which way do we suppose the current to flow? (**414.**) What was FABRONI's theory?

whenever a metal is attacked by an acid, the former is positively and the latter negatively electrified.

According to this view, which is now universally adopted, the acidulated cloths, that VOLTA regarded as simply conductors, are in fact the principal cause of the development of electricity.

The Carbon Pile.

415. The CARBON PILE was invented by BUNSEN, about twenty years ago, and is often called the *Bunsen Pile.*

Each *couple* of BUNSEN'S pile consists of four pieces, which are shown both separately and united in Fig. 285. These

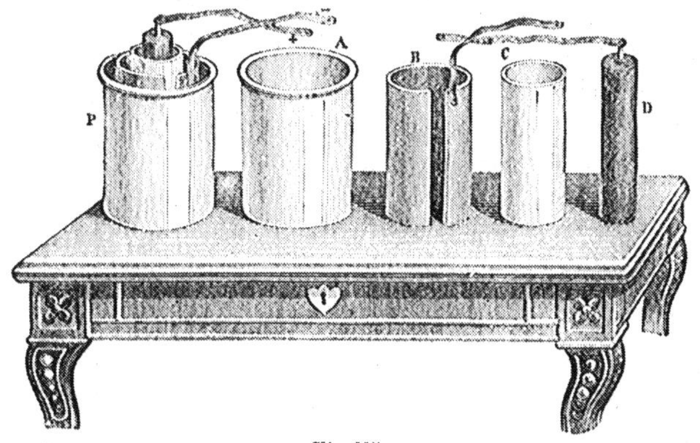

Fig. 285.

parts are: 1. An earthen vessel, *A*, containing dilute sul phuric acid; 2. a zinc cylinder, *B*, open at one side and having a strip of copper soldered to its upper extremity; 3. a vessel, *C*, of porous earthen-ware, containing nitric acid; 4. a cylinder of carbon or coke, which is well calcined, and a good conductor of electricity. At the top of this

How is the action of the pile explained according to this theory? (**415.**) Who invented the Carbon Pile? When? Describe one of the couples of BUNSEN'S pile, in detail.

cylinder a stem of copper is inserted, to which is soldered a thin strip of the same metal.

The completed couple, represented at *P*, is formed by putting the cylinder *B* into *A*, then putting *C* into *B*, and finally, introducing the cylinder *D* into the cylinder *C*. On bringing the slips of copper in contact, a current of electricity is developed, flowing from the carbon to the zinc.

In this case there is a double chemical action. Water is decomposed in the vessel *A*, giving its oxygen to the zinc, forming oxyde of zinc, which is taken up by the sulphuric acid, producing sulphate of zinc. This remains in solution. The hydrogen of the water passes through the porous cell, *C*, and uniting with a part of the oxygen of the nitric acid, decomposes it, reproducing water, and also forming nitrous acid, which escapes in fumes. This double action develops a large amount of electricity, that flows from the carbon, which is the positive, towards the zinc, which is the negative pole of the couple.

Any number of couples may be united by attaching the copper slip of the zinc cylinder in one couple to that of the carbon in the next couple, and so on throughout the combination. The remaining two slips, which will be at the extreme ends of the combination, may be united by a conductor.

Such a combination is called a *galvanic battery*, or sometimes a *voltaic battery*. A galvanic battery has been constructed, containing as many as eight hundred couples. Fig. 286 represents a battery of twenty couples.

II.—APPLICATIONS OF GALVANIC ELECTRICITY.

Effects of the Galvanic Battery.

416. The EFFECTS OF THE GALVANIC BATTERY may, for convenience of study, be divided into *physiological*, *heat-*

Explain the action of a couple of BUNSEN'S *pile.* How are the couples connected? What is such a combination called? (**416.**) What are the principal effects of the galvanic battery?

ing, *illuminating*, *chemical*, and *magnetic*. They are all due to the recombination of the two electricities, as in machine electricity, but they are more remarkable and more energetic, because of their continuous action.

Physiological Effects.

417. The PHYSIOLOGICAL EFFECTS of galvanic electricity are a succession of shocks producing violent muscular contractions, not only in living, but in dead animals, as shown in the case of GALVANI'S frog.

When we touch but one of the poles of a galvanic battery, no shock is felt, but if we take both electrodes in our hands, as in Fig. 286, we feel a sensation similar to a shock from a Leyden jar, with

Fig. 286.

To what are they due? (**417.**) What are the physiological effects? *How may a shock be obtained from a battery?*

this difference, the latter is instantaneous, whilst that from the galvanic battery is continuous. The action of the battery keeps up a continuous supply of the two fluids, which supplies the place of that lost by recombination in passing through the body of the experimenter.

The effect of galvanic electricity upon the bodies of dead animals is peculiarly striking. It produces violent contractions of the muscles, causing motions similar to those of the living being. On the occasion of performing some experiments upon the body of a criminal who had been executed, in England, a violent and convulsive respiration was produced, the eyes opened, the lips moved, and the face, no longer under the control of the will, assumed expressions so strange and horrifying that one of the assistants fainted from terror, and only recovered his natural state of mind after several days.

Heating Effects.

418. When a current of galvanic electricity is passed through a conductor, it becomes heated, and often to such a degree as to produce fusion or even vaporization. When a powerful current is passed through a wire, it soon becomes incandescent, and then melts or is dispersed in vapor. Small wires burn with splendid brilliancy. Silver burns with a greenish light, and much smoke arising from the vaporization of the metal. Gold burns with a bluish white light. Platinum, which is infusible in the most intense heat of our furnaces, melts into spherical globules with a dazzling light.

With a battery of 600 couples, Despretz fused nearly half a pound of platinum in a few minutes. Carbon is the only body which has not been fused by galvanic electricity. Despretz, however, by passing a current through small rods of pure carbon, succeeded in softening them so much that they could be bent and made to adhere, which indicates an approach to fusion.

What effect has galvanic electricity on dead animals? (**418.**) Describe the heating effects of the battery. Despretz' experiments.

Illuminating Effects.

419. The heating effects just described are accompanied with a disengagement of more or less light, but to obtain the most brilliant electrical light possible, calcined carbon points are employed. They are at first placed in contact, one being connected with the positive, and the other with the negative pole of a powerful galvanic battery, as shown in Fig. 287. The points immediately become incandescent,

Fig. 287.

(**419.**) Describe the illuminating effects.

emitting a light of dazzling brightness. If the points are slightly separated, the current still continues to pass between them, and the light takes the form of a luminous arch, called the *voltaic arch.* In this experiment the point connected with the positive pole wastes away, whilst the other increases in size; hence we conclude, that particles of carbon are transported from the former to the latter; this explains how the current continues to pass in spite of the interval which separates them.

The intensity of the *electrical light* is very great. A battery of 48 small couples furnishes a light equal to that of 572 wax candles; with 100 couples a light is produced so intense as to dazzle the eyes, and with 600 couples the intensity is such as to render it as impossible to look at it, as it is to look at the sun.

In 1844, FOUCAULT first made use of the electrical light instead of that of the sun, to illuminate the solar microscope. Since then, many attempts have been made, and with some success, to apply it to purposes of general illumination. Fig. 287 represents an apparatus employed for the purpose of illumination. The battery is contained in the interior of a cast-iron pedestal, upon which is erected a column of the same material. At the top of the column are two carbon points, one connected with each pole of the battery by copper wires, insulated by gutta percha coverings.

Chemical Effects.

420. The most important chemical effects produced by galvanic electricity, are the decomposition of bodies traversed by it, and the transportation of their elements.

In order to understand the chemical effects, we must explain the meaning of certain terms employed in chemistry.

Oxydes are *generally* compounds of oxygen with the metals. Thus, iron rust is an oxyde of iron, that is, it is a compound of oxygen and iron; vermilion is an oxyde of lead; potash is an oxyde of potassium, and so on.

What is the voltaic arch? *Illustrate by example. Describe* FOUCAULT'S *experiment. Describe the apparatus for illumination.* (**420.**) What are the most important chemical effects? *What are oxydes?*

Acids are *generally* compounds of oxygen with some non-metallic body. Thus, sulphuric acid is a compound of oxygen and sulphur; nitric acid is a compound of oxygen and nitrogen; carbonic acid is a compound of oxygen and carbon, and so on.

Salts are *generally* compounds of an acid with an oxyde. Thus, sulphate of potash is a compound of sulphuric acid and potash, nitrate of copper is a compound of nitric acid with an oxyde of copper, and so on.

In these definitions, we say *generally*, because the definitions given are only intended for illustration, and it is not thought best to enter into a detailed account of the various substances described. That belongs to Chemistry.

Decomposition of Water.

421. A galvanic current was first employed to decompose water in the year 1800, by CARLISLE and NICHOLSON.

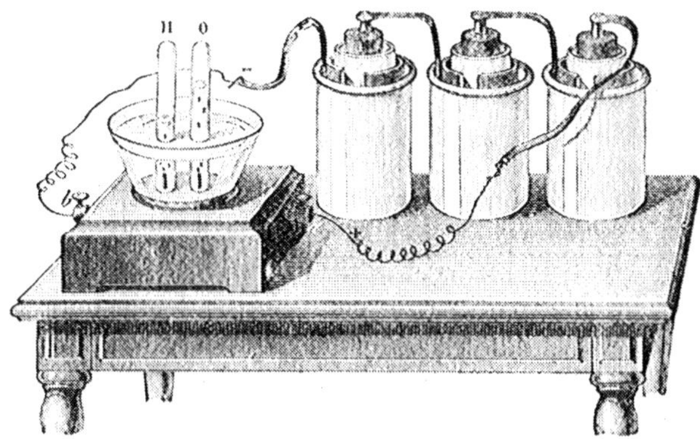

Fig. 288.

To repeat the experiment, we may employ the apparatus shown in Fig. 288. It consists of a glass dish with a wooden bottom. Rising from the bottom are two platinum wires,

Acids? Salts? (**421.**) When was water first decomposed? How may the experiment be repeated?

which pass through the wooden stand and terminate in the tubes, *a* and *b*. These wires serve as electrodes.

The glass vessel is partially filled with water, to which a small quantity of sulphuric acid is added to improve its conducting power. Two narrow bell-glasses, *H* and *O*, are filled with water and inverted over the two platinum wires. The tube, *a*, is then connected with the positive pole of the battery, and the tube, *b*, with the negative pole. A current is set up from one wire to the other through the water, and decomposition begins, as is shown by bubbles of gas rising in the two bell-glasses.

By testing the gases thus obtained, we find that in the glass, *O*, corresponding to the positive pole of the battery, is pure oxygen, whilst that in the glass, *H*, corresponding to the negative pole, is pure hydrogen. We see also that the volume of hydrogen is twice that of the oxygen. This experiment shows that water is composed of oxygen and hydrogen, mixed in the proportion of one volume of the former to two of the latter.

Decomposition of Oxydes and Salts.

422. Oxydes and salts may in like manner be decomposed by a current of electricity. In decomposing oxydes, the oxygen is transported to the positive electrode, and the metal to the negative electrode. In the decomposition of acids there is a like transfer of elements, the oxygen going to the positive, and the other component to the negative electrode.

In decomposing salts, there may be several cases. Sometimes there is a simple resolution of the salt into an acid and an oxyde, in which case the acid goes to the positive and the oxyde to the negative electrode. Sometimes, besides the separation into acid and oxyde, the latter is decomposed;

Explain in detail. (**422.**) How may oxydes and salts be decomposed? Explain the different cases of decomposition of salts.

in this case the oxygen and the free acid go to the positive, and the metal alone goes to the negative electrode. This action is utilized in the process of electrotyping.

DAVY, at the beginning of the present century, used the battery to decompose potash, soda, lime, baryta, magnesia, alumina, &c., and thus demonstrated that all of these substances, previously regarded as simple bodies, were in reality compound. They consist of oxygen united with metals, which are called respectively, potassium, sodium, calcium, barium, magnesium, aluminium, &c.

Application of Electricity to Galvanoplasty.

423. GALVANOPLASTY, or ELECTROTYPING, is the operation of copying medals, statues, and the like, in metal, by the aid of galvanic electricity.

Such copies were formerly made by the process of casting; now they are in many cases more elegantly produced by galvanoplasty. This process was discovered simultaneously by SPENCER, of London, and JACOBI, of St. Petersburg, in 1838, the year preceding the discovery of the daguerreotype process.

Method of Electrotyping.

424. The first step is the preparation of a mould of the object, upon the accuracy of which depends the success of the entire operation. Wax, plaster, or gutta percha may be used, but the latter material is now considered the best. At ordinary temperatures, gutta percha is hard, but on being heated, it becomes soft and ductile. To form the mould, the gutta percha is warmed by putting it into a vessel of warm water and allowing it to remain till it is of the proper softness; it is then placed upon the object to be

---

Explain DAVY'S experiments. (**423.**) What is Galvanoplasty? When discovered? By whom? (**424.**) Explain the operation of electrotyping, in detail.

copied, and pressed with the fingers till it touches every point of the surface of the object, when it is left to harden by cooling. After hardening, it is removed and is ready for use. As gutta percha is apt to adhere to certain bodies, precaution should be taken to cover them with a thin layer of powdered plumbago or graphite. This may be laid on with a soft brush, and if properly applied, it effectually prevents the adhesion of the mould.

The second step is to deposit the metal in the mould. As gutta percha is a non-conductor of electricity, it is necessary to cover it with a conducting substance. This is done by laying on a coating of plumbago in the same manner as in forming the mould. The mould thus prepared is then made ready for the bath by attaching to it the proper suspending wires, as shown in Fig. 289.

Fig. 289.

Fig. 290.

Fig. 290 represents one face of a medal to be copied, and Fig. 289 represents the gutta percha mould prepared for receiving the metallic deposit. For making the deposit, which we shall suppose to be of copper, a DANIEL'S battery of two or three couples is usually employed.

A couple of DANIEL'S battery differs from one of BUNSEN'S in the following particulars. The carbon cylinder is replaced by one of zinc, denoted by *Z*, in Fig. 291, and the zinc cylinder is replaced by one of copper, denoted by *C*, in the same figure. The outer vessel is of glass, and is filled with a solution of sulphate of copper (blue vitriol), which is kept saturated by some crystals of the sulphate

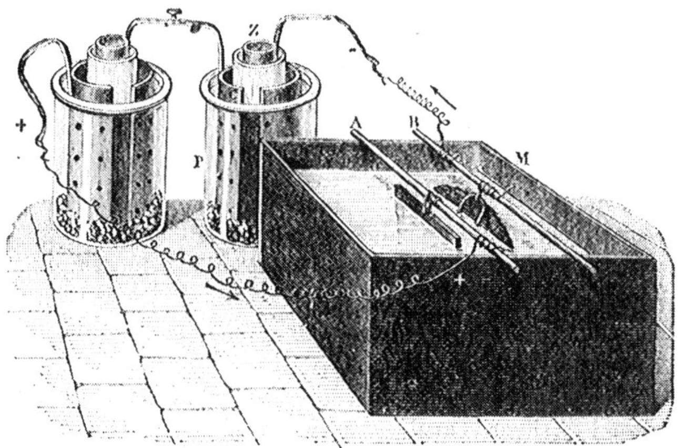

Fig. 291.

placed at the bottom of the vessel. The porous vessel is filled with dilute sulphuric acid. When this battery is in action, water is decomposed: the oxygen goes to the zinc, forming oxyde of zinc, which is dissolved by the sulphuric acid, giving sulphate of zinc. The hydrogen of the water goes to the sulphate of copper in *P*, and decomposes it. The result of these decompositions and recompositions is to keep up a current of electricity, as shown by the arrows, which will continue as long as the vessel, *P*, is kept full of the saturated solution of sulphate of zinc.

Fig. 291 shows the method of depositing the metal upon

What kind of a battery is used for depositing copper? *Explain one of* DANIEL'S *couples*. Explain Fig. 291.

the mould. *M* is a vessel filled with a solution of sulphate of copper; *A* and *B* are metallic rods communicating with the two poles of the battery; the mould is suspended from the rod, *B*, and facing it is a plate of pure copper suspended from the rod, *A*; these constitute the electrodes, the mould being the negative one.

The current which is set up through the solution of copper between the electrodes, decomposes the sulphate into sulphuric acid, oxygen, and pure copper. The sulphuric acid and oxygen go to the copper plate, and uniting with it, produce sulphate of copper; the pure copper goes to the negative electrode, that is, to the mould, and is there deposited. After about two days the coating of copper becomes thick enough to be removed from the mould, and it then presents a fac-simile of the object to be copied. In copying medals, each face is copied separately, and the two are united by means of some fusible metal placed between them

Electro-gilding and Electro-plating.

425. The process of covering bodies with thin coatings of gold or silver is analogous to that of electrotyping. The perfection of the process consists in making the coating of gold or silver, not only of uniform thickness, but also closely adherent.

The object to be gilded or silvered is first heated upon a charcoal fire to remove all fatty matter; it is next plunged into dilute sulphuric acid, and then rubbed with a hard brush to remove any oxyde that may exist upon the surface; it is next plunged into common nitric acid, and then into nitric acid into which a small quantity of salt and soot has been thrown; it is then washed in pure water and carefully dried in sawdust, and is ready for use.

Explain the chemical changes which take place. (**425.**) What is the process of electro-silvering and electro-gilding? How is the object cleaned?

The method of silvering, or electro-plating, is shown in Fig. 292. The object to be silvered is suspended in a bath of a silver solution by a metallic rod which connects with the negative pole of a BUNSEN'S battery. Immediately below it is a plate of pure silver, which is connected with the positive pole of the battery. The object to be silvered and the silver plate, *a*, constitute the electrodes, *a* being the positive one. The explanation of the process is analogous to that in the preceding article.

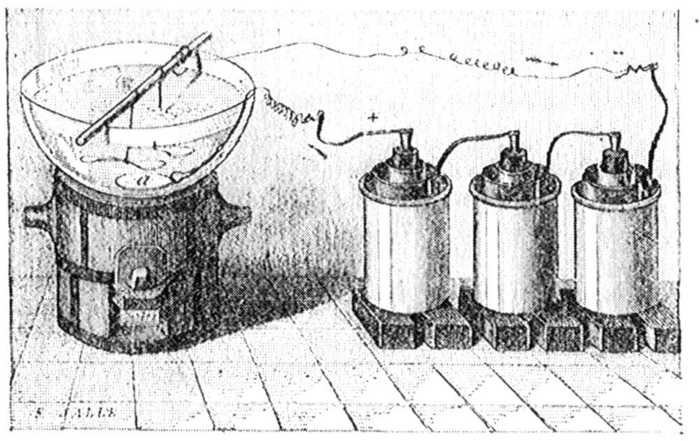

Fig. 292.

The salt of silver generally employed is a double cyanuret of silver and potassium. The thickness of the coating deposited will depend upon the power of the battery and upon the time of immersion.

The process of gilding is the same as that of silvering, except that we use a cyanuret of gold and potassium, and a plate of gold at *a*, instead of a silver one.

Explain the process of silvering as shown in Fig 292. *What salt of silver is employed?* What is the process of gilding? What salt of gold is used?

The history of electro-plating and electro-gilding is briefly as follows: In 1803, BRUGNATELLI first gilded a silver medal by suspending it in a solution of gold from the negative pole of a battery, but proceeded no further. In 1840, DE LA RIVE, of Geneva, discovered a process of gilding metals with a battery, but by his process much gold was wasted, and the work was unsatisfactory. In the same year, ELKINGTON, an Englishman, discovered the process of gilding by means of the cyanuret of gold and potassium. A few months later, RUOLZ succeeded in silvering and platinizing metals by the methods now in general use. The arts of electro-plating, electro-gilding, and electrotyping are now of general application, and afford occupation to thousands of artisans.

Give an outline of the history of electro-plating and electro-gilding.

CHAPTER X.

ELECTRO-MAGNETISM.

I.—FUNDAMENTAL PRINCIPLES

Relation between Magnetism and Electricity.

426. It was observed at an early period that the magnetic and electrical fluids had many analogous properties. In each case fluids of the same name repel, whilst those of an opposite name attract. It was also observed that a stroke of lightning often reversed the poles of a magnetic needle, and sometimes completely destroyed its magnetism. The two have also points of dissimilarity. Magnetic fluids are not transmitted like electrical fluids through conductors. A magnet does not, like an electrified body, return to a neutral state when brought into communication with the earth. Magnetism can only be developed in a few, whereas electricity may be developed in all bodies.

Between these analogies and dissimilarities nothing positive could be affirmed with respect to the identity of magnetism and electricity, until, in 1819, Œrsted made a discovery which showed that these physical agents are most intimately allied, if not identical.

(**426.**) What early observations were made on the relation of the phenomena of electricity and magnetism? What dissimilarities were noticed?

Action of an Electrical Current upon a Magnet.

427. ŒRSTED discovered the fact that an electrical current has a directive power over the magnetic needle, tending always to direct it at right angles to its own direction.

This action may be shown by the apparatus represented in Fig. 293. If a wire be placed parallel to and pretty near a magnetic needle, and then a current of electricity be passed through it, the needle will turn around, and after a

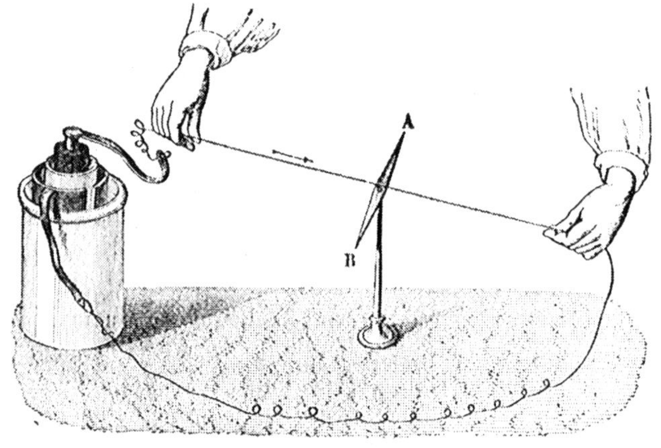

Fig. 293.

few oscillations will come to rest in a position sensibly at right angles to the current. That it does not take a position absolutely perpendicular to that of the current, is because of the directive force of the earth, which partially counteracts that of the current.

The direction towards which the austral pole, that is, the north end of the needle, will turn, depends upon the direction of the cur-

(**427.**) What discovery was made by ŒRSTED? Explain the action of the electrical current on the needle. *Which way does the north end turn?*

rent. If that flows from south to north, and above the needle, the needle deviates towards the west; if it flows towards the south, and above the needle, the latter deviates towards the east. When the current flows below the needle, the phenomena are reversed.

Ampere's Law.

428. AMPERE, to whom the discovery of the greater portion of electro-magnetic phenomena is due, gave a simple expression to the law which governs the action of a current upon a magnet. He supposes an observer lying down upon the wire along which the current flows, the current entering at the head and going out at the feet. Then, if he turn his face towards the needle, the austral pole will in all cases be deviated towards his right hand.

Action of Magnets upon Currents, and of Currents upon Currents.

429. AMPERE established the following principles:

1. Magnets exercise a directive force upon currents.

To illustrate this, we bend a copper wire into a circular form, and then dip its extremities, which should be pointed with steel, into cups of mercury, one above the other, as shown in Fig. 294. These cups communicate with the two poles of a battery, by means of which a current may be generated, flowing as indicated by the arrows. Now if a bar magnet be brought near this current, the axis being in the plane of the current, we shall see the hoop turn about the steel points in the cups, and come to rest, with its plane perpendicular to the axis of the magnet. This experiment, which is due to AMPERE, is the reverse of that made by ERSTED.

2. The earth, which acts like a huge magnet upon a magnetic needle, acts in the same manner upon movable cur-

(**428.**) Explain AMPERE's law. (**429.**) What is AMPERE's first principle? *How illustrated?* His second principle?

rents; that is, it directs them so that they are perpendicular to the magnetic meridian.

This may be shown by the apparatus of Fig. 294. If the communication with the battery be cut off, and the hoop be turned till its plane coincides with the magnetic meridian, it will remain in that

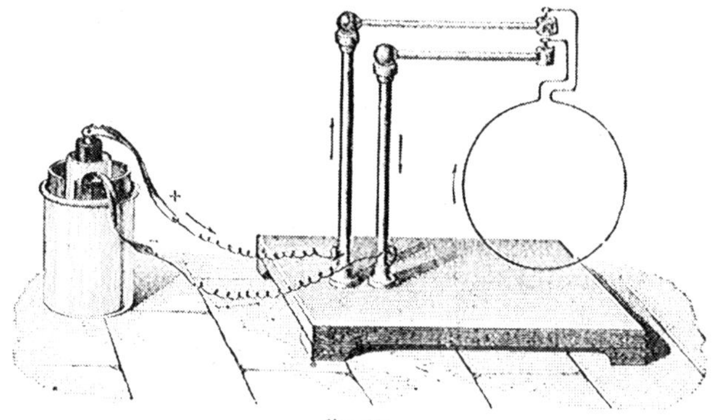

Fig. 294.

position. If now a current be passed through it, we see it turn slowly around the pivots, so as to take a position at right angles to the meridian. It will turn in such a direction that the current in the lower part of the hoop will flow from east to west.

3. Two parallel currents attract each other when they flow in the same direction, and repel each other when they flow in opposite directions.

4. If a wire be coiled into a double helix, as represented in Fig. 295, and then be suspended by its steel points in the cups of mercury (Fig. 294) it will, when a current is passed

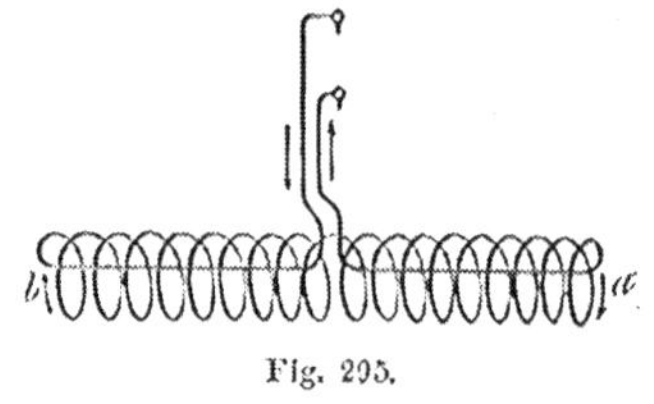

Fig. 295.

How may this be shown? His third principle? His fourth principle?

through it, arrange itself in the meridian like a magnetic needle. When the current takes the direction of the arrows, the end, *a*, becomes an austral pole, and is directed towards the north.

When thus suspended, the helix has all the properties of a magnet, and is subject to the same laws of attraction and repulsion. A helix of the kind described is called a *solenoid.*

Ampere's Theory of Magnetism.

430. From the facts explained in the last article, AMPERE deduced a theory of magnetism. He supposes magnetism to be due to currents of electricity flowing around the ultimate molecules of a magnet, always in the same direction. The currents in the interior of the magnet neutralize each other, and consequently the total effect of all the currents in a magnet, is the same as that of a set of surface currents flowing around the magnet, in such a direction, that if we place the eye at the south end of a magnet, and look in the direction of the axis, the current will flow around in the same direction, as the hands of a watch.

He supposes the directive force of the earth to be due to currents of electricity flowing around it, parallel to the magnetic equator, from east to west. These currents are produced by variations of temperature, which arise from the earth's revolution continually presenting a new portion to the direct action of the sun's rays.

If we conceive all the currents of the magnetic needle to be replaced by a single resultant about its equator, and all of the terrestrial currents to be replaced by a single equatorial current, then that portion of the latter current which lies nearest the magnet, will attract the lower part of the current on the magnet, and repel that on the upper part, thus compelling the magnet to place itself in the meridian.

What is a solenoid? (**430.**) What is AMPERE'S theory o magnetism? To what did he attribute the directive power of the earth? *Explain the action of the terrestrial current upon the magnetic needle.*

AMPERE supposes that natural magnets owe their properties to the long-continued action of electrical currents. We may suppose magnetic bodies to be made up of atoms, having electrical currents flowing around them; that is, of little magnets. These, when they are arranged heterogeneously, will exhibit no magnetic properties. When they are by any action brought into positions in which their similar poles are arranged in the same direction, they become magnets.

Galvanometer.—Galvanic Multiplier.

431. A GALVANOMETER is an instrument for measuring the force of an electrical current. In its simplest form, it consists of a magnetic needle, *ab*, Fig. 296, with a conducting wire passed around it in the direction of its length.

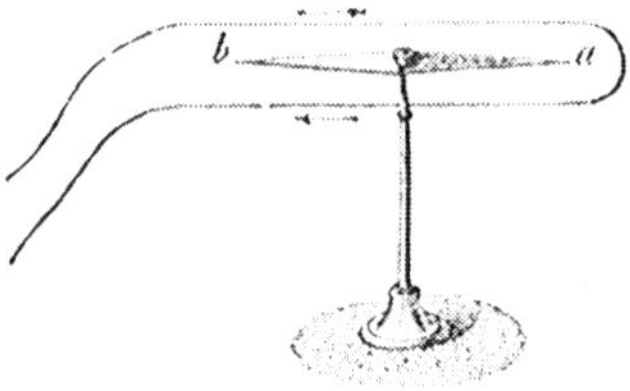

Fig. 296.

When a current of electricity is passed through the wire, its presence will be indicated by a motion of the needle, its force by the amount of deviation of the needle, and the direction of the current will be indicated by the direction towards which the north end of the needle deviates.

The GALVANIC MULTIPLIER is a galvanometer of great sensitiveness, but constructed on the same principles as the one already described.

It is represented in Fig. 297. It consists of a copper stand, *M*, supporting a glass cylinder, as shown in the figure. Under the cylinder is a graduated circle, beneath which is a wooden frame wound with a great number of coils of copper wire. The wire is insulated by being covered

How does AMPERE *explain the formation of natural magnets?* (**431.**) What is a Galvanometer? Describe it. *Its action.* What is a Galvanic Multiplier? Describe it in detail.

with silk. The two ends of the coil communicate with the binding screws, *m* and *n*, by means of which they may be made to communicate with the poles of a magnetic couple. A metallic frame supports a hook, from which is suspended a delicate silken cord, *s*. This cord supports two fine magnetic needles, the one, *ab*, above the graduated circle, and

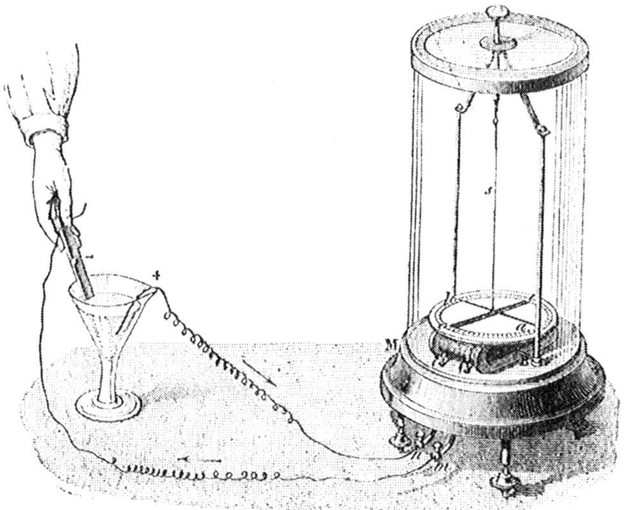

Fig. 297.

the other, *B*, within the coil, only a part of which is visible in the figure. The two needles are so united that one can not turn without the other, and their poles being placed in opposite directions, the action of the earth upon them is completely neutralized. Hence they are free to obey the least force.

Uses of the Galvanic Multiplier.

432. The Multiplier is used to indicate the feeblest currents of electricity. By means of it, BECQUEREL established

(**432.**) What is the use of the Multiplier?

the fact that a current is developed in every chemical action, in the imbibition of liquids, and in many other phenomena. By using a galvanometer with many thousands of turns of wire, the existence of electrical currents in animals and vegetables may be demonstrated.

To show the currents developed by chemical action, as, for example, the action of acids upon metals, two fine platinum wires are introduced into the binders, *m* and *n*. One end of one of the wires is then dipped into a glass of dilute sulphuric acid, and the other is held in contact with a plate of zinc which is also dipped into the dilute acid. The two needles which were before parallel to *oi*, and which we suppose to have been placed in the magnetic meridian, immediately turn round and become perpendicular to the meridian, indicating the instantaneous production of a current. The current in this case takes the direction indicated by the arrows, whence we conclude that the acid is positively and the zinc negatively electrified. This corresponds to what was said on this subject in a preceding article.

Magnetizing by means of an Electrical Current.

433. If a wire be wound around a bar of iron, and a current of electricity be passed through the wire, it is at once converted into a magnet. The method of making the experiment is shown in Fig. 298.

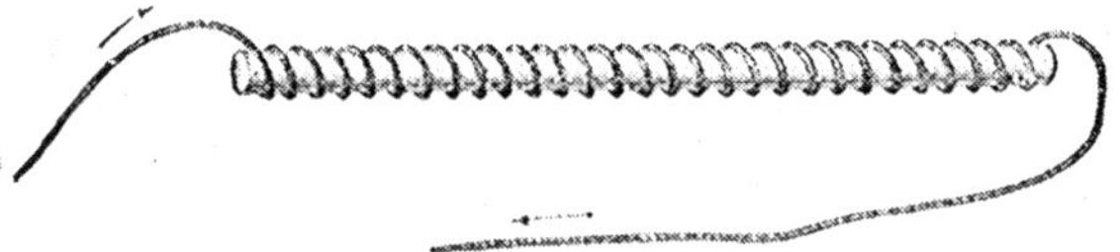

Fig. 298

If the current cease, the iron bar at once loses its magnetism. We may in like manner form a permanent magnet by using a bar of steel instead of a bar of iron.

Illustrate. (**433.**) How is an iron bar converted into a magnet by galvanism? In what way may a bar of steel be converted into a magnet?

The bar of steel may also be magnetized by passing through the wire a spark from a Leyden jar. To do this, one end of the wire is made to touch the external covering of the jar, and the other end is brought into contact with the button of the jar. The steel bar is magnetized instantaneously, thus showing the identity between the electricity of the galvanic current and that of the Leyden jar.

II. — ELECTRO-MAGNETIC TELEGRAPHS. — THE ELECTRO-MOTOR.

The Electro-Magnet.

434. An ELECTRO-MAGNET is a magnet obtained by the use of electricity.

Electro-magnets are generally made of soft iron, bent in the form of a horse-shoe, as shown in Fig. 299. Upon each branch is wound a great number of coils of wire, insulated by being covered with silk. The wire is coiled in different directions upon the two branches, and its extremities are then connected with the poles of a battery.

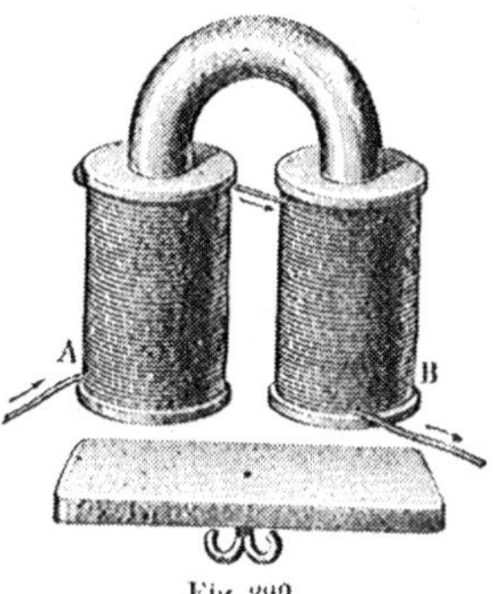

Fig. 299.

In this way magnets may be constructed of immense power, so powerful, in fact, as to support the weight of ten or twelve persons. Fig. 300 represents the method of arranging the details of a magnet which is intended to exhibit a great *sustaining power.*

The plate in contact with the two poles is called an *armature.*

When the instrument is of soft iron, it is magnetized instantaneously by the passage of a current of electricity through the wire, and

In what other way may it be done? What inference is drawn from this fact? (434.) What is an Electro-Magnet? How are they constructed? What is an armature?

Fig. 300.

as instantaneously loses its magnetism when the current is stopped, or *broken*. This property has been utilized in the electro-magnetic telegraph.

The Electrical Telegraph.

435. An ELECTRICAL TELEGRAPH is an apparatus for transmitting intelligence to a distance by means of electrical currents.

In 1820, AMPERE proposed to transmit signals by passing currents over magnetic needles, making use of as many wires and needles as

What is the principal use of the electro-magnet? (**435.**) What is an Electrical Telegraph? *Give an outline of the history of magnetic telegraphs.*

there are letters. In 1837, STEINHEIL, of Munich, actually constructed such a telegraph.

In 1831, Prof. HENRY, now of the Smithsonian Institute, published the results of his researches on the subject of electro-magnetism, and in subsequent years, exhibited experiments illustrative of their application to the transmission of power to a distance, for the purpose of producing telegraphic effects.

In 1837, Prof. MORSE invented a machine for recording signals upon paper, and in 1844, the first working line of telegraph for practical purposes was built from Washington to Baltimore.

Many modifications of the telegraphic apparatus have been made since its first invention. Three principal varieties are now in use, all of which are based upon a common principle, which is very simple.

At the station from which a telegram is dispatched, is an electrical battery, and at the one where it is to be received, is an electro-magnet. The two are connected by a wire running between the stations. When the current is transmitted through the wire, the iron becomes magnetized and attracts an armature of soft iron, which in turn imparts motion to other pieces, by means of which the signals are imparted. When the current ceases, the iron loses its magnetism, and a spring forces the armature back to its primitive position. By successively breaking and restoring the current, the telegram is transmitted.

In one form of the telegraph, the electro-magnet causes a needle to move over a sort of dial, around which are printed the letters of the alphabet. The letter before which the needle stops, is the one to be transmitted. This machine requires as many signals as there are letters in the message. This is the *dial telegraph*.

In another form of the telegraph, there are two electro-magnets, which set in motion two movable arms placed at the extremities of a horizontal black line on a white dial-plate. The relative positions of the hands with reference to the fixed line, serve as conventional signals, nearly in the same way as was customary in the old-fashioned telegraph. This is the *signal telegraph*.

How many kinds are in common use? Explain their general principle. What is the dial telegraph? The signal telegraph?

The dial telegraph is used in France on the lines of railways. The signal telegraph was used in France for ordinary purposes until it was replaced by MORSE'S registering apparatus.

Morse's Registering Telegraph.

436. In MORSE'S telegraph, the telegram is permanently registered upon paper by means of a conventional alphabet.

Fig. 301.

Fig. 301 represents the method of dispatching a telegram, and Fig. 302 represents the method of receiving it. At

Where were these used? (436.) Describe MORSE'S registering telegraph.

each station the apparatus is identical, but it is double; that is, composed of two pieces, the *manipulator* and *receiver*. These pieces are shown more in detail in Figs. 303 and 304. In order to explain the working of this telegraph, let us commence with Fig. 301.

Under the table is shown the battery which furnishes the electrical current. The current is conducted by the wire, *P*, into the manipulator, which will be described hereafter.

Fig. 302.

From thence it goes into a galvanometer, *g*, which indicates by a needle the passage of the current; it next passes

Explain the method of working this telegraph.

through a piece, *M*, that serves as a safeguard, and from thence reaches the wire, *L*, which passes to the station where the message is to be delivered. We see the same wire entering at the top of Fig. 302, whence the current passes through a safeguard, *M*, then into the galvanometer, from which it goes to the electro-magnet of the receiver. After passing through the electro-magnet, it passes through the wire, *T*, and is lost in the earth.

Morse's Manipulator and Receiver.

437. MORSE'S MANIPULATOR is shown in Fig. 303. It consists of a wooden stand, upon which is a metallic lever, *kh*, turning about a horizontal axis. One end of this lever is raised up by a spring, *r*, and the other is traversed by a

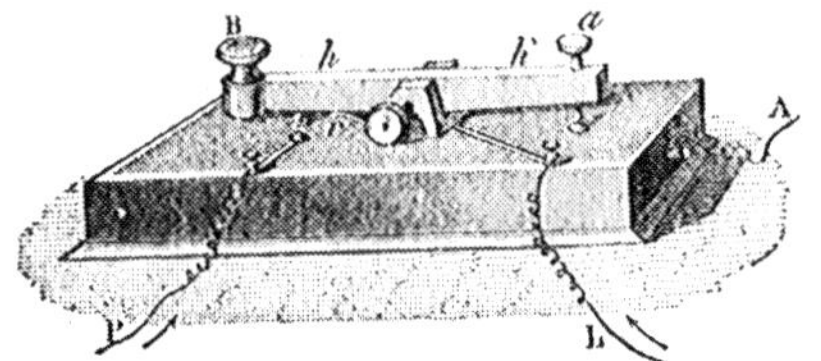

Fig. 303.

stem, *a*, which rests upon a copper button, and this in turn communicates through the stand with the wire, *A*. Fig. 303 represents the manipulator at the instant when it receives a dispatch. The current arrives by the wire, *L*, which is the wire of the line, rises into the lever, *kh*, and descends by the wire, *A*, to the receiver.

When it is desired to transmit a signal, it becomes necessary for the current from the battery, *P*, to enter the manipulator. This is not effected when the latter is disposed as in Fig. 303, for the lever, *kh*, does not touch the button in which the wire, *P*, terminates. By pressing the

(437.) Describe the Manipulator in detail. Its use.

button, *B*, the lever, *kh*, is lowered; a contact is established, the current passes immediately into the wire, *L*, which leads to the other station.

The RECEIVER, Fig. 304, is composed of an electro-magnet, *E*, which, whenever a current is transmitted, acts by attraction upon an armature of soft iron, *m*, fixed at the extremity of a lever, *mn*, and movable about an axis. At its extremity, *n*, the lever carries a point, *x*, which may be made to press against a movable fillet of paper, *ab*. When the current does not pass through the electro-magnet, the point, *x*, does not press against the fillet; but as soon as the current passes, the point is pressed against the paper, and traces upon it either a point, or a line more or less elongated,

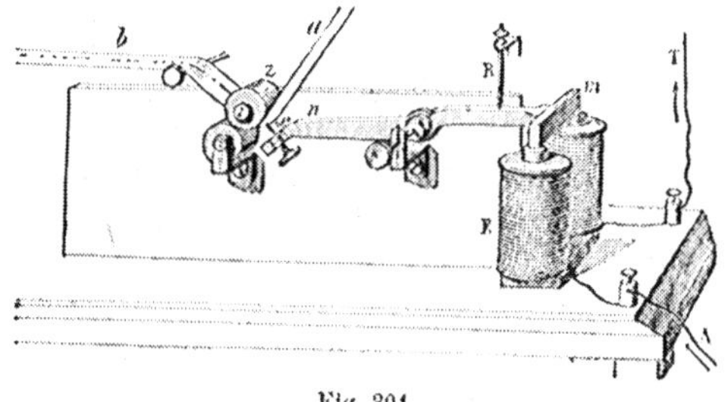

Fig. 304.

according to the length of time during which the current is uninterrupted.

The fillet of paper is kept in motion uniformly by means of a train of clock-work, *V*, which turns the cylinder, *Z* (Fig. 301). The fillet of paper moving uniformly in the direction from *a* to *b*, the operator at the other station, by pressing the button of the manipulator, and maintaining the pressure for greater or lesser periods of time, causes a succession of points and marks to be made upon the fillet at

Describe the Receiver in detail. Its use. How is the fillet of paper kept in motion? How are the letters recorded?

pleasure. These marks are, by convention, made to stand for the letters of the alphabet, as shown in the following table

MORSE'S ALPHABET.

a · —	j — · — ·	s · · ·
b — · · ·	k — · —	t —
c · · ·	l ———	u · · —
d — · ·	m — —	v · · · —
e ·	n — ·	w · ——
f · — ·	o · ·	x · — · ·
g — — ·	p · · · · ·	y · · — · ·
h · · · ·	q · · — ·	z · · · ·
i · ·	r · · ·	

It only remains to explain the PROTECTOR, *M*. Experience has shown that the wires may, from atmospheric influences, accumulate upon themselves sufficient quantities of electricity to prove troublesome to the operators of the telegraph. The piece, *M*, is destined to prevent any injurious action of this kind. It is composed of two toothed pieces of metal, disposed so that the teeth are nearly in contact. The current passes into one of these pieces, whilst the other is in communication with the earth. If, from any atmospheric change, electricity accumulates upon the wires or apparatus, it is given off by the points to the piece which is in communication with the earth, and shocks are thus avoided.

In what has been said, only a single wire, *L*, has been spoken of as running from station to station. It would seem to be necessary, in order to complete the circuit, that a second wire should be employed; such however is not the case. The employment of a second wire is avoided by connecting the two ends of the single wire with the earth. The parts *T*, Figs. 301 and 302, are for this purpose

Explain the Protector. Its use. *Why is it possible to operate a line of telegraph with a single wire?*

prolonged from the instruments till brought into free communication with the earth. The fluid then continues to circulate just as though a return wire had been used.

Velocity of Electricity.—Submarine Cables.

438. It has been found by experiment that the velocity of electricity is such as to carry a current around the earth in about a quarter of a second. For short distances, then, we may regard the transmission as instantaneous.

Since the invention of the telegraph, a complete net-work of lines has been established over both continents. Not only have thousands of miles of wires been stretched on land, but submarine wires have been laid, connecting places separated by hundreds of miles of water. Telegraphic wires connect England and Ireland, England and France, France and Algiers, and so on. Finally, an attempt has been made to connect the two continents, and although it has thus far failed to be successful, there is good reason to anticipate a complete success at no distant day. Signals and messages have been transmitted from Ireland to Newfoundland, and the possibility of the connection has thus been fully demonstrated.

Electro-Magnetic Motor.

439. Many attempts have been made, and with partial success, to employ electro-magnetism as a motor for the propulsion of machinery. JACOBI, of St. Petersburg, constructed an engine of this kind in 1838, which was capable of propelling a boat containing twelve persons. Many other machines have since been constructed, but in all cases the expense of moving them has been so great as to preclude their economical use.

Fig. 305 represents an electro-magnetic machine, constructed according to the design of M. FROMENT. It is composed of four electro-

---

(**438**.) What is the velocity of an electrical current? *Give an account of some of the submarine lines of telegraph.* (**439**.) Has electricity been used as a motor? *Describe* M. FROMENT'S *machine in detail.*

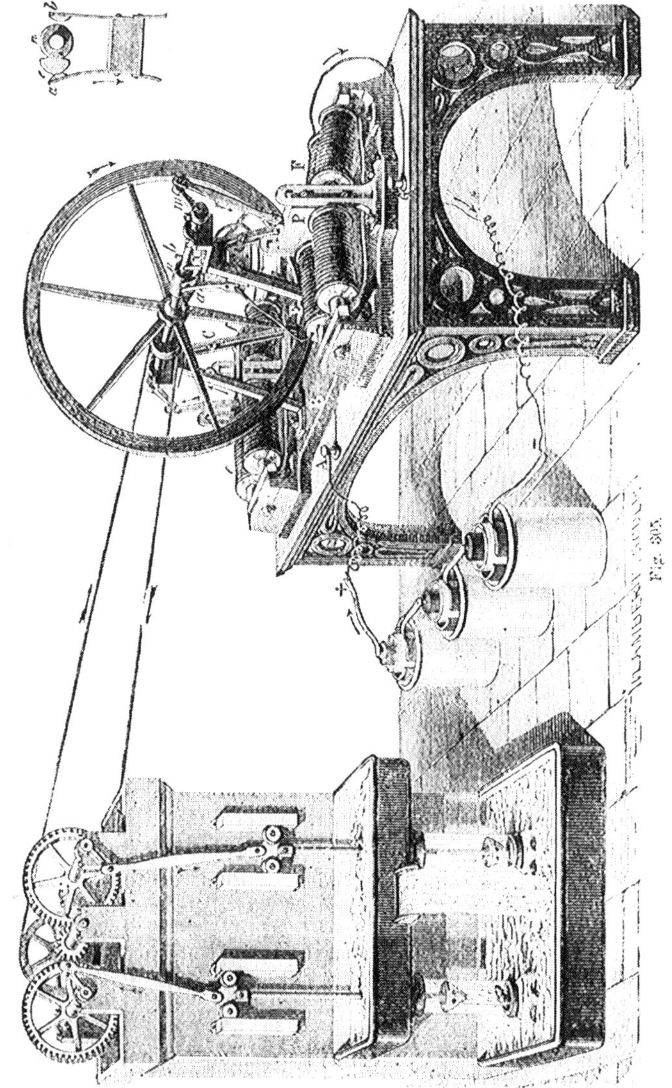

Fig. 305.

magnets, acting in pairs upon two pieces of soft iron, *P*, only one of which is seen in the figure. These pieces, attracted by the electro-magnets, *EF*, transmit the motion by means of a working-beam, to a crank, *m*, fixed at the extremity of a horizontal arbor. The latter bears an iron fly-wheel, which regulates the motion. Finally, the same arbor supports a piece of metal, *n*, of a greater diameter, the use of which will be explained presently.

The current from the battery, *P*, entering *A*, passes into a platform of cast-iron, *B*, then, through different metallic pieces, it reaches the arbor and the piece *n*. From thence the current flows alternately to the electro-magnets, *EF* and *ef*. The manner in which this alternate flow is effected, is shown in Fig. 306, which represents a section of the piece, *n*, and its accessories. Upon the piece, *n*, is a projection, *c*, called a *cam*, which in the course of one revolution touches successively two pallets, *a* and *b*; these transmit to the electro-magnets the current, whose course is indicated by the unfeathered arrows. The feathered arrows in the figure show the direction in which the parts of the machine move.

The current passing alternately into the two pallets, *a* and *b*, and thence into the systems of electro-magnets, *EF* and *ef*, the piece, *P*, is first attracted, and then a similar piece at the other extremity of the arbor of the fly-wheel is attracted, and so on. The result is a continuous rotary motion, which is transmitted by a driving-band to a train of wheels, and so on to the pumps, which it is destined to work.

III.—INDUCTION.—APPLICATION TO MEDICINE.

Induction by Currents.

440. We have seen that the electricity of the machine acts upon bodies by induction. The electricity of the battery acts in a similar manner, but only when the currents begin to flow and when they cease.

To show this, take two copper wires, covered with silk, and wind them side by side upon a bobbin. Then fasten the two ends of the

Its mode of action. (**440.**) Does galvanic electricity act by induction? *How is this shown?*

first wire to the two binders, *m* and *n*, of the galvanometer, Fig. 297. Next connect one end of the second wire with one pole of a feeble galvanic battery. If the other end of the second wire be brought into contact with the second pole of the battery, at the instant of contact, the needle of the galvanometer will indicate the production of a current in the first wire, flowing in an opposite direction to that of the battery. If the contact is kept up, the flow of the induced current ceases, as is shown by the needle of the galvanometer returning to its position of rest. If the current of the battery is broken, the needle of the galvanometer is again deviated, but in a contrary direction, indicating an induced current flowing in the same direction as that of the battery.

The battery current is called the *inducing* current; the other current is called the *induced* one. These currents conform to the following laws:

1. At the instant when the inducing current begins to flow, an induced current is developed flowing in a contrary direction.

2. The inducing current continuing to flow, the induced current ceases.

3. At the instant when the inducing current ceases to flow, an induced current is developed flowing in the same direction.

Properties of Induced Currents.

441. Experiment has shown that induced currents possess all the properties of other electrical currents. Like them, they give sparks, produce violent shocks, decompose water, salts, and the like, and act upon the magnetic needle.

Induced currents are the more powerful, the longer the wires em-

---

What is the direction of the induced current at the instant of closing the circuit? Of breaking it? What is the inducing current? The induced one? What are the laws that govern the induced currents? (441.) What are the properties of induced currents?

ployed. Hence, in practice it is usual to wind the wires upon bobbins, as shown in Fig. 307.

The coil shown in Fig. 307, consists first of a cylinder of pasteboard, upon which is wound about three hundred coils of coarse copper wire. This is the inducing coil. Over it is a finer wire, making several thousand coils. These wires are not only covered with

Fig. 307.

silk, but also with an insulating varnish of gumlac. At the extreme left of the stand on which the coil rests, are two binders in connection with the two poles of a battery. From one of them proceeds a plate of copper, going to a toothed wheel, moved by clockwork, and communicating with one of the ends of the inducing coil;

How are the wires wound? How insulated? How are the battery communications made?

the second binder is in communication with the other end of the same coil. The two ends of the finer wire are also connected with binders, and through them may be connected with any conductor whatever. For the purpose of administering a shock, the binders are provided with wires having copper handles, which are to be grasped, as shown in the figure.

When the instrument is in operation, the current from the battery is continually broken by means of the toothed wheel, and there results a succession of shocks, two at each interruption of the current. The shocks that arise at the beginning of the flow are almost nothing, whilst those which take place at the time of interruption are quite severe.

The principle of induced currents has been applied in constructing instruments for generating large quantities of electricity. The induction coil, as improved by Ruhmkorff, is one of the most remarkable instruments of this class. In some of his larger instruments he uses more than sixty miles of wire in the secondary coil.

Physiological effects of Electrical Currents.

442. Electrical currents have been employed in the treatment of certain diseases, especially those connected with the nervous system. Electricity has a powerful action upon the animal economy, and when judiciously applied possesses considerable curative power.

Fig. 308 represents one of the many forms that have been given to the electrical apparatus, for the purpose of acting upon the human body. It consists of a wooden box, upon which is mounted a copper cylinder, inclosing a bobbin of two wires. The box has a drawer of zinc, in which is a small quantity of salt water. A plate of well calcined carbon, impregnated with nitric acid, is plunged into this solution. In a word, the combination constitutes a modified form of a Bunsen's couple. Two copper slips communicate, the one with the zinc and the other with the carbon, conducting the current to the

How are shocks given? What arrangement is made for continually breaking the current? How may the shocks be increased? (**442.**) What application has been made to medicine? *Explain Fig. 308.*

large wire of the coil, through a piece of machinery for breaking the current. This current-breaker consists of a small plate of soft iron, attracted by an electro-magnet in the centre of the bobbin. It is attracted when the current passes, and immediately interrupts, or breaks it. The induced current is conducted by wires to two sponges

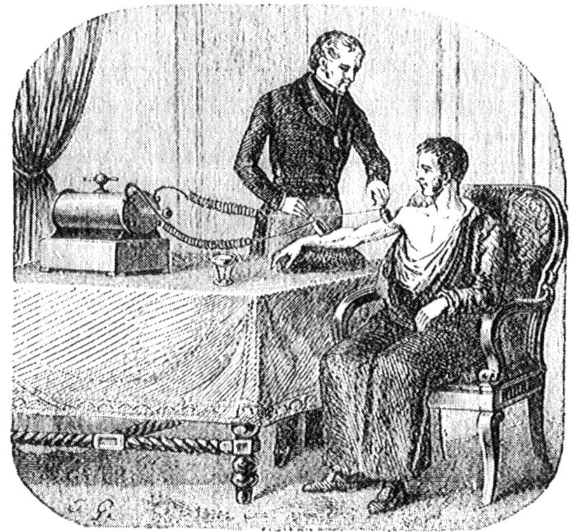

Fig. 308.

saturated with salt water, or fresh, according as it is desired to make a more or less intimate communication with the part through which the shocks are to be passed. Finally, the method of applying the shocks is shown in Fig. 308.

Electrical Fishes.

443. Certain fishes possess the power of imparting a shock that compares in intensity with that of a powerful Leyden jar. Such fishes are called electrical fishes, and are of three kinds, the most

(**443.**) *Describe the electrical fishes.*

interesting of which is the electrical eel of South America. This fish was studied by HUMBOLDT and BONPLAND, who have given a complete description of it.

The shocks given by electrical fishes are due to electricity generated in the body of the fish. MATTEUCI showed that sparks could be obtained from the fish, and also that the galvanometer is affected when one of its wires is brought into connection with the back of the fish, and the other with its belly.

In all cases the shock is voluntary, and serves as a means of defense against enemies.

To what are their shocks due? What observations were made by MATTEUCI?

CHAPTER XI.

APPLICATION OF PHYSICAL PRINCIPLES TO MACHINES.

I.—GENERAL PRINCIPLES.

Definition of a Machine.

444. A MACHINE is a contrivance by means of which a force applied at one point, is made to produce an effect at some other point.

The force applied is called the *power*, and the force to be overcome is called the *resistance.*

Motors.

445. The working of a machine requires a continued application of power. The source of this power is called the MOTOR.

Some of the most important motors are *muscular effort*, as exerted by man or beast, in various kinds of work; the *weight and impulse of water*, as in water-mills; the *impulse of air*, as in wind-mills; the *elastic force of springs*, as in watches; the *expansive force of vapors and gases*, as in steam and hot-air engines. The last is, perhaps, the most useful of the motors mentioned.

Object and Utility of Machines.

446. The object of a machine is to transmit the power furnished by the motor, and to modify its action in such a manner as to cause it to produce a useful effect.

(444.) What is a machine? *The power? The resistance?* **(445.)** What is a motor? *Mention some of the most important.* **(446.)** What is the object of a machine?

In no case does a machine add anything to the power applied to it; on the contrary, it absorbs more or less of this power, according to the nature of the work to be done and the connection existing between the parts.

Some of the circumstances which cause an absorption of power are the rubbing of one part upon another, the stiffness of bands and belts, the resistance of the air, the adhesion of one part to another, and the want of hardness and elasticity in the materials of which the machine is constructed. The resistances arising from these causes are called *hurtful resistances.* They not only absorb much of the power applied, but they also contribute to wear out the machine. The existence of these resistances in every machine requires a continued supply of power to overcome them, in addition to that necessary to perform the useful work. Hence the absurdity of attempting to obtain perpetual motion.

Quantity of Work of a Force.

447. The idea of WORK, in mechanics, implies that a force is continually exerted, and that the point at which it is applied moves through a certain space. Thus, in raising a weight, the work performed depends *first* upon the weight raised, and *secondly* upon the height through which it is raised. The *quantity of work* of a force in any given time, is measured by the intensity *of the force*, expressed in pounds, multiplied by *the distance through which it is exerted*, expressed in feet. This distance is called *the path described.*

Equilibrium of a Machine.

448. A machine is in EQUILIBRIUM when the power and resistance exactly balance each other.

In determining the circumstances of equilibrium, it is customary to neglect the hurtful resistances in the first approximation, and then to

---

Can a machine create power? What are hurtful resistances? Their effect? (**447.**) What is meant by work? Illustrate. *What is the measure of the quantity of work?* (**448.**) When is a machine in equilibrium? *What is the condition of equilibrium when the hurtful resistances are neglected?*

take account of them as corrections. If the hurtful resistances be neglected, it will be found that any machine, working uniformly, is in equilibrium, when in any given time *the quantity of work of the power is equal to that of the resistance.*

II.—ELEMENTARY MACHINES.

Mechanical Powers.

449. The elementary machines are seven in number, viz., the *cord;* the *lever;* the *inclined plane;* the *pulley;* the *wheel and axle;* the *screw;* and the *wedge.* These seven are called *mechanical powers.* The first three are simple elements; the remaining ones are combinations of these three.

The principles of the lever and inclined plane, so far as necessary to an understanding of the principles of Physics, have already been explained in Chapter I. In the following articles those principles are repeated, in connection with a description of the other mechanical powers. In the cuts which follow, the power and resistance are represented by arrow-heads, the former being denoted by the letter P, and the latter by R.

The Cord.

450. CORDS, and BANDS or BELTS, are used for transmitting motion from one point to another, as in the pulley. Chains are often employed for the same purpose, as in the watch.

Cords, belts, and chains, should be as flexible as is consistent with sufficient strength.

The Lever.

451. A LEVER is an inflexible bar, free to turn about an axis. This axis is called the FULCRUM. (See Arts. 30, 31, and 32.) Levers may be either straight or curved. The dis-

---

(**449.**) How many mechanical powers are there? Name them. (**450.**) What is the use of a cord or band in machinery? (**451.**) What is a lever?

tances from the fulcrum to the lines of direction of the power and resistance are called *lever arms.*

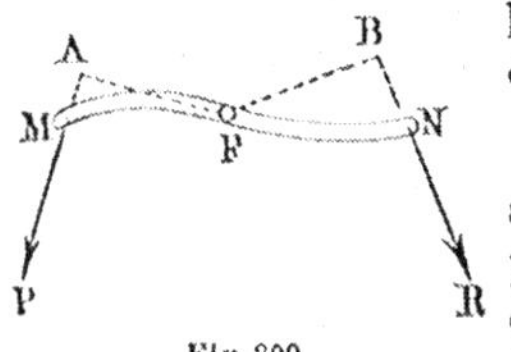

Fig. 309.

In the lever, MN, F is the fulcrum, MP and NR are the lines of direction of the power and resistance, FA is the lever arm of the power, and FB is the lever arm of the resistance.

Levers are divided into *three classes:*

In the *first class* (Fig. 310), the fulcrum is between the power and the resistance. The steelyard is a lever of this class.

In the *second class* (Fig. 311), the resistance is between the power and fulcrum. The rudder of a ship is a lever of this class.

In the *third class* (Fig. 312), the power is between the resistance and the fulcrum. The treadle of a lathe is a lever of this class.

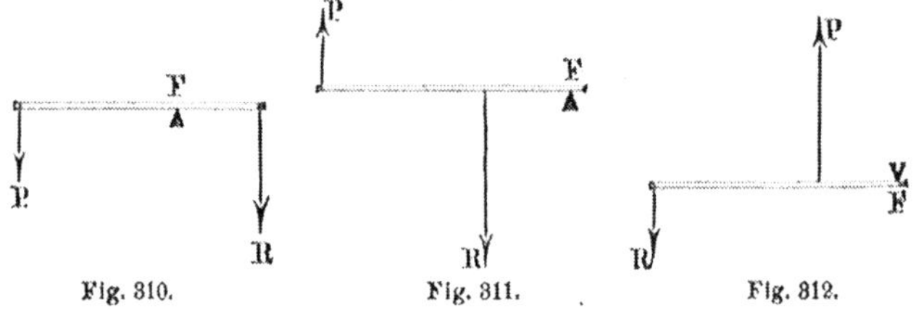

Fig. 310. Fig. 311. Fig. 312.

In all cases the paths described by the points of application of the power and resistance will be proportional to their lever arms, and when in equilibrium, the power will be to the resistance as the lever arm of the resistance is to the lever arm of the power.

Compound Levers.

452. A COMPOUND LEVER is a combination of simple levers, so arranged that the resistance in one, acts as a power in the next, throughout the combination.

Compound levers are used for magnifying small motions, as in showing the expansion of bodies; or to enable a small weight to balance a large one, as in the hay-scale and in other weighing machines. The principle of the compound lever applies in trains of wheelwork.

Fulcrum? Lever arms? *Illustrate. How many classes of levers are there? Describe each class, and illustrate.* (452.) *What is a compound lever?* Its uses?

The Inclined Plane.

453. An INCLINED PLANE is a plane inclined to the horizon. (See Arts. 49, 50, and 51.)

In machinery, the inclined plane is seldom used, except in combination.

The Pulley.

454. A PULLEY is a wheel free to turn about an axis and having a groove around it to receive a cord. The axis turns in a frame called a *block*.

A pulley is said to be *fixed or movable*, according as its block is fixed or movable.

Single Fixed Pulley.

455. In this pulley the block, O, is fixed, and the wheel, AB, turns within it. The effect of the fixed pulley is simply to change the direction of a force.

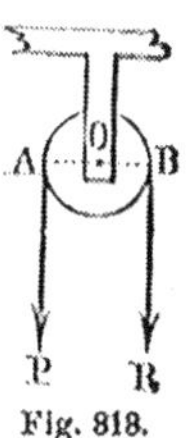

Fig. 313.

Single Movable Pulley.

456. In this pulley the block, O, is movable, and the wheel turns within it.

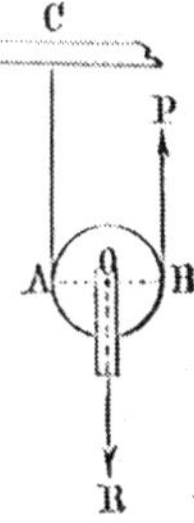

Fig. 314.

Pulleys are combinations of the cord and lever. In the fixed pulley we may regard AB as a lever, whose lever arms are OA and OB, and whose fulcrum is O. In the movable pulley we may regard AB as a lever of the second class, whose fulcrum is A, and whose lever arms are AB and AO.

Pulleys are used for raising weights, for working the rigging of ships, and the like. They are frequently used in combinations.

(453.) What is an inclined plane? (454.) What is a pulley? A block? *When fixed? Movable?* (455.) Describe the single fixed pulley. Its effect. (456.) Describe the single movable pulley. *Of what simple machines are pulleys composed? Illustrate.*

Combinations of Pulleys.

457. Figure 315 represents a combination of three movable pulleys, in which there is a separate cord for each pulley. One end of each cord is attached to a fixed support, the other end being attached to the block of the pulley next in order, except the last one, at the free extremity of which the power is applied. The resistance, R, is applied to the block of the first pulley. This combination is difficult to use, and occupies a great deal of space.

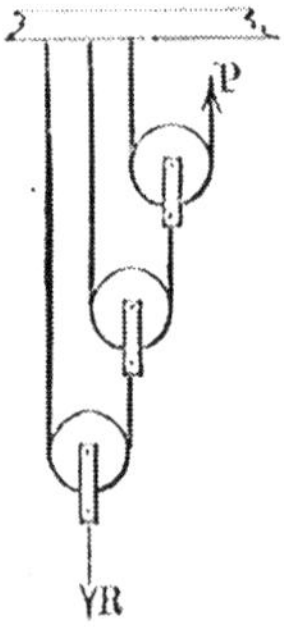

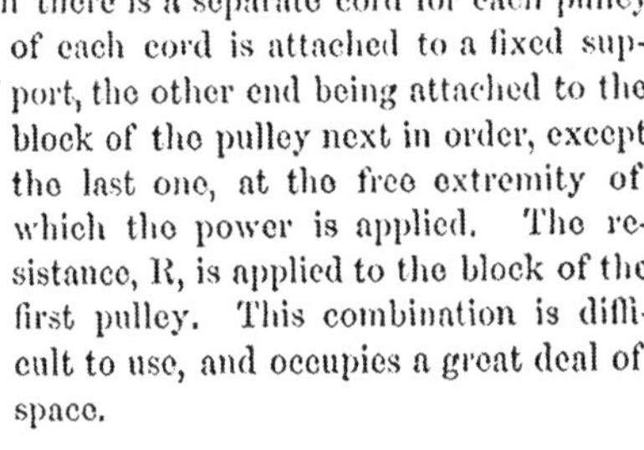

Fig. 315.

Fig. 316.

Figure 316 represents a combination of fixed and movable pulleys, the former in one block and the latter in another. A single cord is employed, having one end made fast to the fixed block, and then passing around the wheels, alternating between those of the movable and those of the fixed block. The power is applied at the free extremity of the cord, and the resistance to the movable block. The pulleys in each block, instead of being one above another, as in the figure, are often placed side by side.

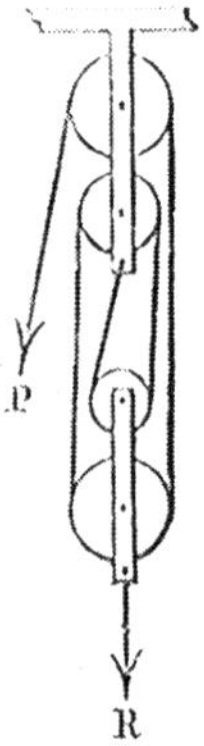

The Wheel and Axle.

458. The WHEEL AND AXLE consists of a wheel, or drum, A, mounted upon an axle, B. The power is applied at one extremity of a cord wrapped around the wheel, and the resistance at one extremity of a second cord wrapped around the axle in a contrary direction. The whole is supported on a suitable frame, by means of pivots projecting from the ends of the axle.

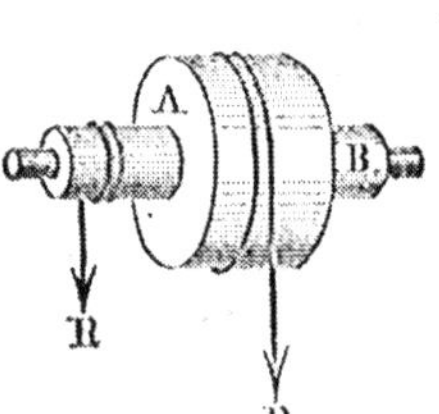

Fig. 317.

(457.) *Describe the combination of pulleys with separate cords. With a single cord.* (458.) What is a wheel and axle?

The principle of the wheel and axle is similar to that of the pulley, and it is, like that machine, a combination of the cord and the lever. In machinery its principal use is in transmitting motion of rotation from one piece to another.

The Windlass and Capstan.

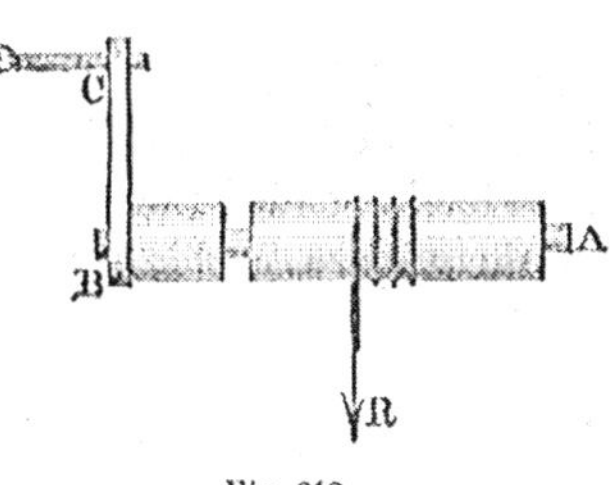

Fig. 318.

459. The WINDLASS consists of an axle, or arbor, AB, and a crank, BCD, by means of which it is turned. The crank consists of an arm, BC, perpendicular to the axle, called the *crank arm*, and a second arm, DC, perpendicular to the first, called the *crank handle*. The power is applied to the crank handle, and the resistance to a rope wrapped around the axle. The windlass is principally used in raising weights.

The capstan differs from the windlass in having its axis vertical, and in being turned by means of levers inserted in holes made in the head of the axle, instead of by a crank. It is much used on shipboard for raising anchors and the like.

The Differential Windlass.

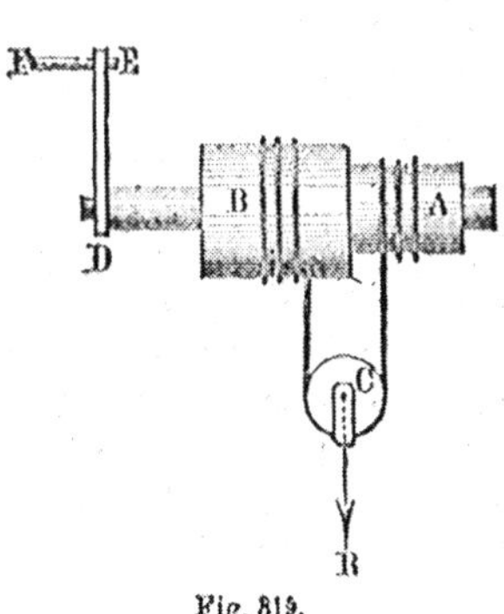

Fig. 319.

460. This differs from the common windlass, in having an axle formed by two drums, A and B, of different diameters. A cord is attached to the larger cylinder and wrapped several times around it, after which it passes under a movable pulley, C, and is then wrapped in a contrary direction around the smaller cylinder. The power is applied to the crank arm, and the resistance to the block of the movable pulley.

Its use in machinery? (**459.**) *What is a windlass? Describe it. What is a crank? What is a capstan?* (**460.**) Describe the differential windlass.

When the cord is wound upon the larger cylinder, it unwinds from the smaller one, but in a less amount, so that the total effect is to raise the weight, R. The power of this machine may be increased by making the two drums nearly equal in diameter.

The Screw.

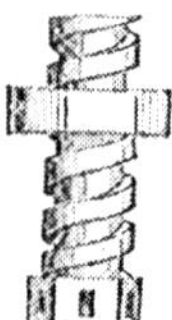

Fig. 820.

461. The SCREW is essentially a combination of inclined planes. It consists of a solid cylinder, enveloped by a spiral projection called the *thread.* The two faces of the thread are nothing more than inclined planes wound around the cylinder of the screw.

The screw works into a solid, fitted to receive it, called the *nut.* The nut may be fixed, the screw turning within it, or the screw may be fixed, the nut turning upon it. Motion is imparted to the one or the other, as the case may be, by means of a lever, at the extremity of which the power is applied.

The endless screw is a screw secured by shoulders, so that it can not move in the direction of its length, and working into a toothed wheel. When the screw is turned, it imparts motion to the wheel, which, in turn, may be made to move a train of wheelwork.

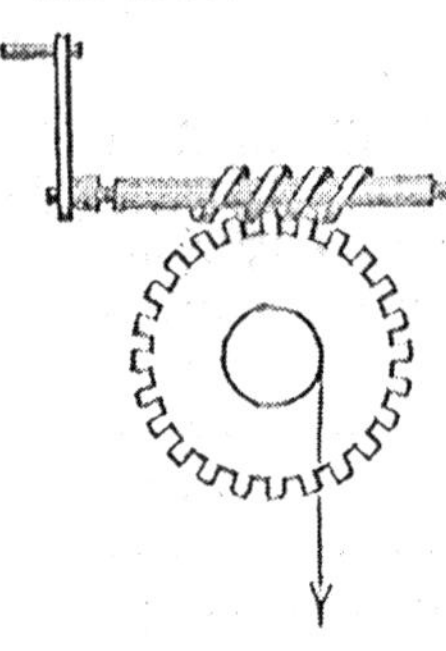

Fig. 821.

Machines of this kind are used in registering the number of turns of an axle, as, for example, the shaft of a steamboat. An endless screw is arranged so as to turn as many times as the shaft, and is connected with a train of light wheelwork. The wheels bear indices, by means of which the number of turns, in any given time, may be read off. This arrangement is extensively used in gas and water-meters, and also in various branches of manufacture.

Its use. (461.) What is a screw? Its thread? Its nut? *Describe the endless screw. Its uses.*

The Wedge.

462. The WEDGE is a solid, bounded by a rectangle, BD, called the *back;* two rectangles, AF and DF, called *faces*, and two triangles, ADE and BCF, called *ends.* The line, EF, in which the faces meet, is called the *edge.*

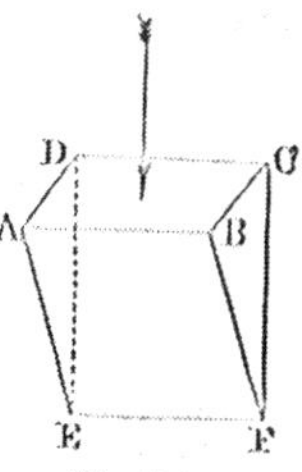

Fig. 322.

The wedge is a combination of two inclined planes, and is used in splitting and cutting instruments. The power is applied to the back, and may consist either of a blow or of a steady pressure. The resistance is applied to the faces.

The power of a wedge may be increased, by increasing the length of its faces, and by diminishing the breadth of its back.

III.—HURTFUL RESISTANCES.

Friction.

463. FRICTION is the resistance which one body experiences in moving upon another when the two bodies are pressed together. This resistance arises from inequalities in the surfaces, the projections of the one sinking into the depressions of the other. To overcome the resistance thus produced, a force must be ap[illegible]d sufficient to break off or bend down the projecting points, [illegible]lse to lift the moving body over the inequalities.

Friction is distinguished as *sliding* and *rolling.* The former arises when one body is drawn upon another; the latter, when one body is rolled upon another. Everything else being equal, the former is greater than the latter.

It has been found by experiment that the sliding friction, between the same two bodies, is proportional to the force with which they

(462.) What is a wedge? Of what is it compounded? Its use? *How may its power be increased?* (463.) What is friction? How is it caused? Distinction between sliding and rolling friction? Which is the greater? *Explain the laws of both sliding and rolling friction.*

are pressed together, and independent of the extent of surface in contact.

It has been found, when a wheel or pivot rolls upon a surface, that the rolling friction is proportional to the pressure, and inversely proportional to the diameter of the wheel or pivot.

In many cases there is a combination of sliding and rolling friction. In the case of a car on a railroad, the friction at the axle is sliding, while that at the track is rolling.

Sliding friction may be greatly diminished by interposing between the rubbing surfaces *unguents*, such as lard, tallow, oil, and various compositions.

For slow motions and great pressures the more consistent unguents, as lard, tallow, and the like, are used; for rapid motions and light pressures, oils are generally employed.

Stiffness of Cords.

464. When a cord is wound upon a wheel or axle, a certain amount of force is required to bend it. The resistance which the cord thus offers to bending is classed as a hurtful resistance. This resistance should be obviated, as far as possible, by selecting bands and cords, which are as flexible as is consistent with due strength.

Atmospheric Resistance.

465. The atmosphere exerts a powerful resistance to the motion of bodies moving through it. It has been found, both by theory and experiment, that this resistance is proportional to the greatest cross section of the body, made by a plane perpendicular to the direction of the motion, and also to the square of the body's velocity. To obviate this resistance, as far as possible, the pieces which have a rapid motion should have as small a cross section as is consistent with due strength.

Explain the two kinds of friction in a car. How is friction diminished? When would you employ consistent and when fluid unguents? (**464.**) How does the stiffness of cords produce resistance? How lessened? (**465.**) Explain the subject of atmospheric resistance. How lessened?

The principle of atmospheric resistance has been employed to regulate the velocity of certain machines. To effect this object, a wheel is arranged bearing *vanes*, which, striking against the air generate a resistance that prevents the velocity from passing a certain limit. It is this principle which renders the *parachute* a safe means of descending from a balloon.

IV.—WHEELWORK.

Trains of Wheels.

466. The power furnished by the motor of a complex machine, is usually transmitted through a succession of pieces, to the working point. These connecting pieces are, in general, wheels and axles, and taken together they form what is called a *train*. A wheel which imparts motion to a succeeding one, is called the *driver*; that to which motion is imparted, is called the *follower*.

Mode of Connection.

467. There are various methods by means of which one wheel may be made to act upon another.

First. By simple contact. The driver, A, being slightly pressed against the follower, B, the friction between the wheels is sufficient to impart a motion of rotation from the former to the latter.

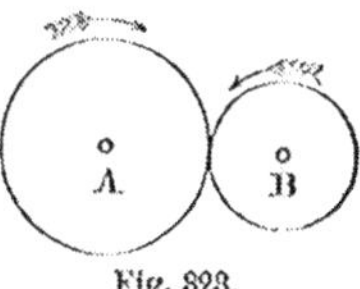

Fig. 323.

To increase the friction and avoid sliding, the surfaces are frequently covered with soft leather. In all cases the motion of the follower is in a contrary sense to that of the driver, as indicated by the arrows.

Secondly. By means of bands or belts. The band is passed around the circumferences of both wheels, and when

What use has been made of atmospheric resistance as a regulator? (466.) What is a train? A driver? A follower? (467.) Explain the mode of connection by simple contact. *How is sliding avoided?*

tightened, a sufficient amount of friction is produced to impart motion from the driver to the follower.

When the band does not cross between the wheels, they both revolve in the same direction, as indicated in Fig. 324. When the band crosses between the wheels, they revolve in opposite directions,

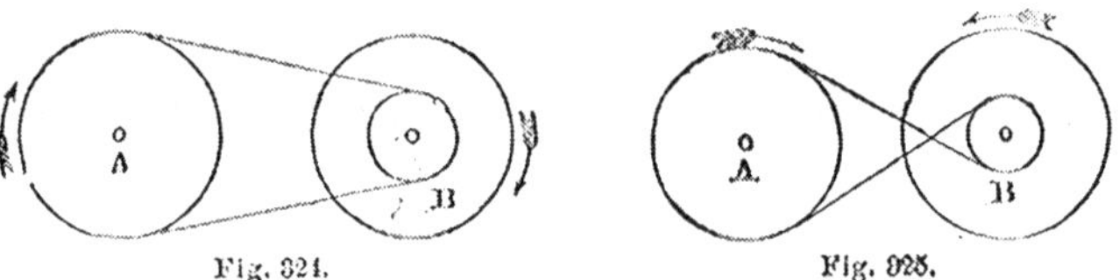

Fig. 324. Fig. 325.

as indicated in Fig. 325. Belts are made of leather, gutta percha, and the like. They are flat and thin, and the drums on which they run should be slightly elevated toward the middle of their thickness. Cords are made of catgut, hempen fibres, or wire, nearly cylindrical. The drums, or pulleys, on which they run should be elevated at the edges. Chains are also used, and in this case the drums should be grooved, and either notched or toothed, so as to fit the links of the chain.

Thirdly. By means of projections on the circumferences of the wheels called *teeth.*

A small wheel, C, mounted on the axle of a large one, B, is called a *pinion*, and its projections are called *leaves*. In the figure the teeth of the wheel A engage with the leaves of the pinion C, and the teeth of the wheel B engage with the leaves of the pinion D. If the wheel A is turned in the direction indicated by the arrow, the wheel B will revolve in a contrary direction, and the wheel F in the same direction. A wheel whose teeth project from its circumference, as shown in Fig. 326, is called a *spur-wheel.*

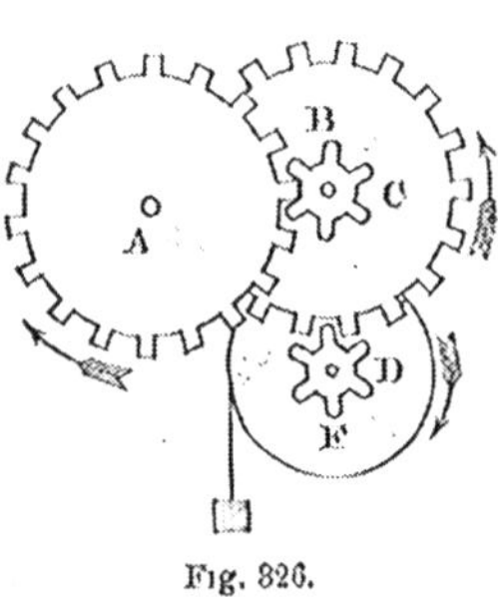

Fig. 326.

Explain the mode of connection by bands and belts. *Explain the two methods of arranging the bands. The kind of wheel required for belts. For cords. For chains. Explain the mode of connection by teeth. What is a pinion? Leaves? Illustrate. What is a spur-wheel?*

In the combination shown in Fig. 326, the axes of the wheels are supposed to be parallel. When the axes are not parallel to each other, motion of rotation may be communicated by means of beveled wheels, as shown in Fig. 13. If the axes are perpendicular to each other, beveled wheels may be used, as in Fig. 13, or the driver may have its teeth set perpendicular to its face and working into a pinion. Such a wheel is called a *crown-wheel.*

When the number of teeth of the driver is greater than that of the follower, the angular velocity of the latter will be greater than that of the former. By a suitable adjustment of the number of teeth on the different wheels, the angular velocity may be multiplied at pleasure.

V.—REGULATORS.

The Governor.

468. The GOVERNOR is a contrivance for regulating the supply of motive force. One form of this contrivance is shown in Fig. 327.

AB is a vertical axis, connected with the machine near its working point, and revolving with a velocity proportional to that of the working point; FE and GD are arms turning with the axis, and bearing heavy balls, D and E, at their extremities; the arms are attached, by hinge joints, at G and F to two bars, CG and CF, and these bars are connected by hinge joints with the axis at C. The arms FE and GD are also connected by hinge joints, with a ring, H, which is free to slide up and down the axis, AB.

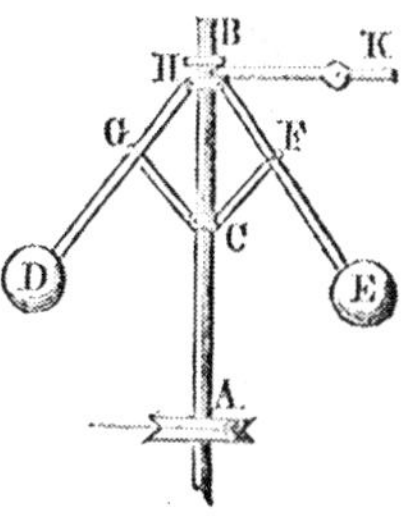

Fig. 327.

When the axis revolves, the centrifugal force developed in the balls causes them to recede from AB, and depresses the ring, H. This causes the lever, BK, to turn about its fulcrum, K, and when the velocity has become sufficiently great, the lever operates to close a

A beveled wheel? A crown wheel? When is the angular motion of the follower greater than that of the driver? (468.) What is a governor? *Describe that shown in the figure, and explain its action.*

valve, and shut off the motive power. When the velocity again diminishes, the balls approach the axis, the ring, H, rises, and the valve is opened. The governor may be adjusted so as to secure any desirable velocity at the working point.

The Fly-Wheel.

469. The FLY-WHEEL is a contrivance for obviating irregularities of motion in a machine.

It usually consists of a heavy rim of iron connected with the axle by radial arms, as shown in the figure. It is mounted on an axle near the working point of the machine, and when the motive power exceeds the amount required to do the work, the excess goes to overcome the inertia of the fly-wheel, with but a slight increase of its angular velocity. On the other hand, when the motive power is less than that required to do the work, the fly-wheel acts by its inertia, giving out the force stored up in it, with but a slight decrease of angular velocity, and thus supplies the deficiency. By a proper adaptation any desired uniformity of motion may be attained.

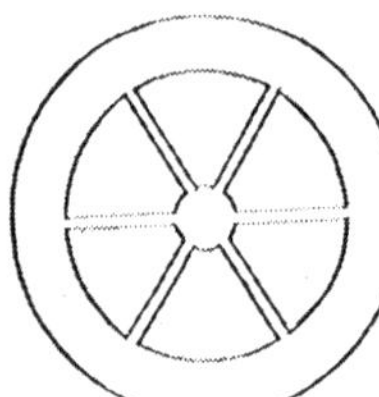

Fig. 323.

VI.—PRIME MOVERS.

Definition of a Prime Mover.

470. A PRIME MOVER is a piece, or combination of pieces, to which the force of the motor is immediately applied, and from which that force is transmitted through the connecting pieces to the working point. The most important prime movers are *water-wheels* and *steam-engines*.

(469.) What is a fly-wheel? *Describe it and its action.* (470.) What is a prime mover? What are the most important ones?

Water-Wheels.

471. A WATER-WHEEL is a wheel put in motion by the action of water. Water-wheels are divided into two classes—*vertical* and *horizontal* wheels. There are three principal varieties of vertical wheels—*overshot*, *undershot*, and *breast wheels.* The most important of the horizontal wheels is the *turbine.*

The *overshot wheel* consists of a cylindrical *drum*, A, terminated at its ends by projecting rings, B, called *crowns.* The space between the crowns is divided into cells, by curved or angular partitions; these cells are called *buckets*, and they are constructed so as to retain the water as long as possible. The water is delivered by a sluice, C, near the top of the wheel, and acting by its weight, it imparts a rotary motion to the wheel.

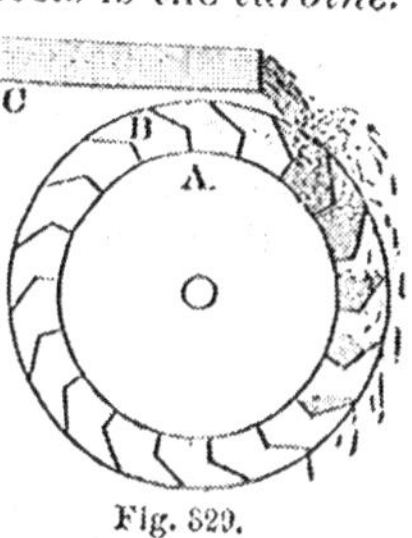

Fig. 829.

The *undershot wheel* resembles the overshot wheel in its general construction, but differs from it in the form of the partitions between the cells, which may be plane or curved. These partitions are called *floats.* In this wheel the water is delivered at the bottom, and striking against the floats with a velocity due to the height of the water in the reservoir, A, rotary motion is produced. In this wheel, the water acts solely by its impulse.

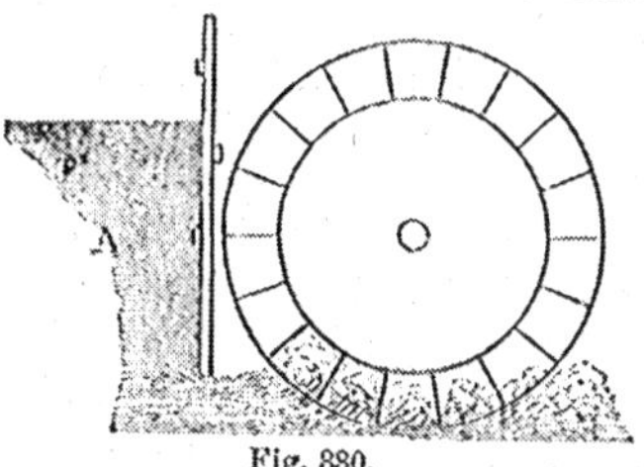

Fig. 830.

The *breast wheel* differs from the overshot wheel in having the water delivered into the buckets at a lower level, and in being provided with a casing, or trough, A, called a *breast*, nearly fitting the periphery of the wheel which revolves within it. In this wheel the water acts both by its weight and its impulse.

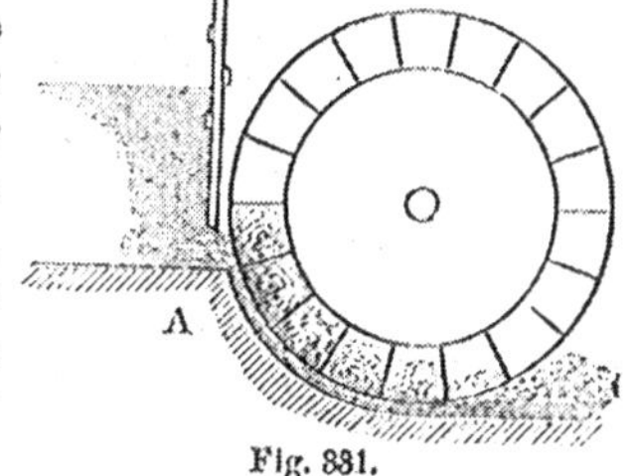

Fig. 831.

(471.) What is a water-wheel? How many classes? The principal varieties of vertical wheels? *Describe the overshot wheel, its drum, crowns, and buckets. Describe the undershot wheel. The breast wheel.*

A description of *turbines*, which are more complex than vertical wheels, does not fall within the scope of the present work.

The Steam-Engine.

472. A STEAM-ENGINE is a combination of pieces for utilizing the expansive force of steam and converting it into a motive power.

In the following pages only a few of the most prominent points are explained. A complete discussion would require more space than can be given within the limits assigned to this treatise. For further details the pupil must consult those works in which the subject of steam is treated as a specialty.

Steam.

473. Let AB represent a glass tube of uniform bore, and C, a piston, fitting it steam-tight, and suppose a little water to be in the tube below the piston. If heat be applied to the bottom of the tube, by means of a spirit-lamp, the water will be converted into steam, and the piston will be driven to the top of the tube. If the lamp be removed, and the tube allowed to cool, the steam will be condensed, and the pressure of the atmosphere will drive the piston back to its original position. By again applying heat and withdrawing it, the operation may be repeated, and so on indefinitely. This simple experiment involves the fundamental idea of the steam-engine.

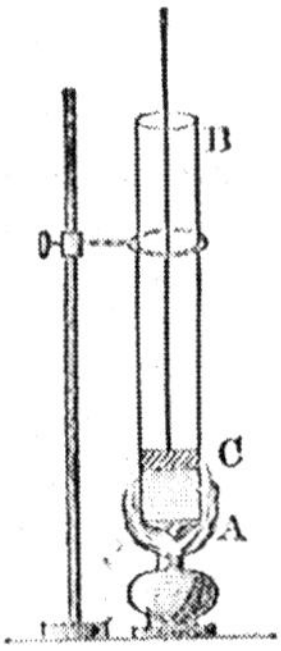

Fig. 332.

Under the ordinary pressure of the atmosphere, a cubic inch of water gives 1,700 cubic inches, or nearly a cubic foot, of steam. In this case the expansive force of the steam is in equilibrium with the pressure of the atmosphere, and it is said to have a *tension* of 15 lbs. to the square inch. If a cubic inch of water be converted into steam, under a pressure of two atmospheres, it will yield but 850 cubic inches of steam, but the *tension* will now be 30 lbs. to the inch.

(472.) What is a steam-engine? (473.) *Describe the experiment illustrating the idea of the steam-engine. How many cubic inches of steam does a cubic inch of water give under a pressure of 15 lbs. to the inch? How many under a pressure of 30 lbs. to the inch?*

In general, the volume of steam yielded by a given volume of water varies inversely as the pressure under which it is generated, and in all cases the *tension* of the steam is equal to this pressure. Hence, the quantity of work arising from the conversion of a given bulk of water into steam is always the same, no matter what may be the pressure under which it is generated. In round numbers, we may say, that the conversion of a cubic inch of water into steam produces a quantity of work sufficient to raise a weight of one ton through a height of one foot.

It is found that the quantity of heat necessary to convert a cubic inch of water at 32° F. into steam is constantly the same, no matter what may be the pressure. Hence, so far as fuel and work are concerned, it is of no consequence what the pressure may be.

It was to utilize the immense expansive force of steam that the steam-engine was devised.

Varieties of Steam-Engine.

474. Steam-engines may be either *condensing* or *non-condensing*. In the *former*, the steam, after having acted upon the piston, is condensed, and the warm water returned to the boiler; in the *latter*, the steam is not condensed, but after having acted upon the piston, is blown off into the air. In condensing engines, steam may be, and generally is, used of a lower tension than 15 lbs. to the inch, in which case the engines are called *low-pressure engines.* In non-condensing engines, steam is always used of a tension greater than 15 lbs. to the inch, and the engines are then called *high-pressure engines.*

Condensing engines are more economical of fuel, but they are heavier and more complex in their construction. Hence they are generally used as stationary engines. Non-condensing engines are used for locomotives, and where fuel is cheap they are often employed as stationary engines.

The efficiency of a steam-engine is measured in terms of a unit

---

How much work does the evaporation of 1 *cubic inch of water produce? What is the rule for the heat required under different pressures?* (**474.**) What is a condensing engine? A non-condensing engine? High-pressure engine? Low-pressure engine? *Advantages of each? What is a horse-power?*

called a *horse-power*, that is, a force which is capable of raising a weight of 33,000 lbs. through a height of one foot in one minute. Thus an engine that can perform a work equivalent to raising 33,000 lbs. through 10 feet in one minute, is said to be an engine of 10 horse-power.

Boilers and their Appendages.

475. The BOILER is a shell of metal, generally of wrought iron, but sometimes of copper, in which steam is generated.

Boilers are made of various shapes. One of the simplest, has the form of a cylinder, with rounded ends; it is set in a furnace, as shown in Fig. 85. Sometimes two smaller cylinders, also with rounded ends, called *heaters*, are placed below the main shell and connected with it by suitable pipes. The object of this arrangement is to increase the heating surface. In the Cornish boiler the cylindrical shell has a large flue passing through it, containing an internal furnace. Sometimes two such flues exist. The tubular boiler has a great number of tubes, or flues, passing through it, for transmitting the flame and heated gases from the furnace.

The principal appendages of the boiler are the *furnace*, or *fireplace*, with its flues and dampers for regulating the draft; the *feed apparatus*, by which water is introduced to supply the place of that converted into steam; the *safety valve*, to guard against danger of explosion (see Art. 229); the *manometer*, for measuring the tension of the steam in the boiler (see Arts. 123 and 124); the *steam-guage*, to indicate the height of the water in the boiler; the *blow-off apparatus*, usually a cock near the bottom of the boiler, which, when opened, permits the pressure of the steam in the boiler to force out the sediment and impurities that collect there; and the *steam-pipe*, to convey the steam from the boiler to the engine proper.

The boiler and its appendages are variously arranged in different engines, the object in all cases being to obtain the greatest amount of steam with a given quantity of fuel. In stationary engines, the furnace is usually made of brick or some other bad conductor of heat, and the flues are so arranged as to bring the flame and heated gases in contact with as large a portion of the boiler as possible. To

(475.) What is a boiler? *Describe some of the forms of boilers. What are the principal appendages of boilers, and what are their uses? What arrangements are made for economizing heat in stationary engines?*

prevent waste of heat, the exposed surface of the boiler is covered with a *jacket* of coarse felt. In locomotive engines, the fire-box is made of boiler-iron, and is so constructed, that it is nearly surrounded by the water in the boiler. An additional heating surface is also obtained by means of flues, running through the boiler, for conveying the flames and heated gases.

To guard against explosion from too great a pressure of steam within the boiler, the safety valve is employed. In addition to this, a fusible safety plug is sometimes used. This consists of a plug of metal, inserted in the boiler, which is capable of being fused at that temperature beyond which there is danger of explosion. If the temperature is raised above this limit, the plug melts and falls out, permitting the water and steam to escape through the hole which it leaves.

Mechanism of the Condensing Engine.

476. The essential parts of a condensing engine are shown in Fig. 333. The figure is only intended to illustrate the principles of the engine, and, for the purpose of illustration, the parts are arranged in such a manner as best to exhibit them at a single view.

The principal parts of the condensing engine are the following:

The *cylinder*, shown on the left, with a portion broken away.

The *piston*, P, which receives the action of the steam, alternately on its upper and lower faces, and is thereby moved up and down in the cylinder.

The *steam-chest*, *b*, into which the steam from the boiler enters through the *steam-pipe* at *o*, and from which it passes through the *steam passages*, alternately to the upper and lower ends of the cylinder.

The *sliding valve*, moved up and down by the rod, *m*, which alternately opens a communication between the steam-chest and the two steam passages leading to the top and bottom of the cylinder.

The *eduction pipe*, U, connecting with the cylinder at *a*, by which the steam, after having acted upon the piston, is conducted into the *condenser*, O.

The *piston-rod*, A, working through a *packing-box*, *d*, which transmits the motion of the piston to the *working-beam*, L.

The *parallel bars*, DD, and the *radial bars*, C,E, which keep the

In locomotive engines? Describe the safety plug. (476.) *Describe the principal parts of a condensing engine and their uses.*

piston-rod from pressing against the side of the packing-box. This arrangement is called *Watt's parallel motion.*

The *connecting rod*, I, which transmits the motion of the working-beam to the *crank arm*, K, and through it imparts a motion of rotation to the shaft of the engine.

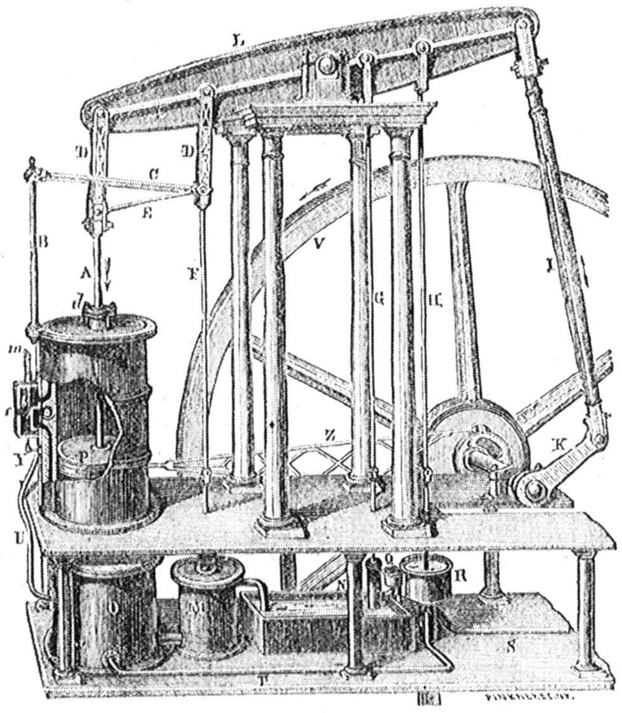

Fig. 353.

The *fly-wheel*, V, which obviates to a certain extent the irregularities of motion in the engine.

The *excentric*, *e*, which, acting like a crank, produces a backward and forward motion in the *connecting rod*, Z. This rod acting on

Describe the principal parts of a condensing engine and their uses.

the *bent lever*, Y, causes the rod *m*, of the sliding valve, to move up and down.

The *cold-water pump*, R, worked by the rod, H, which draws cold water from a reservoir, and forces it through the pipe, T, into the condenser. This pipe, terminating within the condenser in a *rose*, delivers the water in the form of a shower, and condenses the steam.

The *air-pump*, M, worked by the rod, F, which draws the hot water and the air that is mixed with it from the condenser, and forces it into the hot well, N.

The *feed-pump*, Q, worked by the rod, G, which draws the water from the hot well and forces it into the boiler.

To explain the action of the engine, let the position of the parts be as represented in the figure. The steam entering the steam-chest finds the upper passage open, and flowing through it, acts upon the upper face of the piston and drives it to the bottom of the cylinder. The steam below the piston meanwhile flows through the lower passage, and entering the eduction pipe at *a*, is conveyed to the condenser, where it is condensed. When the piston reaches the bottom of the cylinder, the excentric acts upon the bent lever to open the lower and close the upper passage. The steam from the steam-chest now flows through the lower passage, and acting upon the lower face of the piston, forces it to the top of the cylinder. Meantime the steam above the piston, flowing down the upper passage, enters the eduction pipe and is conveyed to the condenser. When the piston reaches the top of the cylinder, the excentric again acts to change the position of the sliding valve, and thus the motion of the piston is continued indefinitely.

The Locomotive.

477. The figure represents a section of a locomotive, the principal parts of which are the following:

The *boiler*, BB, with its *flues*, *pp*, and *safety-valve*, M. The dotted line represents the height of the water in the boiler.

The *fire-box*, A, communicating with the *smoke-box*, C, by means of the *flues*, *pp*. The fire-box has a double wall, the interval being

---

Explain the action of a condensing engine. (477.) *Explain the principal parts of a locomotive engine.*

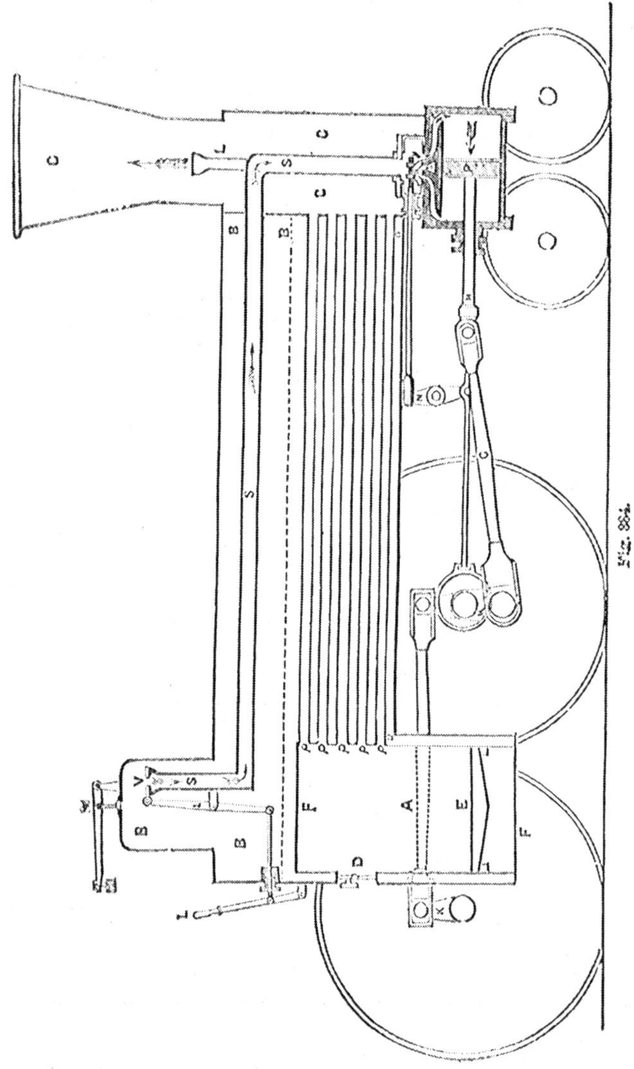

Fig. 334.

filled with water and communicating with the boiler. E is the *grate*, and D the *door* for the supply of fuel.

The *steam-pipe*, SS, conveys the steam from the *steam-dome* to the *steam-chest*. It may be closed by a valve, V, worked by a lever, L.

The *steam-dome* is an elevated portion of the boiler, the object of which is to permit the steam to enter the steam-pipe without any admixture of water, as might be the case were the steam taken from a lower level.

The *cylinder*, the *piston*, P, and the *piston-rod*, R, are similar to the corresponding parts of the condensing engine.

The *blast pipe*, L, through which the steam is blown off after having acted upon the piston, terminates in the *smoke-box*, and the blast of steam from it serves to increase the draft of air through the flues, and thus promotes the combustion of fuel.

The *connecting rod*, G, transmits the motion of the piston to the *crank arm*, by means of which a rotary motion is imparted to the driving wheels of the locomotive.

The manner in which steam acts to impart motion to the piston is the same as in the engine already described.

The Hydraulic Ram.

478. The HYDRAULIC RAM is a machine for raising water by means of *shocks*, caused by sudden stoppages of a stream of water. It consists of a reservoir, B, with a supply pipe, A, and an orifice, D, which may be closed by a spherical valve. Attached to the reservoir is an air-vessel, G, with a valve, E, and a delivery pipe, H.

The stream of water entering the reservoir through the supply pipe, forces the valve D into its place, and closes the orifice. The sudden stoppage of the water causes a shock which forces a portion of water into the air-vessel through the opening, E. The equilibrium being restored, the valve D falls, as does the valve E, and immediately a second shock ensues,

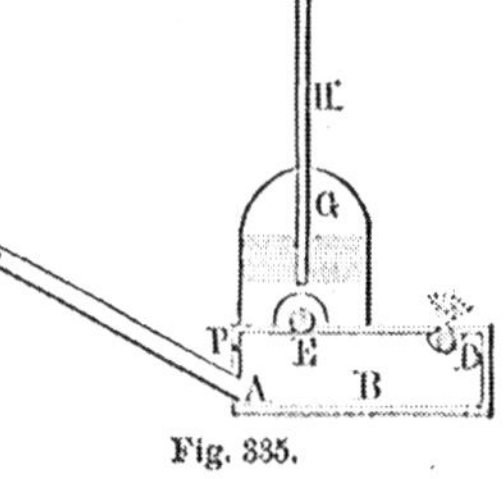

Fig. 335.

Explain the action of a locomotive engine. (478.) *What is the hydraulic ram? Explain its construction and use.*

and a second supply of water is forced into the air-vessel, and so on indefinitely. The water forced into the air-vessel compresses the air in the upper portion of it, until its elastic force becomes sufficient to force a jet of water up the delivery pipe. The delivery once commenced will continue as long as the machine remains in order. Only a small portion of the water which enters the reservoir is raised into the delivery pipe.

INDEX.

CHAPTER I.

CHAPTER II.

CHAPTER III.

CHAPTER IV.

CHAPTER V.

CHAPTER VI.

CHAPTER VII.

CHAPTER VIII.

CHAPTER IX

CHAPTER X.

CHAPTER XI.

www.ingramcontent.com/pod-product-compliance
Lightning Source LLC
LaVergne TN
LVHW021320110826
845150LV00003B/630